World Hunger

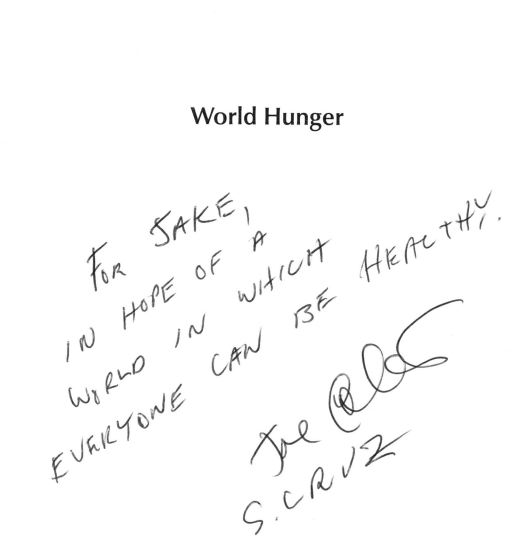

For Jake,
In hope of a
world in which
everyone can be healthy.

Joe Cruz
S. Cruz

World Hunger: 10 Myths

Frances Moore Lappé
and Joseph Collins

Grove Press
New York

Published simultaneously in Canada
Printed in the United States of America

ISBN 978-0-8021-2346-6
eISBN 978-0-8021-9098-7

Grove Press
an imprint of Grove Atlantic
154 West 14th Street
New York, NY 10011

Distributed by Publishers Group West

groveatlantic.com

15 16 17 18   10 9 8 7 6 5 4 3 2 1

To Olivier De Schutter

United Nations Special Rapporteur

on the Right to Food, 2008 to 2014,

for your courageous leadership

". . . all that we are and will and do depends,

in the last analysis, upon what we believe

the Nature of Things to be."

—Aldous Huxley, *The Perennial Philosophy,* 1945

Contents

Contents

Acknowledgments

A work that draws on so many disciplines could not have been written without the contributions of countless knowledgeable and generous people.

For their expertise, dedication, hard work, and willingness to assist us, we are profoundly grateful. Many have been invaluable to us in digging for the most critical source material and helping us to weigh complex questions. Of course, we alone take full responsibility for the book's content.

First, we give special thanks to three superb Research Fellows who worked tirelessly and enthusiastically at the Small Planet Institute: Giulio Caperchi, Rachel Gilbert, and Ashley Higgs. For our chapter on trade, we are indebted especially to Sophia Murphy for her expert assistance. Additionally, SPI researchers Kelly Toups and Jessica Wallach added much-needed help. We give special thanks also to the Institute's manager, Natalie Vaughan-Wynn, who cheerfully orchestrated every step to the finish line.

And the list of those contributing goes on.

We thank the following interns, staff, and volunteers at the Small Planet Institute: Michael Barry, Noel Bielaczyc, Caroline Campbell, Anna Cimini, Allyson Clancy, Nadia Colburn, Lauren Constantino, Bryson

Acknowledgments

Cowan, Ellen Donahue, Vahram Elagöz, Dylan Frazier, Olivia Gool-kasian, Ella Harvey, Tiffany Hawco, Zulakha Iqbal, Olivia Kefauver, Pa Kim, Ria Knapp, Curt Lyon, Jiwon Ma, Jeff Meltzer, Emily Nixon, Crystal Paul, Emma Puka-Beals, Michelle Russell, Freya Sargent, Derek Small-wood, and Emma Walters.

Others with special expertise in the diverse topics covered by our book kindly read drafts of the book or chapters related to their fields, and offered helpful feedback. For their guidance, we thank: Molly Anderson, John C. Berg, Jennifer Clapp, Tim Fessenden, Omar Clark Fisher, Benedikt (Benny) Haerlin, Meghann Jarchow, David H. Kinley III, Anna Lappé, Andre Leu, Mia MacDonald, Peter Mann, Bill Rau, Travis Reynolds, Gyorgy Scrinis, Paul Susman, Brian Tokar, Matthew Vork, and Richard L. Wallace.

Moreover, in our research, we turned many times to experts in specific fields who were willing repeatedly to answer our queries or connect us with other specialists. For their patient and prompt assistance, we thank the following: Abdolreza Abbassian, F. Phillip Abrary, Jeff M. Anhang, Ray Archuleta, Ned Beecher, Chuck Benbrook, Jennifer Blesh, David Briske, Carlo Cafiero, Thomas F. Carroll, Emily Cassidy, M. Jahi Chap-pell, Taarini Chopra, David A. Cleveland, Dana Cordell, Patricia Crease, Olivier De Schutter, Sue Edwards, Sonia Faruqi, Jonathan Foley, Maria Gabitan, Courtney Gallaher, James N. Galloway, Andreas Gattinger, Grace Gershuny, Doug Gurian-Sherman, Hans Herren, Betsy Hartman, Jack Heinemann, Eric Holt Giménez, Pushker Kharecha, Gawain Kripke, Rattan Lal, Allison Leach, Michael Lipton, Fred Magdoff, Clare Mbizule, Nora McKeon, Tracy Misiewicz, Luke Nave, Marina Negroponte, Henry Neufeldt, Kristine Nichols, Meredith Niles, Yacouba Ouedraogo, Ian Paton, Ivette Perfecto, Michel Pimbert, Dr. G. V. Ramanjaneyulu, Talia Raphaely, Jake Ratner, Bill Rau, Chris Reij, William J. Ripple, Peter Rosset, Nadia Scialabba, Jessica Shade, Elson J. Shields, Pete Smith, Jomo Sundaram, Christoph Then, Sapna Elizabeth Thottathil, David Vaccari, John H. Vandermeer, Gaëtan Vanloqueren, Juergen Voegele, Linda Wessel-Beaver, Paul C. West, Timothy A. Wise, Hannah Wittman, Gretchen A. Zdorkowski, and the Landless Workers' Movement.

Additionally, three professors kindly shared chapters with their students, enabling the book to benefit from their feedback. Our thanks to Molly Anderson, Lauret Savoy, and William Ripple.

Acknowledgments

At Grove Press we've been blessed to have the support of editors Patsy Wagner and Allison Malecha, managing editor Amy Vreeland, copyeditor Tom Cherwin, and cover designer Gretchen Mergenthaler.

Finally, Frances wishes to thank her family: Anthony and Clarice Lappé and Anna Lappé and John Marshall, for your steadfast encouragement at every step on this long path; and Richard Rowe, for your careful attention to seemingly endless drafts read aloud, your always insightful comments, and your love that kept me going.

Beyond Guilt, Fear & Despair

Welcome to our continuing exploration.

In 1977, the education-for-action organization we cofounded, Food First—flourishing today with new leadership—released a booklet in which we first took on the "myths of hunger." Two extensive book-length editions followed, and over the years both Food First and we personally have been deeply moved by the many people who've told us that this book changed the course of their lives. These responses, along with the startling realities of hunger in the twenty-first century, have spurred us to create the new book now in your hands.

Please join us on our journey of more than forty years, in which we've sought to understand why there is hunger in a world of plenty, and to discover the most powerful steps we each can take to end it, once and for all. For us, learning had to begin with unlearning. Cutting through the simplistic and scary clichés about hunger, we arrived at some surprising findings:

- The world produces more than enough for everyone to eat well.
- No country is a hopeless basket case. Even countries many people think of as utterly lacking actually have the resources necessary for people to free themselves from hunger.

- Population growth is not the cause of hunger. Rather, hunger and continuing population growth share the same root causes.
- Climate change does not mean hunger is inevitable.
- Increasing a nation's food production may not reduce hunger. Food production per person can increase while more people are nutritionally deprived.
- Our government's foreign aid often hurts rather than helps the hungry. But each of us can play a critical role in ending hunger.
- Unlikely as it may seem, the interests of the vast majority of Americans have much in common with those of the world's hungry.

Our book explains these surprising findings and many more that have freed us from a response to hunger mired in guilt, fear, and hopelessness. But first we must ask the seemingly grade-school question, *Just what is hunger?* Many people assume they know—they've felt it, they've read about it, they've been touched by images of hungry people on television or the Internet. But the greatest obstacle to grasping the causes and solutions to world hunger is that few of us stop to ponder this elemental question.

WHAT IS HUNGER?

Television images haunt us. Stunted, bony bodies. Long lines waiting for a meager bowl of gruel. This is famine hunger in its acute form, the kind no one could miss.

But hunger comes in another form: the day-in-day-out hunger of hundreds of millions of people and nutritional deprivation affecting many more. While chronic hunger doesn't make the evening news, it takes more lives than famine. Every day this largely invisible hunger and its related preventable diseases kill as many as eight thousand children under the age of five. That's roughly three million children each year. Imagine: Every eight days the number of children lost worldwide equals the entire death toll of the Hiroshima bomb.[1]

Statistics like this are staggering. They shock and alarm. But numbers can also numb. They can distance us from what is actually very close to us. So we asked ourselves, What really is hunger?

Is it the gnawing pain in the stomach when we miss a meal? The physical depletion of those suffering chronic undernutrition? The listless stare of a hungry dying child in a televised appeal? Yes, but it is more. We became convinced that as long as we conceive of hunger only in physical measures, we will never truly understand it, or its roots.

What, we asked ourselves, would it mean to think of hunger in terms of universal human emotions, feelings that all of us have experienced at some time in our lives? We'll mention only four such emotions, to give you an idea of what we mean.

A friend of ours, Dr. Charles Clements, is a former Air Force pilot and Vietnam veteran who years ago spent time treating peasants in El Salvador. He wrote of a family he tried to help whose son and daughter had died of fever and diarrhea. "Both had been lost," he writes, "in the years when Camila and her husband had chosen to pay their mortgage, a sum equal to half the value of their crop, rather than keep the money to feed their children. Each year, the choice was always the same. If they paid, their children's lives were endangered. If they didn't, their land could be repossessed."

Being hungry thus means anguish. The anguish of impossible choices. But it is more. . . .

In Nicaragua some years ago, we met Amanda Espinoza, a poor rural woman, who under the long dictatorship of the Somoza dynasty had never had enough to feed her family. She told us that she had endured six stillbirths and watched five of her children die before the age of one.

To Amanda, being hungry means watching people you love die. It is grief.

Throughout the world, the poor are made to blame themselves for their poverty. The day we walked into a home in the Philippine countryside, the first words we heard were an apology for the poverty of the dwelling. Being hungry also means living in humiliation.

Anguish, grief, and humiliation are a part of what hunger means. But increasingly throughout the world, hunger has a fourth dimension.

More recently in Brazil we spent time with a peasant organization known as the Landless Workers' Movement. Since the 1980s, its members have struggled to achieve fair access to farmland to feed their families, for in Brazil a tiny minority controls most of the agricultural land—much gained illegally—while using little of it. Sitting in a countryside meeting room, we were captivated by the enthusiasm of a Catholic nun explaining key details of the movement's upcoming national assembly. Suddenly, a young man on crutches, his foot bandaged, hobbled through the door. Everyone immediately burst into emotional cheers of support. What was this about? we wondered. At break we learned: The man had been seriously wounded in an attack by landowners on the camp where he and his family waited for legal land title. Later we were told that, since the movement's founding, fifteen hundred members have been killed by landowners and corrupt law enforcement officers who've felt threatened by the demand for fair access to land by poor landless people.[2]

Often, then, a fourth dimension of hunger is fear.

Anguish, grief, humiliation, and fear. What if we refused merely to count the hungry and instead tried also to understand hunger in terms of such universal emotions?

How we understand hunger determines what we think its solutions are. If we think of hunger only as numbers—numbers of people with too few calories—the solution also appears to us in numbers: numbers of tons of food aid, or numbers of dollars in economic assistance. But once we begin to understand hunger as real people coping with the most painful of human emotions, we can perceive its roots. We need only ask, When have we experienced any of these emotions ourselves? Hasn't it been when we have felt out of control of our lives—powerless to protect ourselves and those we love?

Hunger has thus become for us the ultimate symbol of powerlessness.

Appreciating that hunger tells us a person has been robbed of the most basic power—the power to protect ourselves and those we love—is a first step. Peeling back the layers of misunderstanding, we must then ask, If powerlessness lies at the very root of hunger, what are hunger's causes?

Certainly, the cause is not scarcity. The world is awash with food, as the chapter on Myth 1 will show. Neither can we blame hunger on natural disasters or climate change. Put most simply, the root cause of hunger isn't a scarcity of food; it's a *scarcity of democracy*.

Wait a minute! What does democracy have to do with hunger? Well, in our view—everything. Democracy carries within it the principle of accountability. Democratic forums and structures are those in which people have a say in decisions that most affect their well-being. Leadership is kept accountable to the needs of the broad public.

Thus, human dignity lies at democracy's core.

Antidemocratic structures, by contrast, are those in which power is so tightly concentrated that the majority of people are left with no say at all. Leaders are accountable only to the powerful minority.

In the United States, we think of democracy as a strictly political concept. We grow up absorbing the notion that political democracy enables us as citizens to protect certain rights—to vote, to have our civil liberties upheld, and to enjoy fair access to opportunities for safe, satisfying lives. We take pride in knowing that we've achieved voting rights and the rule of law—however imperfect—when many societies haven't.

But we hope that reading this book will stretch people's hearts and minds as writing it has stretched ours. We hope it will help us to conceive of democracy as more than a political structure—that we will come to see democracy as a set of values of inclusion, fairness, and mutual accountability applying to all arenas of life, including economics, in all societies. And we hope that in so doing, we will realize the many ways each of us can contribute to the end of hunger.

ON POWER AND POWERLESSNESS

To begin to make these connections, let us look briefly at how the opposite of democracy—antidemocratic decision making—robs people of power over their lives on many levels. Here and throughout this book we focus on realities in much of the Global South.* However, similar dynamics show up in many industrial countries as well.

First, within the family. Women are responsible for growing much of the world's food—60 to 80 percent in the Global South.[3] The resources women have to grow staple foods largely determine their family's nutritional well-being. But many women are losing their special role in food production, as a result of continuing privatization of land ownership and a focus on cash crops, especially for export. Agricultural credit for producing

---

* "Global South" is shorthand for what many call "developing countries," or "poor countries," or "less industrialized countries." Although "Global South" is hardly geographically accurate, we use it in most cases because it is in common use and relatively neutral. Most important, it doesn't risk misleading readers to assume that the designated countries lack resources or that their populations are uniformly poor.

cash crops goes overwhelmingly to men, allowing food crops to languish. This dynamic within the family, often reinforced by government policies, helps explain the crisis of "nutritional deprivation" defined in Myth 1.[4]

Second, at the village level. In most countries, a consistent pattern emerges: Fewer and fewer people control more and more farm- and pastureland. Increasing numbers of people have none at all.[5] Today, global "land grabs" are speeding this trend as foreign capital takes control of huge swaths of land in the Global South. "When people lose access to their land, they also lose their means to obtain food, their communities, and their cultures," writes Fred Magdoff.[6]

Third, at the national level. Wherever people have been made hungry, power is in the hands of those unaccountable to their people. Antidemocratic governments answering only to elites, including large landholders and monopolistic agricultural industries, lavish them with credit, subsidies, and other assistance. Often brutally, such governments fight any reform that would make control over food-producing resources more equitable.

These levels are not hard to see. But there is another, wider arena shaping all of our lives today, a fourth level on which democracy is scarce: the international arena of commerce and finance. A handful of corporations dominate world trade in the agricultural commodities that are the lifeblood of economies on which many of the world's poorest people depend. Historically, efforts by governments in the Global South to bargain for higher prices have repeatedly failed in the face of the preeminent power of the giant trading corporations and the trade policies of the industrial countries and international governing bodies they dominate. Myths 6 and 7 explore these realities.

In attempting to encapsulate the antidemocratic roots of hunger, we have traveled from the level of the family to that of international commerce and finance. Let us complete the circle by returning to the family.

As economic decisions are made by those unaccountable to their citizens, insecurity deepens for millions of people. Economic pressures tear family bonds asunder as men are forced to leave home in search of work. More and more women shoulder family responsibilities alone. Although most nations fail to report on this trend, woefully inadequate World Bank data suggest that worldwide as many as one-quarter of all households are now headed by women.[7] On top of the weight of poverty, they confront barriers of discrimination. The breakdown of the traditional family structure does not bring liberation for women; it often simply means greater hardship.

Given this pattern, it should come as no surprise that most of the hungry in the world are women and the children they care for. Most of those who die from hunger every year are children.[8]

In our effort to grasp the roots of hunger, we have identified the problem: the ever-greater scarcity not of food but of democracy—democracy understood to include power over the life-and-death question of access to food.

PROGRESS AGAINST HUNGER?

As we begin to probe the roots of and solutions to hunger, let's also take in the key messages we receive in the media about humanity's progress toward ending hunger.

Some are quite upbeat.

The number of hungry people in what the Food and Agriculture Organization of the United Nations (FAO) calls "developing countries" has dropped from almost a billion in 1990 to about 800 million in 2015, and the percentage of their people who are hungry has fallen fast, down 45 percent since 1990.[9]

The United Nations Millennium Development Goals set its hunger-reduction target at cutting in half the percentage of hungry people in developing countries by 2015, and now the UN is celebrating.

We didn't quite hit the target, we're told, but we came really close. Seventy-two countries have cut in half the proportion of hungry people since 1990. While the FAO's 2015 annual hunger report frames its message as "uneven progress," the UN News Center strikes a different note picked up by wider media: Overall, humanity is on the right track. If we continue what the FAO calls "inclusive growth" with "social protection," and we reduce civil conflict while dealing better with climate and other natural disasters, we can triumph. "The near-achievement of the MDG hunger targets," declared FAO Director General José Graziano da Silva, "shows us that we can indeed eliminate the scourge of hunger in our lifetime."[10]

These messages are mostly about progress.

But 800 million is a lot of people. And if our primary concern is reducing suffering, then it's the number of people, not the percentage, experiencing hunger, that matters most. Just try telling hungry people they constitute a smaller proportion of a population and see if they feel better. Measured by a reduction in *number*, hunger in the developing world has decreased by just one-fifth since 1990.[11]

Yes, it is progress—modest progress. And how much of humanity has participated in it?

Advances against both poverty and hunger have been tightly concentrated. Without progress in China—more than half of which was achieved in the 1990s—what are officially called "developing regions" would have seen only a 6 percent drop since 1990 in the number of hungry people. It's worth noting that China's reforms in the 1990s making land available to small farmers are credited with much of this progress, as we note in Myth 5, but are not promoted today by those most celebrating a global advance against hunger.[12]

Additionally, missing in this picture of progress is a deeply troubling sign of setback: Inequality is increasing in most regions of the world.[13] By 2016, Oxfam reports, "the top 1 percent will have more wealth than the remaining 99 percent of people."[14]

Returning specifically to hunger, let's examine the yardstick the FAO uses to arrive at its annual number of chronically hungry people, today the 800 million noted above. Most important to understand is that *the measure captures only calorie, not nutrient, deficiency.*

And the calorie deficiency registered is only the direst. The FAO's assistant director-general, coordinator for economic and social development, Jomo Kwame Sundaram, has underscored that this measure "is a very, very strict definition" of hunger, making clear that "the number of hungry people is probably higher" than the FAO's official count.[15]

Nevertheless, the FAO's annual estimate is received by the public and policymakers alike as a "total." But it is not. It is actually a "partial," as we explain more fully in Myth 1. There we also put forth what for us is a more meaningful approach to grasping the true scope of the crisis.

Finally, in this big-picture look, let's not miss the fact that hunger is not limited to low-income countries. Even by the FAO's limited measure, fifteen million people in the "developed countries" suffer severe, long-term calorie deprivation.[16] In the United States, a much broader view of food deprivation includes those who are often unsure of where their next meal is coming from. By that yardstick, we get a disturbing result: One in six Americans is afflicted.[17]

BRINGING THE DEMOCRACY DIAGNOSIS HOME

Having begun to explore the meaning of hunger and the tricky question, "How many go hungry?" let's return to the link between hunger and democracy.

We've suggested that to grasp hunger's roots we have to grapple with the meaning of democracy itself. On this, in recent decades we've sensed a radical change in our own country. What does the right to vote mean, many ask, if the rules of political participation allow a tiny economic minority to steer public policies to serve its narrow interests?

Nearly 90 percent of Americans believe corporations have too much power in our political system.[18]

In 2014 two political scientists confirmed what most Americans feel. In the United States, "average citizens and mass-based interest groups have little or no independent influence," concluded an extensive data analysis by Princeton and Northwestern professors. In sharp contrast, they found that "business interests have substantial independent impacts on U.S. government policy."[19]

When their findings were released, citizens and commentators lit up the media with the question: "Are we now an oligarchy, too?"

We hope this book will encourage all of those fearing the betrayal of America's democratic promise to see new possibilities. As we probe the roots of world hunger, we learn that for genuine democracy to thrive, or even survive, its practice must evolve toward what we call "Living Democracy." By this we mean democracy practiced as much more than a particular political structure. After all, numerous countries have all the trappings of democracy—constitutions, multiple parties, voting—yet so many of their citizens endure great deprivation. Think of India, Kenya, or Guatemala.

The heart and soul of democracy are about voice—who has it and who doesn't. Thus, Living Democracy is a way of life in which everyone has a voice as the principles of inclusion, mutual accountability, and fairness expand to economic and social relationships. They include, for example, access to land, food, jobs, and income.

> Living Democracy is a way of life in which everyone has a voice as the principles of inclusion, mutual accountability, and fairness expand to economic and social relationships.

In this book, you will meet citizens, including some of the world's poorest, stepping up to show us what economic and social citizenship looks like. With their help, we can come to perceive our political rights in a deeper context, one including economic rights. In the United States, this step would mean picking up the torch handed to us more than seventy years ago by President Franklin Delano Roosevelt in his call for a Second Bill of Rights, including "freedom from want."[20] Perhaps there is no more basic right than the right to nutritious food.

What we hope to show in this book is that only as we deepen our concept of democracy to include true accountability to those most affected by decisions—in economic as well as political arenas of life—can people transcend the anguish, grief, humiliation, and fear arising from powerlessness.

HOW WE THINK ABOUT HUNGER MATTERS

Now to the barriers in our own heads.

Especially in troubled times, all of us seek ways to make sense of the world. We grasp for organizing principles to help us interpret the endlessly confusing rush of world events. It's a natural human process—perhaps as natural as eating itself. But living effectively depends on how well our organizing principles reflect reality.

Unfortunately, the principles around which many of us have come to organize our thinking about world hunger block our grasp of real solutions. This entire book is structured around ten such organizing principles. We call them "myths" to suggest that the views they embody may not be totally false. Many have some validity. It is as organizing principles that they fail. Not only do they prevent us from seeing solutions; they obfuscate our own legitimate interests as well as those of hungry people. Some fail us because they describe but don't explain, some are so partial that they lead us down blind alleys, and some simply aren't true.

What we set out to do in this book is to probe the underlying assumptions people have about world hunger's causes and cures. For we've come to believe that *the way people think about hunger is the greatest obstacle to ending it.*

After you read our book, we hope you'll find that you no longer have to block out bad news about hunger but can face it squarely, because a more realistic framework of understanding—to be repeatedly tested against your own experience—will enable you to make real choices, choices that can contribute to ending this spreading, but needless, human suffering.

Our book may shake your most dearly held beliefs or it may confirm your deepest intuitions and experiences. Whichever, we hope it leaves you with the conviction that until humanity has solved the most basic human problem—how to ensure that every one of us has food for life— we cannot consider ourselves fully human.

myth 1

Too Little Food, Too Many People

MYTH: Food-producing resources are already stretched to their limits, and in many places there's just not enough to go around. More people inevitably means less for each of us. So continuing population growth, which could lead to several billion more people by mid-century, is a major crisis. To end hunger today and to have any hope of preventing ever-greater hunger in the future, we must stop population growth.

OUR RESPONSE: "Too many people pressing on too few resources" is perhaps the most common and intuitive explanation for continuing hunger. But sometimes our intuitions just don't line up with the evidence. The world produces more than enough food today. And, given the striking decline in population growth in recent decades, there's every reason to believe it is possible to halt population growth before we overshoot the Earth's capacity.

Let's begin by probing more deeply the extent of hunger that many assume to be evidence of too little food for too many mouths. How we measure hunger turns out to be trickier than we'd long assumed.

In our opening essay we noted that the UN Food and Agriculture Organization (FAO) defines hunger only in terms of calorie deficiency, and reports about 800 million hungry people.[1] In this widely used measure, the FAO explained to us, those who lack calories for many months at a stretch—say, between harvests or jobs—do not register if their calorie supply averaged over a year is minimally sufficient. Yet, medical authorities tell us that even short-term calorie deficiency can have devastating effects, especially on children and anyone weakened by disease.[2]

Appreciating the inadequacy of this single measure, in 2013 the FAO began to emphasize a "suite of food security indicators" that includes not only the adequacy of available protein and calorie supplies but also stunting and factors such as grain-import dependency and access to safe water and sanitation that signal vulnerability to hunger.[3] The FAO also added an assessment tool called Voices of the Hungry, drawing on self-reported experiences of food insecurity.[4] We applaud these efforts to gain a truer understanding of the depth of hunger.

Still, only *one* hunger measure—that of calorie deficiency—reaches the broad public, even as this measure increasingly fails to capture nutritional well-being.

Why do we say "increasingly"?

Because the quality of food in many parts of the world is degrading, so more of us can be suffering from lack of nutrients even when our calories are more than sufficient. For example, take India, where one in seven people is "hungry" by the current calorie measure, yet at the same time four in five infants and toddlers and half of all women suffer from iron deficiency, with potentially deadly consequences.[5]

From 1990 to 2010, unhealthy eating patterns outpaced dietary improvements in most parts of the world, including the poorer regions, reports a 2015 *Lancet* study. As a consequence, "most of the key causes" of noncommunicable diseases are diet-related and predicted by 2020 to account for nearly 75 percent of all deaths worldwide, the study emphasizes.[6] By 2008 nearly four-fifths of deaths from cancer, heart disease, and other noncommunicable diseases were not in the Global North, long associated with these largely diet-related ailments, but rather in "low- and lower middle-income countries," according to the World Health Organization (WHO).[7]

In these alarming trends, *The Lancet*'s study implicates "transnational marketing and investment."

This widening disconnect between calories and nutrients has another devastating outcome: Worldwide, roughly one in eight people is now obese, and thus at risk for heart disease and diabetes among other ailments.[8] Almost two-thirds of obese people live in the Global South.[9]

These realities hit us when a doctor working in a rural Indian clinic serving two thousand impoverished farmers each month described a major change in his practice over the last few decades: "My patients get enough calories, but now 60 percent suffer diabetes and heart conditions."[10]

Clearly, the world urgently needs a more meaningful primary indicator of the nutritional crisis than one based on calories alone—a measure of what we call in this book "nutritional deprivation" that captures both calorie and nutrient deficiencies. Since we don't yet have one, let's review the indicators we do have and then see where we stand.

In addition to the calorie-deficiency measure, arriving at about 800 million people worldwide in 2014, another is "stunting," estimated by the WHO in collaboration with UNICEF and the World Bank.[11] In children under five, stunting is diagnosed when a child's height is significantly below the median compared with the "reference population."[12] To most ears, "stunting" merely suggests being unusually short; but it actually indicates a set of medical problems including a depressed immune system and impeded cognitive development.[13]

One-quarter of the world's children are stunted, report these agencies, with many factors conspiring to cause the problem, including too little food and nutritionally poor food for pregnant women and children, along with other deprivations.[14] New research underscores that poor sanitation also contributes to poor nutrition, and thus perhaps to as much as one-half or more of stunting, even when a child is well fed, because repeated bacterial infection associated with unsafe water interferes with nutrient absorption.[15]

Stunting remains "disturbingly high," notes the FAO. Without China, the global decline in stunting since 1990 would be significantly less than the decline in calorie deficiency—to us more evidence of a widening gap between calories and nutrition.[16]

Evidence grows that the consequences of stunting commonly last a lifetime, including cognitive impairment and a weakened immune system, as noted; and, for females, reproductive problems. All show up in reduced educational and economic achievement. Thus, we believe, because stunting typically brings lifelong harm, individuals designated as stunted during childhood should be counted throughout their lives among those suffering the consequences of nutritional deprivation.

By this reasoning stunting affects not just one-quarter of our children but one-quarter of our whole population, or 1.8 billion people. We know this approach breaks with conventional wisdom, but we ask you to weigh it seriously.

One might counter by observing that not every child diagnosed as stunted experiences significant harm as an adult, so isn't applying the same percentage to a whole population bound to overstate the problem? Unfortunately, no. Because stunting afflicted prior generations as well, this measure actually undercounts many adults born when stunting was even more common. Those in their 30s today, for example, were themselves under five years old at a time when stunting was much more widespread than it is today.[17]

Beyond calorie deficiency and stunting, are there any additional indicators that might help us to grasp the magnitude of the nutritional crisis?

A third is WHO's estimate that *two billion* of us have a deficit in at least one nutrient essential for health—a deficit often causing great harm. Vitamin A deficiency, for example, means blindness for as many as half a million children each year, and iron deficiency is linked to one in five maternal deaths.[18]

So taking into consideration these three indicators, with considerable overlap—calorie deficiency at about 800 million, stunting at 1.8 billion, and nutrient deficiency at 2 billion—arguably at least one-quarter of the Earth's 7.3 billion people suffer from nutritional deprivation. That's roughly twice as many as are "hungry" measured by calorie deficiency.[19]

What is "nutritional deprivation"?
Capturing both calorie and nutrient deficiency, nutritional deprivation means being so deprived of healthy food, and the safe water needed to absorb its nutrients, that one's health suffers. "Being deprived" refers to the result of inequities in power relationships blocking people's access to food and to sanitation.

We urge the UN system to prioritize the critical work it's begun to develop a comprehensive measure of nutritional deprivation.

We've chosen "nutritional deprivation" to define the crisis this book addresses, mindful that it isn't a common term. With this background, we can now clarify its meaning. Here and throughout our book nutritional deprivation means being so deprived of healthy food—and the safe water needed to absorb its nutrients—that one's health suffers. It thus captures both calorie and nutrient deficiency. "Being deprived" in this definition refers to the result of inequities in power relationships, such as those we touched on in the opening essay, that block people's access to food and to santitation. It therefore conveys a social malady—not simply being in a state of deficiency but the widespread harm caused by being actively deprived.

The implication of all of this?

We'll say it again: *In a world of abundant food resources, at least one in four of us suffers from nutritional deprivation*, yet humanity still lacks a comprehensive measure of this crisis.

In our opening essay, we described hunger as painful, universal human emotions that arise from feelings of powerlessness to protect ourselves and our loved ones, and here we stress that hunger must be measured not only as calorie and nutrient deficiency but also as the resultant, ongoing impairment and disease. Hunger means all of this, and it affects all of us.

Because the word "hunger" carries such powerful emotion, we will continue to use it. We hope that you do, too. Still, we want to be clear that for us hunger means not only calorie deficiency but the much broader, and often more devastating, dimensions captured in "nutritional deprivation." In this sense, "hunger" is no longer understood primarily as an uncomfortable, even painful experience; it is a condition creating great and often lasting harms that we can all be part of ending.

Now let's tackle head-on the premise that scarcity explains the widespread misery of not being able to secure a healthy diet. Does scarcity hold up as an explanation in light of the facts?

BEHIND THE SCARCITY SCARE

Global population more than doubled between 1961 and 2013, but world food production grew even faster. So today there's about 50 percent more food produced for each of us.[20] In fact, the world produces enough food to provide every human being with nearly 2,900 calories a day.[21] That's enough to make many people chubby!

> Between 1961 and 2013, world food production per person grew by about 50 percent.
> —calculated from UN Food and Agriculture Organization

Plus, those 2,900 calories are just from the "leftovers"—what's left after we've diverted about half of the world's grain and most soy protein into feed for livestock and nonfood uses.[22] Worldwide, 9 percent of major crops is now used to produce ethanol—what we call "agrofuel" to remind us of its agricultural roots—and for other industrial purposes.[23]

Nor do the 2,900-and-climbing calories for each of us include much of the breathtakingly large amount of food we waste each year, about one-third of all edible parts of food, amounting to 1.3 billion tons in 2009.[24] As a result, we lose one out of every four calories produced.[25] Consumers in industrialized countries waste almost as much food as the net food production of sub-Saharan Africa.[26]

Beyond the vast abundance represented by these numbers are the uncounted but sizable quantities of food that 1.6 billion people living in or near forests secure for themselves from herbs, animals, fruits, nuts, and berries.[27] A sense of the richness that's not counted in the world's food supply is suggested in a finding of the National Academy of Sciences that "most of Africa's edible native fruits are wild—rarely cultivated or maintained or improved."[28]

While we hear from longtime food analyst Lester Brown that scarcity is the "new norm," the UN agency responsible for forecasting our future food supply, the FAO—even after taking into account expected population growth—forecasts global calories available per person in 2050 to be even slightly higher than the generous supply we have today.[29]

Abundance, not scarcity, best describes the world's food supply.

But Don't Price Spikes Prove Scarcity?

On average, global food prices in 2014 were 45 percent higher than a decade ago, after adjustment for inflation, a huge increase in a short time.[30] And they are predicted to rise further as climate change affects agriculture.

But are shortages really the cause?

From time to time, the world experiences price spikes in grains and other agricultural commodities—accompanied by experts blaming food "shortages." The most recent and deadly price spikes occurred in 2008 and 2011. But such spikes often do not reflect a real shortage of food: Over the decade that included this food-price crisis, global per capita agricultural production continued, with one tiny dip, its steady growth of the previous decade.[31] Rather, these spikes are "bubbles" generated in large measure by commodity speculators whose gambles transform small declines in forecasted supply into *much* higher prices.[32]

Unfortunately, for impoverished people increasingly dependent on imports, international price swings bring harsh consequences, as we examine in Myth 6.[33]

Beneath the Big Picture: What About the Hunger Zones?

All well and good for the global picture, you might be thinking, but doesn't such a broad stroke tell us little? What about countries we tend to associate with widespread hunger—those in the Global South, especially in Africa?

Are not food supplies scarce there?

Food output per person in what are called "low income, food deficit" countries increased almost 30 percent between 1990 and 2012.[34] If we look more closely at areas that account for most of the world's hungry people, scarcity cannot explain hunger.

India. Over 190 million Indians do not get enough to eat—that's almost one-quarter of the world's calorie-deficient people.[35] Yet, over the years from 1990 to 2012, food production per person in India has outstripped population growth by about a third, while the number of undernourished Indians—almost one in seven—declined by just 10 percent.[36] India not only exports grain, but in 2012 it had the world's second-largest grain stockpile after China. In that year, India's stockpile alone could have provided one cup of cooked rice to every Indian every day plus almost 50 loaves of bread for everyone that year, and bursting granaries often force the government to store wheat outdoors under tarps, exposed to rot and rats.[37]

Despite all this, India is home to 38 percent of the world's stunted children, and stunting brings lifelong impairment and vulnerability to disease.[38] As already noted, new research suggests that poor sanitation, by exposing children to pathogens that interfere with nutrient absorption, likely plays a huge role in this lost potential.[39]

Scarcity of food, however, is not to blame.

Africa. When most people in industrial countries think of hunger, no doubt images of Africa come to mind first. Yet food production on the

African continent outstripped population growth between 1990 and 2013 by 22 percent, not that far from the global average of 29 percent.[40]

South of the Sahara, since 1990 the number of Africans suffering from long-term, severe calorie deficiency has increased by 22 percent.[41] But during the same period, food production per person rose almost 10 percent, even though the region includes countries with the world's highest population growth rates.[42]

> If food available within sub-Saharan Africa were equitably distributed, *all* Africans could meet their basic caloric needs.
> —African Human Development Report, 2012

Roughly 2,300 calories are available per person every day in sub-Saharan Africa. That's somewhat above the "basic minimum nutritional requirement," of 2,100 calories a day, as defined by the United Nations Development Programme. Thus, if food available within sub-Saharan Africa were equitably distributed, *all* Africans could meet their basic caloric needs.[43]

In rethinking scarcity in Africa, also note that almost a dozen sub-Saharan countries—some with high levels of undernourished people—export more food than they import. The Ivory Coast, for example, uses prime land to grow cocoa and coffee; this makes it a net food exporter, yet 30 percent of its young children are stunted, a proportion higher than the world average.[44]

Despite its production gains, sub-Saharan Africa's food output for local consumption remains far below its potential. This reality of unrealized potential isn't surprising given the range of forces that over centuries have thwarted and distorted the region's agricultural development:

- *Foreign interests take over agricultural lands.* Colonial seizures of land in the nineteenth century continued into the twentieth. They have displaced peoples and pushed agricultural production from good soils into less fertile areas, with the best land dedicated largely to export crops.[45] In new forms, these seizures continue today as China, Saudi Arabia, Kuwait, Qatar, and South Korea, among others, are busy buying up or leasing vast tracts of land to provide food—not for local people but for their own consumers—as well as to produce crops for fuel. While reliable data are hard to come by because the companies involved are secretive, a UN report nevertheless includes this dramatic estimate: Up to two-thirds of all of what are now called "land grabs" are to grow crops for fuel. Since 2000 in Africa, land grabs so far total an area as large as Kenya.[46]

- *Governments underinvest in agriculture—failing to put resources into farm credit, local market roads, and crop storage facilities.* That's true even though agriculture engages at least 70 percent of the African work-force.[47] Tanzania is just one example where "the main reasons farmers do not produce more," observes the FAO, are "difficulty in accessing markets and a lack of infrastructure."[48] Forty African governments recently committed publicly to devote 10 percent of their budgets to agriculture, but by 2014 only nine of them had achieved this goal.[49]
- *Government resources tend to back agricultural exports more than staple foods of small farmers.* The colonial era's focus of public resources—from research to credit—on export crops has continued after independence.[50]
- *Foreign aid policies have often reinforced this emphasis on exports.* Much official foreign assistance has bypassed Africa's small farmers and pastoralists in favor of expensive, large-scale projects, back-ing export-oriented, elite-controlled production.[51] (We explore U.S. foreign aid in Myth 8.)
- *Low prices for farmers stifle production.* With an eye to preventing urban unrest and meeting the desire of the better-off for meat and dairy products, African governments have often maintained low prices for food and feed. One result is that their own small farmers earn so little that their capacity and incentive to produce are under-mined.[52] Plus, some countries, including the United States, "dump" their food surpluses in African markets—that is, sell them below their cost of production. The net effect has been to depress local produc-tion.[53] (More on this also in Myth 8.)
- *Corporations based in the industrial countries have shifted urban tastes.* Thirty years ago, for example, only a small minority of urban dwellers in sub-Saharan Africa ate wheat. Today bread is a staple for many, and bread and other wheat products account for a large portion of the region's grain imports;[54] U.S. food aid and advertising by global food companies ("He'll be smart. He'll go far. He'll eat bread.") have helped mold African tastes to what the industrial countries have to sell.[55]

Thus, beneath scarcity as the diagnosis of Africa's hunger problem lie many human-made and therefore reversible causes. Throughout our book, we share highlights of Africans' progress in overcoming this legacy.

Lessons from Home

In reflecting on the relationship between hunger and scarcity, we should also never overlook the experience of the United States. In 2006, the U.S. government chose to abandon the word "hunger" and replace it with "food insecurity" in the official count of the food-deprived. The U.S. Department of Agriculture defines food security as access "at all times to enough food for an active, healthy life."[56] Thus "food insecurity" is the lack of such access, affecting, as noted, one of every six Americans.[57] But would anyone argue that there is not enough food in the United States?

Surely not.

The United States is the world's leading agricultural exporter. For U.S. farmers, "overproduction"—which knocks down prices—is a persistent worry.[58] Plus, over a third of this country's enormous corn harvest, used for fuel, feeds no one.[59]

In the United States, just as in the Global South, hunger is an outrage precisely because it is profoundly needless.

Behind the headlines, the media images, the superficial clichés, we can learn to see that hunger is real; scarcity is not. Scarcity is a human creation.

With this clarifying evidence of food sufficiency along with vast, wasted potential, let's now turn to the other side of the equation: the number of people who need to eat. After virtually every public talk on hunger we've given over more than forty years, there's been one question we've had to be ready to answer: "What about population growth—isn't it the *real* problem?"

Clearly, many people who appreciate that there is more than enough food today still worry that, if population continues to grow, very soon there will not be.

So let's ask:

WHAT *IS* THE POPULATION PROBLEM?

As we examine the relationship between population and hunger, let's first register the obvious but often-overlooked absence of any link between population density and the extent of hunger.

Scanning the globe, we see population density in the European Union at about twice the world average; but the region has the least hunger. Now consider two regions that are home to most of the world's hungry people: India and sub-Saharan Africa. India's population per square kilometer of 416 is many times the world average of 54, while sub-Saharan Africa's density of 39 per square kilometer is considerably below the average.[60] Now imagine this comparison: Bangladesh's density is equivalent to half the entire U.S. population living in an area the size of Alabama, *yet* the total calories in its food supply could meet the needs of every citizen.[61]

Of course, in localities where people have been pushed off their land and forced to settle on fewer acres of less fertile land, the number of people per unit of land is likely to contribute to hunger. But in no way does such local injustice explain global hunger.

Yet we all must take seriously the continuing growth of the human population. For who would look forward to our species so dominating the planet that other forms of life were squeezed out, and all wilderness was subdued, and the mere struggle to feed and warm ourselves would keep us from more satisfying pursuits? Plus, of course, the size of the human population is one of the key variables in dealing with climate change.

The population question is so vital that we can't afford to be the least bit fuzzy in our thinking. So here we will focus on the most critical questions: Is human population growth "out of control"? And what are we learning about the link between halting population growth and ending hunger?

In the early 1950s, the global total fertility rate was 5. That's the average number of children a woman would bear if she were to live out her childbearing years and have children in line with the current age-specific fertility rates. This total fertility rate of 5 was well more than double the "replacement rate"—the point at which a population begins to level off and stops growing over time.[62]

Then, the 1968 best seller *The Population Bomb* by Paul Ehrlich delivered this frightening verdict: "The battle to feed all of humanity is over. In the 1970s the world will undergo famines—hundreds of millions of people are going to starve to death."[63] A few years later, ecologist Garrett Hardin called for a "lifeboat ethic," in which we must let some starve if the majority is to survive.[64]

People got really worried.

And so what has happened?

Food per person, as we've seen, kept climbing while at the same time, by the mid-1990s, the global fertility rate had dropped from 5 to 3. By 2010, globally, it reached 2.5. (The replacement rate is now 2.1.) More specifically:

- In the "more developed" regions as a whole, fertility rates—with major exceptions, including the United States—had dropped to 1.7 by 2010, well below the replacement rate.
- In Asia and Latin America, fertility has fallen steadily from around 5 in the mid-1970s to about 2.3 in 2010.[65]
- In Africa, the rate of fertility decline has been considerably slower, falling from more than 6.7 in the mid-1970s to 4.7 in 2012.[66] That's about where South Asia and Latin America stood forty years earlier, just as their accelerated fertility declines began.[67]

TOWARD A "DEMOGRAPHIC TRANSITION" FOR A SUSTAINABLE WORLD

All of this lines up neatly with the concept of "demographic transition," first observed in what are today's industrial countries over the two centuries preceding 1950. It works like this: As public health and living standards improve, mortality falls and population grows fast. But over time, fertility rates drop, and overall population growth slows, then stops.[68]

Demographers have observed a similar pattern in countries in the Global South as well. The two most populous countries on the planet, China and India, have experienced dramatic declines in their fertility rates. From the 1950s to 2010, China dropped from 6.1 to 1.7, and India from 5.9 to 2.5.[69]

Thus, the population transition in the Global South as a whole, again with exceptions, has occurred much faster than it did in the Global North.

And what about the future?

Can we get to replacement level fertility, while healing the Earth from our current damaging practices, without overwhelming food-growing resources?

Here's what the United Nations lays out: According to its "medium" projection, global population will grow to 9.6 billion people by 2050, or about a third more of us than in 2015. At that point the world fertility rate is predicted to be 2.2. But even at that level, our population would add another billion-plus people by 2100. By then, while estimates vary, the medium projection suggests we'd have reached an average fertility rate of roughly replacement level—2.0 births per woman.[70]

This big picture is vitally important to absorb, but when we think only in terms of "world population," we miss a lot. For example: Already almost half the world's people live in countries where fertility rates are below replacement levels.[71]

Even more dramatic: If the UN projections pan out, just *eight* countries, six of them in sub-Saharan Africa, will account for more than half of all population growth worldwide to 2100. Those eight are Nigeria, India, Tanzania, the Democratic Republic of the Congo, Niger, Uganda, Ethiopia, and the United States.[72] (The U.S. population increase is expected largely to reflect immigration.) And, within India, the population growth is occurring, not throughout the country, but primarily in nine states that are home to only about half the country's population. Already, eleven Indian states are at or below replacement level.[73]

> Already, almost half the world's people live in countries where fertility rates are below replacement levels.

Thus, we see that more than half of the increase in world population is actually occurring in just seven countries, plus nine states in India—together representing just a fifth of the world's population.[74] Beyond this group, the picture among low-income countries is extremely mixed but, overall, encouraging: Among 156 countries categorized as "less developed" and with at least ninety thousand people:[75]

- 32 have reached below-replacement-level fertility.
- Only 31 still have high fertility rates—5 or above—but these fast-growing countries constitute only 9 percent of the world's population.[76]
- Of the remaining 93 countries among those "less developed," only 11 show increasing fertility, while 88 percent—82 countries—show declining fertility rates.[77]

Combined, these trends lead to projections that all regions except sub-Saharan Africa will reach replacement level fertility by 2050.[78]

Globally, the United Nations predicts a doubling of the number of countries with below-replacement-level fertility by the middle of this century. If this prediction unfolds, by 2050 more than three-fourths of people on our planet, or 7.1 billion, will live in countries with below-replacement-level fertility rates.[79]

Taking all this in, we suddenly see not a "world" population crisis but, rather, a challenge in specific areas of our world—areas with high fertility rates where people are experiencing poverty, hunger, the oppression of girls and women, and other human rights violations. Think of the implications.

Roughly four in ten pregnancies worldwide in 2012 were unintended and almost a quarter of all births were unplanned. Three-quarters of these births occurred in Africa and Asia.[80] Thus, working toward a world where all families have access to contraception and the knowledge and power to avoid these births can move us toward a stable world population that our Earth can support.

Is this not further, powerful evidence that the population challenge is not about stifling an innate drive to reproduce—a daunting task!—but rather about achieving what people truly want: an end to poverty and powerlessness, and therefore the opportunity for real choice.

UN Predictions, Which Choices, Our Choices?

The UN projection taking us to 9.6 billion by mid-century is the one typically referenced in the media. It is the in-between estimate, neither the most pessimistic nor the most hopeful. In the next chapter we explore how humanity can ensure that food supplies continue to be ample if this projection unfolds.

But nothing here is set in stone. Note that population projections before have turned out to be too high because human beings changed things. Case in point? The medium estimate of 9.6 billion forecast for mid-century is a lot lower than the medium estimate of 11.2 billion projected by the United Nations forty years ago.[81]

Now, let's look at the United Nations' "low" projection—8.3 billion people at mid-century—and what it might take to get there.

Getting there is desirable for a slew of reasons. It would make more likely the possibility of halting humanity's destruction of the natural world and that core human rights, including access to food and reproductive choice, would be fulfilled. The good news is that we have clear evidence of what it takes to halt population growth, because it's already worked in most of the world. So the question is only whether humanity will step up and apply these clear lessons to the root causes of the remaining population growth.

Proven Pathways to Bringing Population into Balance with the Earth

To act wisely, as so many people want to, we need to identify the specific social conditions linked to movement toward ending population growth.

What are they?

A short answer is the expansion of basic human rights. They include education (particularly for women) and economic opportunity, as well as access to food, health care, and contraception.[82]

One health factor associated with fertility is the rate of child death. Understandably, women who experience the death of a child will on average give birth to more children.[83] And progress in avoiding the tragic loss of a child can be speedy: In Myth 6 we note the experience of a large Brazilian city that cut child deaths by 72 percent in only a dozen years.[84]

That women's education is associated with improved child survival also makes intuitive sense. For this and other reasons, it's no surprise that female education is associated with lower fertility rates, as indicated by a 2011 survey in Uganda: Women with no formal education have on average 4.5 children, whereas those with even a few years of primary school have three. Women with one or two years of secondary school average 1.9 children, and with more years of schooling, the number falls further.[85]

All of these advances interact, of course.

The relative powerlessness of women subordinated within the family and society is, arguably, one factor in high birthrates. Men make reproductive decisions in many cultures, research finds.[86] But, it's important also to consider the wider context in which such imbalance exists: Within societies characterized by extreme inequality, many men who hold power over women are themselves part of subordinated groups. As long as society's economic and political rules deny men sources of self-esteem and decent income through productive work, it is likely some will cling tenaciously to their superior status vis-à-vis women, and one manifestation will be more births than women would themselves choose.[87]

Thus, it is not surprising that greater gender equality is associated with lower fertility rates.[88]

"Schools for Husbands"

Among gender-equity concerns are the protection of women's health and the responsibility for the use of contraceptives. On these counts, consider what's happening in Niger, a West African nation ranking near the bottom of the UN Gender Equality Index, with both its fertility rate and its rate of maternal death among the world's highest.[89]

Just eight years ago, only 5 percent of couples there used modern contraceptives.

Since then, villagers have created 137 groups they call *Écoles des Maris* (Schools for Husbands) that are already making a difference. Members are chosen by a local social-benefit organization because they are trusted by their communities and their wives use local health services. The men meet twice monthly to discuss problems and solutions, then "become guides and role models within their own family and among others in the community," reports the UN Population Fund.[90]

The "schools" were launched in 2008 in response to the Population Fund's survey identifying "men's dominance and attitudes to be one of the major obstacles to women taking advantage of reproductive health care."

Since the *Écoles des Maris* began, the "use of family planning services has tripled [and] the number of childbirths attended by skilled health personnel has doubled." Still, few couples in Niger use birth control. But change is under way: "When the schools started, only 5 percent of women in Niger reported using contraception. Now that figure is up to 13 percent."[91]

The People-to-People Effect

The Schools for Husbands tap a force for change identified by collaborating researchers in southern India and Bangladesh. They came up with another key variable in lowering fertility after discovering that certain fertility rates didn't line up with widely recognized economic and educational-status fertility determinants.

People accept contraceptive use, the researchers found, based on the "sum total of . . . interaction within the community." They tracked a number of indicators mentioned above and noted the "sudden decline in fertility in [the Indian states of] Tamil Nadu and later in Andhra Pradesh without significant improvement in female literacy or decline in IMR [infant mortality rate]." Strongly associated with lower fertility, however, was what these researchers call "diffusion."[92]

So what is *diffusion*?

It is the people-to-people effect. Fertility rates fall when norms change as women participate in groups of all sorts, including micro-credit groups, or self-help groups more generally. Meeting together, the women absorb one another's experience and exchange useful information about contraception. In the early 2000s, notes the report, the southern Indian state of Andhra Pradesh (now divided into two states) had four hundred thousand micro-credit groups, almost 40 percent of all such groups in

> The "Diffusion" Effect
> What matters most in our reproductive behavior may be the influence of others in our close social circles.
> —drawn from the work of the Centre for Economic and Social Studies and Bangladesh Institute of Development Studies

India.[93] Today, its fertility rate is 1.8 and its poverty rate is less than 10 percent, both far below India's overall averages.[94]

In other words, what matters most to our reproductive behavior, these researchers posit, may be the influence of others in close social circles. Through social connections, women gain a greater sense of their own power over the most intimate matters of reproduction. This finding reinforces observations that self-empowerment through learning in groups can be transformative. Research on small self-help groups of women in Bangladesh, for example, suggests that members' lower fertility rate reflects not just greater economic security but the influences of being in a group, which helps to bring independence from control by family members.[95]

The population question is indeed complex. Even in a few low-income societies with great economic inequality, one can find low fertility rates. Examples in Latin America include Brazil and Costa Rica, both of which are at below-replacement-level fertility; and Argentina and Mexico, where rates are relatively low.

But lest we leave our readers imagining that the lowering of birthrates in the Global South invariably represents a positive transition—including progress for women—consider the experience of China and India.

When Pressure to Lower Birthrates Puts Women at Risk

First, China. Its "one-child" policy, most stringently enforced from 1979 to the late '80s, continues in some form to this day.[96] At its height, women were forced to abort second pregnancies, and those who gave birth to females, who were less favored, were sometimes scorned by their families. Because female fetuses were disproportionately aborted, today China faces a highly uneven sex ratio at birth of 118 boys for every 100 girls.[97]

Then consider India. Although it is home to 17 percent of the world's people, more than a third of female sterilizations worldwide occur there. In fact, female sterilization accounts for two-thirds of the country's contraceptive use.[98]

Sterilization can be a good choice—when it is a *choice*.

In India, however, "choice" gets murky when at least some government health workers must meet sterilization quotas or risk having their salaries cut, and when very poor women are paid or offered gifts if they agree to sterilization.[99]

In 2014, at least thirteen women died and many others were sickened in India's "sterilization camps." At one, the doctor performed eighty-three operations in six hours. It was "nearly impossible to limit the number because recruiters generated such a crowd," noted a doctor quoted by the *New York Times* editorial board. Each woman was paid 1,400 rupees, roughly $22, to undergo the procedure.[100]

"I did it out of desperation," a twenty-five-year-old mother of three told a Bloomberg News reporter in 2013, as she lay recovering from the procedure on a concrete floor.[101]

Interestingly, in the 1970s it was a public outcry over forced male sterilization that led to a spotlight being put on female sterilizations.[102] So, while studies report that male sterilization by vasectomies is both less risky and less costly, in India today only 1 percent of men are sterilized, compared with 37 percent of women.[103]

On all sides of the population question, there is a lot to learn. But one clear pattern stands out:

Neither population density nor population growth is the cause of hunger. Rather, the two often occur together because they have similar roots in extreme power inequalities. So, let's now turn to a few examples of the rapid slowing of population growth as societies evolve to correct those imbalances by securing basic economic and social rights.

Lessons from Country Successes

Some of the earliest and most spectacular fertility declines occurred in the context of broad-based improvements in nutrition, health, and education. Let's look at some of those examples.

Sri Lanka. Significant decline in the fertility rate began from the time of Sri Lanka's independence in 1948 to 1978. Over those thirty years, the government supported citizens' access to basic foods, notably rice, through a combination of free food, rationed food, and subsidized prices. Since then, continued fertility decline has been associated with increased availability of contraception and the rise in age of marriage, linked to longer female schooling. In the early 1980s, the fertility rate was 3.2, but by the late 2010s it stood at 2.3, just above replacement level.[104]

Cuba. From the start of the revolution in 1959 till the acute economic crisis of 1989, rationing and price ceilings on staples kept basic food affordable and available to the Cuban people. All citizens were guaranteed enough rice, beans, oil, sugar, meat, and other food to provide them with 1,900 calories a day.[105] As health care and education also became available to all, Cuba's fertility rate fell from almost 4.2 in 1955 to 1.5 in 2010.[106]

Two States in India—Himachal Pradesh and Kerala. In the small northern state of Himachal Pradesh, 90 percent of the population is rural; yet its per capita income and education levels are among India's highest. "Strong community involvement," notes a World Bank study, with high levels of female participation and local accountability, have helped the state's health and other services reach even "far-flung villages."

Among the roots of the state's success are "land reforms in the 1950s and the 1970s [that] laid the early foundations for social inclusion," adds the report. Today, almost eight in ten rural households in Himachal Pradesh own some land.[107] And its fertility rate? On a par with Denmark's, at 1.7.[108]

At the opposite end of the country is Kerala, also a small state and one of India's most densely populated; yet judged by measures of well-being, including infant mortality and life expectancy, Kerala is superior to most low-income countries as well as to India as a whole.[109]

The state's 1969 land reform law "abolished tenancy in both rice land and house compound plots," notes Kerala analyst Richard Franke, who adds that the reform's major consequences include the "abolition of landlord and tenant classes, reduction in inequality of land ownership and income as measured by the Gini index, and reduction in caste inequality."[110] Kerala's female literacy rate is over 90 percent, and its fertility rate of 1.6 is below that of the United States.[111]

Some critics downplay the state's progress, noting that it is facilitated by large remittances sent home by Kerala citizens working abroad. What makes Kerala impressive to us, however, is what its citizens have chosen to do with their resources.[112]

In addition, Thailand and Costa Rica deserve our attention. In both, health and other social indicators offer clues as to why they experienced early declines and now have fertility rates below replacement. Infant death rates are relatively low, especially in Costa Rica, and life expectancy is high—for women, it is eighty-two.[113] Perhaps most important, in Costa Rica an unusually high proportion of women are educated, and in Thailand proportionately more women work outside the home than in most countries in the Global South.[114]

The experiences of these countries confirm that advances in nutrition, health care, and health outcomes, as well as education and employment for women, typically are associated with a decline—sometimes quite rapid—in fertility rates.

THE CHALLENGE AHEAD

In this chapter, we've outlined what we believe are critical points too often muddled in discussions of food availability and population:

- Today, the world produces more than enough for everyone to thrive, yet roughly one-quarter of the world's people suffer nutritional deprivation; and at the same time we waste a quarter of all calories.
- In no country does population density explain hunger.
- Continuing population growth is a critical challenge, but it is not the root cause of today's hunger. Most often it is—like hunger—an outcome of inequities that deprive the majority, especially poor women, of basic human rights to food, security, economic opportunity, health care, and education.
- Fertility and population growth rates have dropped dramatically. More than half of all remaining growth is occurring in just seven countries plus nine states in India. Together they represent one-fifth of the world's population.

- Countries with high fertility rates—of 5 or more—are home to only 9 percent of the world's total population; and most of the "less developed" countries show declining fertility rates.
- All of these trends lead to the (medium) projection that by 2050 the world as a whole will reach a fertility rate of 2.2, on the path to a replacement level of 2.1.
- Fertility rates decline as women gain access to education, health care, and employment. They also fall as women participate in social groups in which they acquire a range of attitudes, skills, and knowledge from each other and gain resources that enable them to make more independent choices.

The biggest lesson we take from these points is this: Precisely because population growth is such a critical problem, we cannot waste time with approaches that do not work. To be serious about bringing human population into balance with the natural world and with our food-producing resources, we must address the unfair structures of economic and political power—from the local to the global level—that lie at the root of the crisis.

To attack high birthrates without attacking the causes of poverty, hunger, and the disproportionate powerlessness of women is fruitless.

It is a tragic diversion our small planet can ill afford.

myth 2

Climate Change
Makes Hunger Inevitable

MYTH: There may well be plenty of food today, but climate change is a game changer. More frequent extreme weather and rising temperatures threaten agriculture. Climate change not only makes ongoing hunger virtually impossible to overcome but also means more and more episodes of outright famine.

OUR RESPONSE: The hard reality of climate change is certainly no myth. It's already destroying lives and unraveling ecosystems. In no way, however, does this challenge make hunger and famine inevitable. It turns out that within our inefficient, inequitable, and climate-harming food system is vast scope for both increasing the food supply and addressing the roots of hunger. And here's some really encouraging news: Changes in food and farming that best address global climate change are precisely those that most benefit the world's hungry people, the environment, and everyone's health.

Let's start with what we know about climate change: "Nobody on this planet is going to be untouched by the impacts of climate change," warned Rajendra Pachauri, former chair of the Intergovernmental Panel

35

on Climate Change (IPCC), upon the release of its 2014 report.[1] As climate change brings us more chaotic weather and rising temperatures, farming and fishing face new challenges.

- The IPCC expects that crop productivity could decline by as much as about 13 percent after 2030 to the end of the century.[2] Heat and water stress, the panel reports, are already threatening crops.[3]
- "[S]evere drought conditions by the late half of this century" are expected "over many densely populated areas," write scientists in *Nature Climate Change*, including "Europe, the eastern USA, Southeast Asia and Brazil."[4]
- South Asia and southern Africa are likely to experience the greatest impact, predicts the IPCC, with much of Africa's cropped area expected to be outside normal temperatures by 2050.[5]
- In response to warmer temperatures, crop pests are expected to expand into new regions.[6]
- Scientists warn that elevated atmospheric carbon dioxide can reduce nutrients such as zinc and iron in some widely eaten crops.[7]
- Climate change is making oceans more acidic, and this effect could reduce the supply of seafood on which a billion people in the Global South depend.[8]

Seeing these food-related trends, the World Food Program (WFP) expects "the number of malnourished children to increase by 24 million by 2050, or about one-fifth more than without climate change."[9]

As humanity continues to increase greenhouse gas emissions while undermining the Earth's capacity to absorb carbon, these expert observations form a powerful call to immediate action. *But they are a far cry from a verdict that hunger and famine are inevitable.* We alive today will largely determine whether such adversity can be turned into opportunity—not just limiting misery and dislocation but creating more life-supporting societies.

If we let ourselves be ruled by fear and respond unreflectively to dire predictions, we could miss positive possibilities lying within the huge climate puzzle that we explore in this chapter. Here are three:

Within our food system is vast room for improvement. Precisely because today's food system is so inefficient, destructive, and inequitable, there is vast room to improve food availability before we test Earth's actual capacity to feed us. Achieving greater efficiency and equity would mean not only enough food for all, but also healthier food, even as we respond to warmer, more erratic weather.

The food system has a unique capacity to help rebalance the carbon cycle. Notice that we cast our challenge as one of "rebalancing," not just reducing, to remind us that our task is twofold: cutting emissions *and* storing more carbon in the soil in order to reach a balance of atmospheric carbon supportive of life. Today, we're pushing this cycle out of balance.

Our entire food and agricultural system—from land to landfill—is itself, climatically, a big troublemaker, estimated to account for up to 29 percent of total greenhouse gas emissions.[10] Most startlingly, emissions from food and agriculture are growing so fast that, if they continue at this rate, in thirty-five years they alone could nearly reach the safe target set for *all* greenhouse gas emissions.[11] That projection should get us moving!

At the same time, remade to better serve us, our food system would not only reduce emissions but also store more carbon, becoming not a climatic curse but a cure.

Strategies addressing climate change can directly help reduce hunger. Many changes necessary for the food and farming system to contribute to climate solutions also hold special power to reduce hunger. Why? These farming practices are low-cost and especially benefit small-scale farmers and farmworkers, who are the majority of hungry people.[12] Here, and especially in Myths 4 and 10, we highlight how farmers are transforming their practices to yield more and better food while at the same time reducing greenhouse gas emissions, becoming less vulnerable to extreme weather, protecting diverse species, and storing more carbon.

These points suggest a bundle of positive synergies. Before we explore them, let us lay out the main elements of the chapter before you.

First, in "Famine Isn't Fate" we argue that the tragedy of famine, typically blamed on weather extremes, actually lies in human-made vulnerability. Second, in "The Hidden Food Supply" we examine the untapped and unappreciated potential mentioned above. Third, in "From Curse to Cure" we lay out the impressive changes in farming and eating that humanity can make—and has begun making—to contribute to meeting the climate-change challenge.

From there, we show—primarily through stories—how people the world over are "Reducing Vulnerability to Both Climate Change and Hunger," and then close the chapter with "Reflections on Vulnerability and Human Agency."

In these explorations, we share what we find surprising as well as motivating. Now let's begin by examining the most frightening element of this myth—that climate change makes famine inevitable.

FAMINE ISN'T FATE

Throughout history, famines have been blamed on nature. Take the infamous Irish Potato Famine, which between 1845 and 1852 killed about a million people and forced perhaps as many as two million more to emigrate. It was caused by a virulent potato blight—or was it? That same blight devastated potato crops across Europe, while mass starvation occurred only in Ireland.[13]

So why only Ireland?

A letter to the British prime minister from an observer of events in Ireland in 1846 sheds some light: "For 46 years the people of Ireland have been feeding those of England." They "exported their wheat and their beef in profusion, [while] their own food became gradually deteriorated . . . until the mass of the peasantry was exclusively thrown on the potato."[14] Impoverishment under British rule had made the people of Ireland more vulnerable to the blight.

Vulnerability is largely under human control. A much more recent tragedy brings this point home. In 2011, a severe drought hit the Horn of Africa, including Djibouti, Kenya, Somalia, and Ethiopia. More than thirteen million in the region were heavily affected.[15] In Somalia roughly 260,000 people perished, half of them under the age of six.[16] Another 300,000 Somalis were forced to flee their homes in search of food and security—many swelling refugee camps in Kenya and Ethiopia.[17]

A tragic mix of events—from civil conflict to extreme weather and fast-rising food prices—triggered the crisis, says the UN Food and Agriculture Organization (FAO).[18]

The worst drought in sixty years in some areas followed a decade of periodically poor rainfall throughout the region. Then, in the two years immediately before 2011, rains in the eastern areas utterly failed.[19] Most scientists blame in part the recurring meteorological phenomenon known as La Niña.[20] There's no consensus on whether climate change is worsening La Niña's impact, but climate scientists do agree that a warming planet makes extreme weather more likely.[21]

Once hunger became severe, civil conflict made the government's weak response even less effective. Islamist militants in south Somalia limited aid; and despite an "early warning" system, international help was slow to respond.[22]

These woes led the price of food in the region to rise—it went up by almost half between 2009 and 2011, a leap about twice the world average.[23] In Somalia, it was a lot worse: Local grain prices in several areas rose more than fourfold.[24]

Seeing starvation, many people naturally assumed that food production in the region had fallen precipitously. It had not.

Certainly it's true that for some time the famine-stricken countries had been producing fewer calories per person, compared with other regions in Africa; yet, with the exception of Somalia, food supplies in 2011 met minimum energy requirements in each of the affected countries, reports the FAO.[25] In fact, the famine hit after a decade in which calories per person in the Horn of Africa had increased modestly.[26] Even more counterintuitive is the fact that during the famine, when Kenya was importing basic foods, its export of green beans destined for Western supermarkets increased threefold.[27]

Additionally, record-breaking harvests elsewhere in Africa could have been tapped to address the famine. At the time, Zambia had such a surfeit of corn that much of it ended up rotting for lack of storage.[28]

Thus, even in this life-shattering famine, drought-reduced supply was an exacerbating factor, not the cause.

Famine also reflects lack of preparedness. In years of adequate food supply, governments accountable to citizens' well-being could have established at least minimum stores of food to release in such tough times.

Consider the difference that preparation made for people in Ethiopia compared with Somalia during the famine. In Ethiopia, what aid agencies call "prepositioned food supplies" made rapid response possible;

and Ethiopia's safety net programs—such as one compensating low-income workers for their labor in public projects—also helped reduce drought-related suffering.[29]

Globally, commonsense preparedness is equally critical, and more so as climate change intensifies. Arguing for increased reserves, veteran food and environment analyst Lester Brown points out that "more extreme weather events" arriving with climate change mean that "[i]f stocks equal to 70 days of grain consumption were sufficient 40 years ago, then today we should plan on stocks equal to at least 110 days of consumption."[30] But in 2015 world grain stocks are well below this level.[31]

Lack of food can't be the reason that world stocks are low, given the magnitude of the global food supply documented in the previous chapter. Throughout this book, we explore the real reasons.

Reflecting on Famine

The 2011 tragedy in the Horn of Africa echoes earlier famines blamed on nature.

In 1974, a famine in Bangladesh took tens of thousands of lives even though there had been an increase per person in food grain availability, reported Nobel Prize economist Amartya Sen.[32] Rumors of shortfall spread anyway, prompting well-to-do farmers and merchants to hoard food.

As two longtime observers of rural Bangladesh commented on the famine, "While to most people scarcity means suffering, to others it means profit."[33] One peasant described what happened in her village: "A lot of people died of starvation here. The rich farmers were hoarding rice and not letting any of the poor peasants see it. There may not have been a lot of food, but if it had been shared, no one would have died."[34]

For those living in countries like the United States, it's easy to assume that famine deaths "over there" result from natural catastrophes, so it's worth pausing to take in this news item from the *Chicago Tribune*, January 22, 1994: "Man Dies: Found in Unheated Home." The article called it "the fourth fatality of [the] week's cold wave." Surely, the writer did not believe that weather was the cause of the four deaths.[35] In the United States, roughly seven hundred people, most of them homeless, die from cold exposure every year.[36] Here it's clear that weather is the trigger, not the cause.

In no way, of course, do we downplay climate-change impacts already showing up in more frequent and severe droughts and floods, nor the additional changes predicted in the beginning of this chapter. Nevertheless, whether in Ireland, Kenya, Bangladesh, or Chicago, let's keep our sights clear: Those who die of famine have been made vulnerable to weather extremes by human-made structures of power denying them access to adequate food, shelter, and protection.

This perspective can help us gain clarity as we now turn from famine to the relationship between climate change and day-in-and-day-out nutritional deprivation. But first note the imbalance between famine and chronic hunger. In 2004, "humanitarian emergencies," notes Christopher Barrett in *Science*, amounted to "8 percent of hunger-related deaths worldwide." Meanwhile, "92 percent were associated with chronic or recurring hunger and malnutrition."[37]

THE HIDDEN FOOD SUPPLY

So what do we know so far about what we can do in the coming decades to ensure sufficient, healthy food for all people?

From IPPC and the FAO, we learn that we'll probably need to increase food production almost 60 percent by mid-century.[38] This projection feels downright alarming—especially if getting there would entail destroying yet more forests, along with putting even greater pressure on diminishing available water, and more.

So here's an obvious question: If over this period world population is expected to grow by a third, why, according to the climate change panel's midline projection, do we need so much more food?

Part of the answer is an unquestioned assumption behind these projections: that "demand" for animal foods fueled by "prosperity" is an inexorable process. Yet this demand actually reflects the increasing market power of the minority at the top of the economic ladder to subsidize with public funds and personally afford grain-fed animal foods—especially beef—thereby outbidding the majority in the global marketplace and at the same time, as we'll explain, shrinking the total food supply.

Whether we accept this scenario as a given is key to our well-being in a climate-challenged world. Livestock is a critical piece of the climate puzzle, and we'll return to its role later in this chapter, after we explore another aspect of the hidden food supply potentially available to us.

Trash or Treasure?

The assertion that our only choice is to greatly increase production also skirts this big, immediate, largely ignored bird in the hand: food waste. As climate change handicaps food growing in many areas, waste should grab our attention, for what is it but a margin we can use to compensate for any losses without further disrupting the Earth?

In Myth 1 we noted that one-third of all food produced worldwide never reaches our mouths. In the United States, it's a jaw-dropping 40 percent.[39]

And why such vast, but largely hidden, waste?

In the Global South, much waste stems directly from the poverty of small farmers. Of course, impoverished farmers must conserve all they can. The problem is that they often can't afford proper storage, and this problem results in losses of at least 12 percent and up to 50 percent for fruits and vegetables.[40] Another source of waste is a dearth of rural markets, and roads to reach them, sometimes making it futile—especially given inadequate storage—for poor farmers to harvest additional crops once family needs are met.[41]

And in the Global North, food waste is everywhere. One image says a lot: In a U.K. carrot-processing factory featured in a UN report about common sources of food waste, a machine equipped with special sensors tosses into a livestock feed container any carrot that's not arrow-straight![42]

Try this thought experiment. Posit that by mid-century we cut food waste worldwide by half: Instead of the current estimated 33 percent wasted, we get food waste down to, say, about 15 percent. The food made available by this success alone equals roughly enough to feed a billion people.[43] So even with no other gains, our progress against waste would have more than compensated for the 13 percent decline in agricultural productivity the IPCC suggests is possible.

So how do we do it?

Solutions to food waste abound, and some directly address hunger: Affordable loans to help poor farmers secure low-cost crop storage containers and public investment in market roads are obvious paths. Important, too, are community granaries where farmers can store grain safely for the lean season. In northern Cameroon, on the edge of the arid Sahel region, the WFP has helped establish 410 village granaries.[44] Processing helps prevent waste, too. Solar energy dryers for mangoes, for example, are helping reduce losses in western Africa, where more than a hundred thousand tons of the fruit go bad each year.[45]

Another step is to eliminate waste created when good food—those crooked carrots?—is discarded for minor irregularities. In France, what's playfully called the "Ugly Fruit" movement encourages markets not to toss out their blemished produce but to sell it at a lower price. The first stores to try it offered discounts of 30 percent, and shoppers ate it up![46]

On the theme of "matchmakers pair lonely leftovers," the *New York Times* reported in 2014 that in Germany anyone can leave or pick up free food at about a hundred food-sharing sites. And at Foodsharing.de, members can connect online to share food instead of tossing it.[47]

In this war on waste, though, the U.K. may have taken the lead. Its "Love Food Hate Waste" campaign, now active in ten cities, includes "Save More" educational packets offering tools and tips about both wasting less food and saving money. Since 2007, the "Love Food Hate Waste" campaign boasts that its efforts have cut U.K. food and beverage waste by 21 percent.[48]

And then there's the United States, where the average family of four throws out $1,600 of food each year. So the EPA—with partners ranging from schools and restaurants to grocery stores—launched the Food Recovery Challenge initiative. In 2013, participants saved 370,000 tons of food from landfills and incineration—that's more than two pounds for every American. Even Disneyland joined in.[49]

Overall, cutting food waste by half seems like an eminently achievable goal.

Of course, compensating for climate-caused crop shortfalls isn't enough to meet the challenges ahead. Our population is still growing, with the UN's median projection leading to more than three billion additional people by the end of the century. Let's not forget, however, that if humanity really steps up to spread proven solutions to elevated fertility, covered in Myth 1, we could radically alter that outcome. In fact, the UN suggests it would be possible to arrive at the turn of the century with a world population lower than today's.[50]

Fortunately, there is much more we can do beyond cutting literal waste to make sure that good food is available for all as we work to slow population growth and come into equilibrium with the Earth. We'll suggest several approaches that not only protect against further environmental harm but can lead to a healthy planet.

Certainly, we can cut the enormous but much less visible *built-in* waste.

Livestock, a Food-Factory-in-Reverse

In ensuring our food supply, now take in the great room for improvement captured in these striking numbers: Worldwide, three-fourths of all agricultural land, including pasture, is used to produce animal products.[51] And from all this, what do we get? Just 17 percent of our calories.[52]

To understand why humans get so little, consider livestock's "take":

> Worldwide, ¾ of all agricultural land, including pasture, is used to produce animal products—from which we get just 17% of our calories.
>
> —drawn from *Trade and Development Review*, 2013; Herrero and Thornton, *Proceedings of the National Academy of Sciences*, 2013

About half the world's calories from crops don't go to people. Instead they go primarily to feed livestock—which consume a third of the world's grain and 85 percent of soy—and into agrofuel production and other industrial purposes.[53] All this leaves only about half of all crop calories—and a shrinking share—for people to eat directly.

Let this sink in. Then note also that not all livestock are created efficiency-equal. Beef comes in last: Of the calories that cattle eat in feed, humans get a measly 3 percent in the beef we eat. The accompanying table reveals the big differences in the capacity of livestock to convert what they eat into calories we eat. Dairy is about thirteen times more efficient than beef, chicken about four times.[54]

> **Shrinking Calories**
> Percent of feed calories eaten by livestock that end up as calories humans eat:
>
> Dairy = 40%
> Eggs = 22%
> Chicken = 12%
> Pork = 10%
> Beef = 3%
>
> —Cassidy et al., *Environmental Research Letters*, 2013

Given this extreme inefficiency, it shouldn't surprise us that livestock-centric U.S. agriculture—viewed by many as the pinnacle of efficiency—actually feeds fewer people per hectacre, 5.4, than either Chinese, 8.4, or Indian, 5.9, agriculture, both of which are less meat-focused.[55]

What a jarring contrast to the assumptions of so many Americans about the superiority of U.S. agriculture!

But, of course, livestock were not always shrinking our food supply, or humans could never have made it to seven billion. Throughout human evolution animals have converted grass and other things we don't eat into high-grade protein we do eat—a big boon for humans. But over time we have remade livestock into nutrition disposals.

This we can change.

We can tap the huge potential gain to be realized by shifting even modestly away from heavily grain-fed-meat-centered diets. (We revisit this theme and its nutritional implications in "Eating with the Earth" below.)

The food-supply implications of such a shift are striking. Worldwide, converting just half of crops fed to livestock into crops for humans could yield enough food for two billion people.[56]

> Worldwide, converting just half of crops fed to livestock into crops for humans could yield enough food for two billion people.

And here at home, if Americans on average were to eat one-quarter less meat and stop diverting cropland into agrofuel production (ethanol)—which offers no reduction in greenhouse gas emissions compared with oil—more than one in five U.S. cropped acres would be released.[57] That's enough to supply food for more than 231 million people.[58]

Looking at the impact of what we eat from another angle, what would be the result of shifting modestly from beef to other animal foods?

Replacing even a fifth of the beef eaten globally with the more efficient pork or poultry would reduce the *total* agricultural area needed for all to eat in 2030 by about the same amount—a fifth—according to scientists at Sweden's Chalmers University of Technology.[59]

That freed-up land could be used to grow food for people, or it could be used to regrow forests and other carbon-sequestering vegetation, as we discuss below. (Carbon sequestration refers to the capture and storing of carbon by plants and soil.) Scientists estimate that enabling cropland to transition to secondary forest can increase soil carbon stocks per unit of land by more than half.[60]

Let us be clear that these examples are "thought experiments" to indicate the reality of plenty and the possibilities open to us. We know such changes can happen not by individual choice alone, but only if we step up together to change policies governing our food systems.

Narrowing the Yield Gap

There's yet another overlooked angle on how humanity can step up to ensure supply, especially for the most vulnerable, even as climate change challenges agriculture.

Worldwide, many farmers' yields are far below their crops' proven potential—bad for them and a big waste for all of us. Climate change isn't the main reason. It's that so many farmers are resource-poor and lack economic and political power. They're often working degraded soils without access to the training, affordable credit, and tools needed to build soil fertility while conserving water, using the low-cost practices described in Myth 4.

Breakthroughs among such farmers are happening, but today huge "yield gaps" remain between what is and what's already proved possible.[61]

The gaps are so enormous that if they were to shrink for sixteen major crops, achieving an average yield of even 75 percent of known potential, the global food supply would expand enough for roughly two billion people, on the basis of data reported by University of Minnesota scientists in *Nature*.[62] Other scientists at the same university report in *Science* that yields of seventeen major crops are so low in many regions that getting them up to even *half* of what's attainable would add food sufficient for about 850 million people.[63]

And the great news? We have every reason to believe that such yield improvements are achievable using the ecologically sound farming practices we describe in Myth 4.

Thus, as we face the challenges of climate change, at least three wide avenues offer possibilities for tapping the "hidden food supply" available to us: reducing food waste, shifting a portion of today's meat-centered meals to plant-centered meals, and narrowing the gap between current crop yields and what we know is possible. Taking advantage of these three avenues, we can work to ensure a food supply for all in the face of climate change—*without* expanding cropped land. We can also, we hope, restore some land to natural vegetation, all while offering health benefits.

As we'll now explain, these changes bring climate benefits, too.

FROM CURSE TO CURE

So far we've explored ways that humanity can unleash the potential food supply now hidden from view, in order to ensure enough food for all in the face of climate change. But there is much more to the connection between food and climate.

Few yet realize that just over a third of human-caused global greenhouse gas emissions—taking into account the different potencies of gases—is not from carbon dioxide released by burning fossil fuels.[64] Of this third, the largest piece is from agriculture, estimated at nearly 20 percent of all human-caused greenhouse gas emissions, and thus a much greater climate-changer than transportation.[65] (And this estimate may be too conservative, as the UN Environmental Program [UNEP] estimates the world's agriculture emissions to be one-third higher.)[66] Of the nearly one-fifth contributed by agriculture, the most significant emissions are:

- *Carbon dioxide* released in deforestation and subsequent burning, mostly in order to grow feed, and from decaying plants.
- *Methane* released by ruminant livestock, mainly via their belching (ruminants, such as cows, goats, and sheep, have a specialized stomach, the "rumen," where feed is fermented); and, in addition, the methane released by manure and in rice paddy cultivation.
- Plus *nitrous oxide*, largely released by manure and manufactured fertilizers.

Note that these last two—methane and nitrous oxide—pack a particularly dangerous punch: Over a hundred-year period, methane per unit emitted is thirty-four times worse as a climate-heater than carbon dioxide. For nitrous oxide, make that about 300 times worse.[67]

Here we emphasize the often-overlooked climate impact of producing food; yet eating increasingly also involves transportation, processing, packaging, refrigeration, storage, wholesale and retail operations, and waste—all of which emit greenhouses gases. So, from land to landfill, what is the food system's climate impact?

Unfortunately, a dearth of "comparable data" makes it impossible to answer this vital question with confidence, an FAO officer tells us.[68] Nonetheless, a trio of international environmental scientists used what data are available to arrive at an estimate: The total food system's contribution to global greenhouse gas emissions could be as high as 29 percent. CGIAR, a global partnership of fifteen food-security research centers, also relies on this estimate.[69]

However, as we explore below and throughout this book, producing food doesn't have to contribute to climate chaos. Indeed, the way we grow food can actually help *solve* the climate crisis.

A Harvest of Greenhouse Gases

Worldwide, we've moved toward an industrial model of agriculture that destroys forests and uses manufactured fertilizer and other inputs, resulting in two to three times more carbon dioxide released per unit of land than an organic farming model.[70] We explore these contrasting models in the next two chapters.

And keep in mind that food-system emissions include those arising from what never even makes it into our mouths. In fact, we waste so much food that, if it were a country, food waste would be the world's third-largest greenhouse gas emitter after China and the United States.[71]

Additionally, as noted above, agriculture contributes to climate change when its expansion destroys forests. Burning forests emits carbon dioxide, and clearing forests can release soil carbon if that land is cultivated or becomes degraded. Since the beginning of the industrial era, human action has removed a third of the earth's carbon-trapping forest cover. Imagine a forested area roughly the size of South America destroyed largely to grow crops.[72]

> **Food Waste = Climate Killer**
>
> Globally, food waste—including greenhouse gases generated in producing all the food we *don't* eat—is so enormous that if it were a country it would represent the world's third-largest greenhouse gas emitter after China and the United States.
>
> —drawn from FAO, "Food Wastage Footprint," 2013

And it continues.

At the current rate, by century's end we will lose additional forests covering an area nearly one and a half times the size of India, and with it, significant capacity for taking up carbon.

Turning native forest to cropland can reduce soil carbon by more than 40 percent.[73]

In all, we're creating a mode of industrial farming that is storing less and less carbon. Soils managed with synthetic pesticides and fertilizers hold about 30 percent less organic carbon than do organically managed soils.[74] (A note to clarify the different uses of "organic" here. "Organic carbon" simply means carbon associated with living matter, while "organically managed" refers to a set of farming practices we describe in Myth 4.)

Livestock's Climate Liability

So far, we've highlighted the wasted potential built into the extreme inefficiency of grain-fed livestock. Now, let's explore livestock's role in the climate crisis.

Overall, livestock are responsible for at least 14.5 percent of all human-generated greenhouse gas emissions, reports the FAO—not only in their raising, but also via transport, processing, refrigeration, and more. Livestock-related emissions are considered "human-generated" because livestock are bred and raised by people. One peer-reviewed study arrived at a larger estimate: that livestock production contributes about one-fifth of global greenhouse gas emissions.[75] Much more research is needed on these knotty questions.

> Global GHG impact of livestock populations:
> Goats 4%
> Pigs 5%
> Sheep 9%
> Dairy 17%
> Beef 54%
>
> —drawn from Caro et al., *Climatic Change*, 2014

In any case, what's clear is that every bite of hamburger connects us to a vastly bigger greenhouse gas impact than does, say, a mouthful of bean chili or a veggie burger.

Of course, not all livestock are equally guilty. The contribution to climate change depends on the animal's size and number, as well as whether it is a ruminant and therefore emits methane.

Taking all this into account and looking at the planet's livestock populations, the greenhouse gas contribution of dairy is 17 percent, sheep 9 percent, pigs 5 percent, and goats 4 percent—whereas beef accounts for 54 percent of livestock's total greenhouse gas emissions.[76]

Measured differently—by GHG emissions *per unit of food*—the climate change "hoofprint" also varies dramatically by type of livestock, as shown in the "Eaters' Guide" table (page 56). Producing a pound of lamb or beef averages from about twenty to almost fifty times greater climate impact than does producing high-protein plant foods.[77] These estimates take into account "both direct and indirect environmental effects from 'farm to fork' for ruminants, including gas released in their digestion, manure, feed, fertilizer, processing, transportation and land-use," write ecologist William Ripple and colleagues in the journal *Nature Climate Change*.[78]

And then there is the water impact of livestock on our warming planet.

In many areas climate change is making water more precious, and irrigation claims nearly 70 percent of freshwater used by humans. Much of it is for producing livestock, which are big water guzzlers.[79]

In drought-plagued California, for example, meat and dairy account for almost half of the state's entire water footprint.[80] Nearly a fifth of its irrigation water goes to one feed crop, alfalfa.[81] So even as water scarcity worsens, every year 100 billion gallons of California water in the form of alfalfa go to China for meat production there.[82] More than half of water used in the Colorado River basin, reaching six states including California, "is dedicated to feeding cattle and horses," reports the Pacific Institute.[83]

In all, raising a pound of beef uses almost fifty times more water than growing a pound of vegetables, about forty times more than potatoes and other root crops, and about nine times more than grain.[84] Or consider this: One could bathe daily for more than a month with the water used to produce a pound of beef!

> Livestock production is the "single greatest threat to overall biodiversity."
> —Machovina and Feeley,
> *Science*, 2014

But livestock's often unappreciated impact is greater still. Largely through the destruction of natural habitats, human expansion of livestock production is the "single greatest threat to overall biodiversity," concludes a study in *Science*, and is causing the extinction of plant and animal species worldwide.[85] For example, below we note deforestation in Brazil, in part to expand beef production. In the next chapter we take up the threat to a sustainable food supply posed by this shrinking biodiversity.

Putting these pieces together, we see that both the extent to which humanity depends on livestock and the type of livestock we choose to eat really matter as we work toward a world in which we can avoid climate catastrophe and everyone can eat well.

Solutions at Hand, Restoring the Balance

Now to the positive, as we work to restore the carbon balance both by cutting emissions and by increasing sequestration.

First of all, shifting to ecological farming practices itself reduces greenhouse gas emissions, as we explain in Myth 4. Take rice, which provides one-fifth of the world's calories.[86] As noted, rice paddies are big contributors of the powerful climate-changing gas methane and are responsible for 10 percent of total agricultural emissions—a problem expected to worsen as climate warms.[87] But breakthroughs in rice production, including the "System of Rice Intensification," significantly cut the release of methane—mainly because they do not flood the paddies. At the same time, they are increasing rice yields.[88] Also reducing methane emissions are cattle manure "digesters"—a simple technology that captures methane for renewable fuel.[89]

A second step is enhancing carbon absorption.

A surprise to a lot of people is that soil, which sequesters carbon and keeps it out of the atmosphere, holds three times as much carbon as the atmosphere does.[90] To appreciate the value of soil, keep in mind that our atmosphere is already overloaded with carbon dioxide; and its absorption by oceans is slowing. Already, climate change has made oceans more acidic, and thus increasingly inhospitable to sea life.[91]

Only soil can sequester carbon in ways that actually promote climate and human health.

Ecological farming practices contribute to soil's capacity for storing even more carbon. One such practice is composting, in which farmers apply to their fields decaying organic material made from plant and animal wastes such as leaves, stalks, kitchen scraps, and manure. Just one ton of organic material can result in storing almost 600 pounds of carbon dioxide, reports a study using Environmental Protection Agency data.[92]

Other effective practices include keeping the soil covered at all times, applying manure, planting multiple crops in the same field, and, especially, integrating trees on farms, a practice known as "agroforestry."[93]

Agroforestry might seem a rarity for anyone who imagines that a "farm" means vast stretches of a single crop as far as the eye can see.

> **Agroecological Practices Can Help Meet Our Climate Challenge**
>
> The EU's adoption of agroforestry, along with several organic farming practices, holds what scientists call the "technical" potential to sequester carbon equal to 37% of the EU's 2007 annual carbon-dioxide-equivalent emissions.
>
> —drawn from Aertsens et al., *Land Use Policy*, 2013

But mixing trees with crops is hardly rare. A quarter of the world's agricultural land has more than 20 percent tree cover, reports the World Agroforestry Center.[94] (And note that in agroforestry systems, trees aren't stealing resources from crops; rather, they can increase crop productivity.)[95]

And agroforestry's sequestration potential is impressive: If within the European Union, for example, agroforestry and several other ecological farming practices spread to their full potential, the shift has the "technical" potential to annually sequester carbon equal to 37 percent of all EU's carbon-dioxide-equivalent emissions in 2007, according to a study in *Land Use Policy*.[96] That's huge, and agroforestry would contribute most of the gain. Globally, among a number of ecological farming practices, "agroforestry by far has the highest sequestration potentials for all world regions," concludes a World Bank study.[97]

For each unit of land, though, one practice called "biochar" tops all others in sequestering carbon, according to the same study. Biochar is a form of charcoal generated by the careful, controlled smoldering of organic material. An ancient technology, recently rediscovered, it's especially great for poor farmers because biochar can be made from farm waste—in Africa, for example, from cassava stems, oil palm branches, and common weeds. The potential of biochar varies widely by soil type and other variables, so scientists acknowledge great uncertainties.[98]

Nonetheless, biochar has been shown to increase the soil's capacity to store carbon as well as enhance crop yields significantly.[99] Its special carbon-holding power seems to lie in its porous structure, whose surfaces protect carbon from degradation. Some scientists estimate that biochar's long-term carbon-sequestering capacity is huge.[100]

The Big Picture of Possibilities for Earth's Carbon Storing

Considering all of these possibilities, and others, what's the Earth's likely big-picture potential to store carbon and keep it out of the atmosphere? We can:

- Spread agroecology and agroforestry, described in Myth 4.
- Restore wetlands and protect carbon-holding peat lands— waterlogged organic soils layered with decaying plant material, such as bogs.
- Reestablish forests that we have destroyed.

- And, in grassland ecosystems, improve grazing practices, sow leguminous grasses and other forage (that act as natural fertilizer), and manage fires.

These and other biomass-related steps to sequester carbon could be further enhanced by changing how we treat the soil. We can:

- Halt the spread of deserts, thus enhancing soil organic and inorganic carbon.

> **Earth-Based Potential to Meet the Carbon Crisis**
>
> Using known improvements in how we treat the Earth, the attainable potential to sequester carbon equals *a fifth of human-caused global carbon-dioxide-equivalent emissions annually.*
>
> —drawn from the work of Rattan Lal, Ohio State University

- Restore salt-affected soils and those degraded by nutrient depletion, soil-structure degradation, waterlogging, acidification, and more.
- And, in farming, forgo plowing while extending the use of biochar and cover crops, as well as recycling nutrients using mulch and manure; and integrate livestock with cropland and forestry.

Together, these advances could potentially sequester carbon equal to a fifth of global carbon-dioxide-equivalent emissions annually, reports a leading world authority, Professor Rattan Lal of Ohio State University.[101] Whereas the EU study above captures technical—or theoretical—potential, Lal describes this big-picture potential as "attainable."

Note, however, that there's a limit to how much more carbon the Earth can sequester. The estimates for reaching that limit range from just a few decades to well over a hundred years.[102] The hopeful news is that this time frame is precisely when we will most need help from carbon sequestration as we end reliance on fossil fuel.

While grateful for the guidance of Lal and others on these complex questions, we are also dismayed by how much is still unknown and how little productive debate is under way to resolve serious divides among experts. Given the urgent need for common direction, we strongly encourage scientists to pursue this research with transparent communication among those with differing assessments. And, as citizens, we urge more public funding for their work.

So here we are. Even without all the data points, this we know: We've thrown natural cycles out of whack, but in the ways noted above and others we're discovering powerful means by which to rebalance the Earth.

Lightening Livestock's Load on the Planet

From this big picture of possibilities, let's now look specifically at how we can reduce the heavy burden of livestock. Above, we came down hard on livestock for its role in shrinking our food supply, depleting our groundwater, and fueling climate change.

Is there any hope?

In a 2013 "TED Talk," viewed by millions, biologist Allan Savory claimed that his "holistic" grazing method alone—largely by enabling more soil-carbon sequestration—could bring atmospheric carbon concentration down to preindustrial levels.[103]

Actually reverse climate change? Wow! Who would not want to believe such great news?

But others see the potential quite differently. One is Rattan Lal, cited above. According to his estimates, the *total* technical potential of soil sequestration—of which improved grazing practices are just one contributor—would take us, not all the way, but in fact less than 20 percent of the way to preindustrial carbon dioxide levels—or to about half of what's needed to return to even the highest level of atmospheric carbon considered safe.[104] (Note that in this case Lal is referring to "technical" potential, which is more generous than what scientists term "attainable.") Other scientists in peer-reviewed journals strongly challenge Savory's assumptions as well.[105]

Our big concern is that debate over Savory's extreme claim might distract us from many things that are clear. One is that improving livestock management is urgent, and would bring about multiple benefits.

Pastoralists, who rely on livestock for their livelihood and are also many of the world's hungry people, are learning to improve their management practices and thereby livestock's productivity. Simultaneously, they are reducing soil degradation, hunger, and greenhouse gases.

In Namibia, for example, pastoralists using Savory's holistic management methods broke with long-standing tradition and began working together—planning and coordinating their herding. Each night they corral animals, preventing them from returning to recently grazed plants, reports a joint account by the FAO and the International Federation of Organic Agriculture Movements. "The herders only come back to a grazed area when the perennial grasses are fully recovered," and, as a result, they've been able to make progress on four fronts: reducing soil degradation, improving soil-surface cover, conserving water, and enhancing biodiversity.[106] All are critically important to sustaining local people and to fighting climate change.

Namibia's story is encouraging evidence that the "tragedy of the commons"—the notion that shared property is always abused—is not inevitable. Throughout history we humans have shown ourselves capable of making rules and enforcing them together so that our commons can flourish.[107]

Namibia's experience underscores the benefits of improved livestock management. But removing livestock can also lead to ecological restoration. Below, the story of China's Loess Plateau is one example.

Evidence also comes from observation of Oregon's Hart Mountain National Antelope Refuge after grazing livestock were removed. There, regeneration of vegetation—bringing about enhanced biodiversity and carbon sequestration—occurred rapidly. Scientists sampling sixty-four sites in the refuge discovered that the area of bare soil had decreased by 90 percent, soil erosion had dropped measurably, and trees and other vegetation had spread. All this regreening, and in just over two decades.[108]

For us the takeaway here is the wisdom of remaining skeptical of claims of single, sweeping solutions, doggedly pursuing the evidence, and demanding that more resources be devoted to independent research on the best ways of righting the carbon balance.

To Graze or Not to Graze

But we'd be remiss if we moved on from a discussion of livestock without entering another hot debate.

In producing meat, some report our highly polluting and inhumane "concentrated animal feeding operations" (CAFOs) produce fewer greenhouse gas emissions per unit of protein.[109] It may be that CAFOs are so speedy in pumping out a meat product that each bite is linked to fewer greenhouse gases. One study suggests that a reason each CAFO cow may cause less climate harm is that it uses less land, even after we add in the vast acres used to grow feed.[110]

Or not.

Some make the opposite case: that the grazing of livestock means fewer greenhouse gases emitted compared with livestock produced largely by grain feeding in CAFOs.[111] Much depends, no doubt, on whether grazing is managed well so that soil remains covered.

For us, however, a concern is that this greenhouse gas debate—CAFOs versus grazing—could divert attention from what is clear about CAFOs: Even if they were proved to be less damaging in terms of greenhouse gas emissions relative to grazed livestock, CAFOs' pollution, grossly inefficient conversion of feed to food, disease threat, and cruelty—reviewed in the next chapter—make them a dreadful choice.[112]

Moreover, on top of concerns about climate harm and the waste of resources built into livestock's inefficiency, the vast increase in meat production worldwide—that's been growing at two and a half times the rate population has since 1961—is drawing down precious water supplies and destroying biodiversity.[113]

Putting all these pieces together, here's how it looks to us: Humanity's turn toward grain-fed, meat-centered diets is a dangerous aberration indeed.

A course correction is in order, and, fortunately, it is imminently doable.

Eating with the Earth

We can all contribute to food-supply and climate-change solutions by reducing production and consumption of meat, especially beef. Shifting toward plant foods and less-carbon-intense animal foods is a great choice on many counts. Pork, chicken, and eggs carry one-fourth to one-twentieth the climate cost of beef. And the GHG impact of vegetable protein food is even less, at just one-twentieth to one-fiftieth of beef.[114]

Eaters' Guide to Greenhouse Gases
kg CO_2 equivalent per kg of food
Vegetable protein food: 1.2
Eggs: 3
Poultry: 3
Pork: 6
Beef: 24 to 58

—drawn from Ripple et al., *Nature Climate Change*, 2014

As we move toward more plant foods and away from animal foods, each of us can contribute to continuing food-supply adequacy for all and cut greenhouse gas emissions—potentially enabling more carbon storage—while benefiting our own health.[115]

In fact, fruits and vegetables are so beneficial to our health that if Americans ate what the nation's dietary guidelines recommend, we could save $17 billion each year in medical costs, estimates the Union of Concerned Scientists.[116] And in 2015, we were pleased to see that experts advising the next official USDA dietary guidelines describe healthy "dietary patterns" as emphasizing plant foods and eating less meat, particularly beef, and tie this shift to "positive environmental" outcomes.[117]

And, in making this turn, we can let go of any worry about replacing all the protein we now eat: On average, Americans eat almost 70 percent more protein than recommended.[118] (For perspective, the protein in one eight-ounce steak exceeds the total recommended daily for both men and women.)[119] And, since our bodies cannot store protein, the extra ends up being used for fuel and, if we eat too much, turned to fat.[120] Plus, a big bonus: Medical experts stress the health benefits of a shift toward more beans, lentils, grains, nuts, fruits, and veggies.[121]

To succeed, though, in moving toward a planet-friendly diet, citizens need to demand an end to public policies that subsidize the meat industry.[122] In so doing, billions of dollars would be freed up to use to encourage production of affordable, diverse, healthy plant foods and to invest in smart public outreach about the multiple benefits of embracing more plant-centered diets. (Imagine the possibilities: Right now we devote 51 million acres to hay for feed but only four million to vegetables.)[123]

Such foundational change, we realize, requires democratically accountable policymakers—a theme we take up in Myths 9 and 10.

Americans are often dubious about our society's capacity to change. But should we be? After all, we cut the percent of Americans who smoke by half in forty years after warning labels were enforced and TV and radio ads banned.[124] To tackle tobacco, a combination of public policy encouragement and personal choices worked, and presumably meat isn't as addictive as cigarettes! Plus, Americans have already shown a willingness to reduce their beef intake, cutting it by nearly 40 percent since the mid-1970s.[125] (This is great, but keep in mind that during the same period, chicken consumption doubled, and U.S. consumers are still among the top beef eaters in the world.)[126]

In all, there's a place for each of us in this big, positive step toward healthier, climate-friendly farming and eating.

Re-covering the Earth, Ending Hunger

Now, let's turn to solutions in which solutions to hunger and climate change are intersecting.

Courageous citizens and officials are working to halt forest destruction, about three-fourths of it caused by expanding crops and grazing.[127] And some countries are restoring and reforesting their land.

Costa Rica has moved from high rates of deforestation in the 1970s to almost zero today, and a quarter of its land is now officially protected.[128] Brazil—long a notorious forest destroyer—became a leader in slowing the destruction, and by 2012 had cut its rate of deforestation by two-thirds compared with its 1996–2005 average. Brazil's progress has since reversed.[129] It will take citizen pressure to reverse this reverse.

In China, where many see only unmitigated ecological disaster, farmers on its north-central Loess Plateau have reclaimed an area two-thirds the size of Massachusetts. For decades "experts" had tried restoration schemes for the area, sometimes described as a "moonscape" and known for dire poverty and hunger. Its massive erosion had been silting up the Yellow River for millennia. All attempts at restoration had failed.

Then, over fifteen years beginning in the 1990s, farmers created thriving, food-secure, more ecologically sustainable communities, benefiting three million people. They did it, not alone, but in a lesson-filled partnership.

For two years, the World Bank's Agriculture Global Practice director Juergen Voegele and his Chinese counterpart traveled to the poorest and most eroded villages of the Loess Plateau, everywhere asking one question: "What worked in the past?" Basically, Voegele told us, the answer was, "Nothing." Then one day in Shageduo village the two men happened to glance down a slope and spotted one lush, green, wooded area.[130] Oh, Voegele thought, there must be some special, unseen water source there.

But no.

Fifteen years earlier the village leader had tried creating a walnut orchard, but grazing sheep and goats soon destroyed the trees. What to do? Faced with a choice between walnut trees and sheep, villagers chose trees. And everything changed. Animals were kept in pens, and plant cover was reestablished naturally.[131] From this village's experience, the idea for the Loess Plateau Rehabilitation Project—a World Bank and Chinese government partnership—took off.

Penning livestock to prevent destructive grazing was one key to the stunning restoration of vegetation. Constructing wider terraces on the mountainsides also enabled the planting of orchards and improved crop yields.

But, Voegele stressed, success was possible only because of changes in the relationships among people. Farmers gained trust in government because, for the first time, government entered into written compacts with them. It guaranteed farmers' rights to their plots of land if they contributed their labor, mainly in widening the terraces. And farmers whose cash income improved agreed to repay the government a percentage of their gain. For the first time, assured that they would be on the same plot for years to come, farmers were willing to take up long-term, sustainable agricultural practices—penning livestock, maintaining terraces, planting, and protecting trees.[132]

It worked. Ultimately, observing the transformation in the project area, villagers banned grazing throughout much of the Loess Plateau, and planted forest trees, orchards, and drought-resistant shrubs on almost 600,000 acres.[133] Critics point out that in some areas many of the newly planted trees have not survived because they were unsuited to the local terrain.[134] On another 140,000 acres, however, villagers agreed not to farm in order to enable natural regrowth, which avoids the risk involved in planting new trees.[135]

Their efforts have begun to reduce erosion that had been washing into the Yellow River as much as a billion tons of silt every year. The project's new vegetation also enables carbon sequestration estimated to be the equivalent of taking three hundred thousand cars off the road annually.[136]

In sum, says a World Bank report, "the ecological balance was restored in a vast area considered by many to be beyond help." All this while crop yields increased 60 percent, fruit orchards were established, and farmers' incomes more than doubled. New schools, roads, and local enterprises are now improving village life.[137]

As one country inspires another, similar approaches to food security as part of "re-covering" the Earth are taking hold in some surprising places.

Since its independence in 1962, Rwanda has lost almost two-thirds of its forests. The country's 1994 genocide destroyed not only lives but landscapes and livelihoods as well. Then, Rwandan forestry officials visited the Loess Project, and, Voegele told us, "It gave them the vision that they, too, can change their landscape."[138]

Rwanda has recently expanded its forest cover, and the government has committed to increasing its current forested area 50 percent by 2020. Rwanda connects its fight to conquer hunger and poverty with its goal of reversing completely the degradation of the country's soil, water, land, and forest resources by 2035.[139]

In 2011, India set a ten-year goal of restoring forests on about 11 million acres and generating new forest on more than 12 million other acres. Among the country's tree-planting goals is to employ jobless youth to plant 2 billion trees along roadways.[140]

In East Africa, through Kenya's Green Belt Movement, village women have in three decades planted more than fifty million trees.

Earth-wide, the citizens' campaign Plant for the Planet, launched by the United Nations Environmental Program (UNEP) in 2007, encourages and tracks tree planting. By 2015, it had counted more than thirteen billion trees planted. That's nearly two for every person on earth![141]

Most of us don't think "trees" when we think of food and hunger. But whether it's on the Loess Plateau in China or in Green Belt villages in Kenya, trees help conquer hunger as they reduce soil loss and often provide fruit, nuts, and fodder. Even more directly, agroforestry—combining food crops with trees in the same fields—is increasing food supplies in many parts of the world, as we highlight in our story from Niger below.

Such examples, and many more, are restoring not only the Earth but human dignity, as people gain power over their lives. It turns out that enhanced food security, ecological health, and climate mitigation are inseparable. We see that in addressing all dimensions of the climate challenge—whether by regreening land for increased carbon storage or by cutting greenhouse gases—the remaking of our food system offers immense contributions toward solutions.

REDUCING VULNERABILITY TO BOTH CLIMATE CHANGE AND HUNGER

As we argued in Myth 1, natural events are not the cause of hunger and famine but instead the final blow for those already vulnerable. So the key questions are, *What creates vulnerability? And what reduces it?*

Consider the world's 1.5 billion small-scale food producers. Among them are most of the world's hungry people, pushed onto less fertile, drought-prone lands or deprived of land altogether and forced to work as day laborers or sharecroppers. Many are also in debt to moneylenders or landowners who claim much of their harvests. Other vulnerable people are the more than 200 million unemployed, and those so poorly paid that nothing is left to fall back on in hard times.[142]

With this understanding, it shouldn't surprise us that "people in low-income countries are four times more likely to die in natural disasters than those in high-income countries," according to the World Bank.[143]

So what reduces vulnerability?

In part the answer is how we farm.

Agroecology—which we just noted helps to sequester carbon and which we explore in Myth 4—means farming that relies on, and strengthens, synergies among all the domains of life, from bacteria to bees. Agroecology frees farmers from having to buy chemical farm inputs, and thus from risky indebtedness. So farmers become less economically vulnerable.

Agroecology practices increase farms' resilience in many other ways as well. In our water-stretched world of climate change, mulch used in agroecological systems "lowers soil temperatures," as well as "reduces the amount of water lost through evaporation," and "protects the soil from erosion," notes the U.S. Department of Agriculture.[144] Soils on farms not using synthetic fertilizers and pesticides typically have significantly more organic matter in the topsoil, which increases water-holding capacity.[145] All this means that agrological practices heighten farmers' resilience as they face increased rainfall variation expected as our climate warms.

So it is not surprising that scientists find that, compared with farms relying on manufactured fertilizer and other inputs, organic farming systems have higher yields during periods of heavy rains and droughts.[146] Agroforestry trials with corn and native, nitrogen-fixing trees in Malawi and Zambia found an almost fivefold increase in "rain use efficiency," along with significantly lower carbon dioxide emissions, compared with maize grown alone.[147]

Practices such as composting and crop rotation also increase soil organic matter, improving soil structure and lessening risk of erosion. Such practices offer protection in the face of another type of extreme weather: hurricanes.

Here are two telling stories of increased resiliency.

Central America. During the 1970s, hundreds of rural communities in Central America began adopting agroecological farming practices, improving their soil structure to better resist wind and water erosion.[148] Then in 1998 Hurricane Mitch hit hard, in certain regions wiping out virtually the entire banana export crop and causing massive loss of topsoil from cropped hillsides.[149] A collaborative study in Nicaragua, Honduras, and Guatemala by forty civil society organizations in 360 small-farm communities then documented the striking power of agro-ecological farming to reduce loss: Ecological farms retained 20 to 40 percent more topsoil than farms that had not been using ecological practices.[150]

Also promising? After the hurricane more than 90 percent of farmers still using chemical methods said they wanted to switch to ecological practices.[151]

Niger, West Africa. Across the world, on a much larger scale, agro-forestry is reducing the vulnerability of small farmers.

Over the past three decades, poor farmers in southern Niger—among the world's lowest-income countries—have "regreened" an area almost the size of Costa Rica. It's a momentous achievement, reviving a centuries-old practice, not of planting trees but of abetting their natural regeneration. Farmers leave selected tree stumps among the crops in their fields and protect the tree's strongest limbs as they grow.

Their results so far? Two hundred million more trees in southern Niger. Part of a global shift toward agroforestry, the trees help protect the soil and bring increased crop yields, as well as many other benefits.[152]

In an agroforestry success story, Niger's farmers have nurtured the regrowth of 200 million trees on 12.5 million acres, ensuring food security for about 2.5 million people.
—drawn from Reij et al., "Agroenvironmental Transformation in the Sahel," 2009

In the regreened area, "the poorest households often derive significant income from their on-farm trees in the form of fodder, firewood, fruit, and leaves, some of which is sold on the market," Dutch development specialist Chris Reij reported in early 2012. Selling leaves—used in favorite local sauces and more—from a single mature baobab tree in Niger can bring as much as $75, a lot where most people live on less than $2 a day.[153] As noted, some tree species are also natural fertilizers, because they fix airborne nitrogen in their root systems. Trees also help break wind speed, and their shade lowers ground temperature, reducing water evaporation from the soil. In all, mixing trees and crops helps farmers to get by in drought years.

These farmers' efforts have made them less vulnerable to both hunger and climate change. (Also, recall the impressive global estimates of agroforestry's potential contribution to actually addressing climate change.) Such farmer "regreening" has already brought enhanced food security for 2.5 million people, roughly 16 percent of Niger's population.[154]

During the drought and food shortages of 2005, for example, the infant mortality rate was close to zero in villages that protected and managed their trees, but was significant in villages that did not.[155] And, in 2011, a year of poor rainfall and much hunger, Niger researchers found that "a district with high on-farm tree densities" produced a grain surplus.[156] In one such area, chief Moussa Sambo described his village near the capital city, Niamey, as in fact experiencing the greatest prosperity ever, with young men returning to farm. "We stopped the desert," he said, "and everything changed."[157]

Now, as we conclude our reflections on vulnerability, let's return to the famine in East Africa with which we began. How were some people able to reduce their vulnerability there during the dire 2011 crisis?

At the time of the famine, Samuel Loewenberg, a public health investigator at Harvard, traveled through the arid landscape of Wajir in eastern Kenya, seeing only hunger and desperation. In certain regions, health officials were reporting some of the highest rates of childhood malnutrition they'd ever witnessed.[158] Then, entering Kutulo Farm, he wrote of his surprise at encountering "an amazing sight: a green oasis—a farm, a greenhouse, a well, a water pump, a windmill." Running around the farm, he added, "were the first happy, healthy-looking children I had seen."

Loewenberg also approached a women's cooperative founded in the town of Wagberi. The cooperative of thirty women grows kale, cabbage, and peppers. He noticed a well, which he learned had been funded by the European Union. With windmill-powered pumps, the villagers get water to the crops, and have markedly increased the farm's productivity.[159]

Through community action and modest assistance, the people of Wagberi reduced their vulnerability to famine. Their resiliency is another sign that, despite the drought, there was nothing inevitable about the horrific 2011 famine.

REFLECTIONS ON VULNERABILITY AND HUMAN AGENCY

We've argued that both acute famine and chronic hunger result not from nature's vagaries but from human action and human inaction. While climate change increases the urgency of quickly and widely spreading the positive steps described here and throughout our book, in a world of vast food waste and inequity it does not alter this fundamental truth—it is *human* institutions that determine:

Who will have a claim to food. As long as people's only claim to food is through a market dominated by powerful corporations backed by unaccountable governments—and as long as incomes and prices remain volatile—famines will occur, people will go chronically hungry, and many will die *no matter how much food is produced.* For evidence we need look no further than the United States, where people go hungry and even die of hunger-related causes in the midst of an abundance of food. In Myth 6 we highlight one city remaking the rules so that all have a right-based claim to nutritious food.

Who will be chronically vulnerable. Here we've shown a range of actions that are radically reducing the climate-change vulnerability of even very poor people. Throughout the book we chronicle additional ground-level progress that is decreasing vulnerability.

The vulnerability of the agricultural system itself to drought, floods, and other adversities now made more extreme by climate change. As long as our economic system drives resource-depleting, climate-disrupting agricultural practices, our food system, and all of us dependent on it, will become ever more vulnerable. Throughout this book we weave together pieces of another path, one transforming farming systems toward greater inclusion, fairness, and resiliency.

Thus, as with the challenge of human-made climate change now upon us, there is nothing "natural" about either famine or hunger. They are social, not natural, disasters, the result of human choices, not fate.

And every year we delay in taking positive steps ensures more dire impacts.

Fortunately, we now know what to do. One essential step is to replace climate-disrupting energy with green energy. Another is to shift our diets and land use—a step we can make happen relatively quickly and inexpensively, if we act to bring about public policy support. Moreover, every one of us can shift market demand toward climate sanity by taking more of our nourishment from ecological farms and the diverse world of plants, as well as by reducing our own family's food waste.

And why isn't humanity moving quickly on this compelling evidence?

Callousness, we believe, does not explain the weak response to the ongoing tragedy of hunger and to the global climate threat. The bigger problem is that too many of us see private power determining public policies—power exerted by, for example, fossil fuel and agribusiness industries within our political systems—and we feel defeated before we even begin. We see what appears to be immovable power blocking the kind of sensible changes outlined here.

But we can overcome paralyzing defeatism. Here's how.

We recognize that around the world regular citizens *are* assuming their power—from the tree-nurturing farmers in Niger to the lifesaving Wagberi women's cooperative in Kenya to the many more people you will meet in Myth 10. We join them by seizing our own power to influence those around us. With urgency and clarity, we work to assume the power of our choices in the marketplace and our voices as citizens to remake the rules of public life—including those now rewarding the most climate-damaging food and farming practices—so that public decisions indeed serve the common interest. As we contribute to building local-to-global democracies, we create a world in which no human being's claim to food may be denied.

Throughout our book we return to these themes, sharing examples of new power that motivate us.

myth 3

Only Industrial Agriculture & GMOs Can Feed a Hungry World

MYTH: If the world produces enough food today, it's only because modern farming methods—now widespread in the United States and developed countries—have greatly boosted yields and overall agricultural efficiency. So our primary responsibility now is extending modern agriculture to all parts of the world, especially to hungry Africa. But we must do even more. Since we're bumping against the limits of nature's capacity to feed us, we must push ahead with more sophisticated technologies. And that includes genetically modified organisms (GMOs) to help increase yields even further and also help to reduce our use of farm chemicals and water—all enabling us to withstand climate change.

RESPONSE: This myth celebrates what's often called "industrial agriculture," farming that relies on patented seeds and manufactured fertilizers and pesticides, and typically large-scale machinery as well. Certainly, the production success of industrial agriculture is no myth. The myths are that this model of production can end hunger and nourish us all well—now and in the future—and that we have no proven alternative.

Carrying a friendlier nickname—the "Green Revolution"—industrial agriculture was introduced in the developing world in the middle of the last century as the solution to hunger. Today, we often hear it called "conventional agriculture." But, of course, it isn't. Humans have been farming for roughly 10,000 years, but industrial agriculture took hold only about sixty years ago.

Many cite experiments in Mexico in the 1940s as the birth of the Green Revolution.[1] Internationally funded research there on wheat and corn—and later research on rice in the Philippines—led to the development of what became known as "miracle seeds."[2] The new varieties produced much higher yields when accompanied by synthetic fertilizer, pesticides, and, often, irrigation.[3] Overall, the Green Revolution is given much of the credit for more than doubling food production in the Global South between 1960 and the mid-1980s.[4]

And in 1970 Norman Borlaug received the Nobel Peace Prize for developing the new seed varieties, said to have saved a billion lives.

It's worth noting, however, that in India—the poster child of the Green Revolution—the increased output did not derive exclusively from the much-celebrated seeds. Indian wheat output also climbed because wheat acreage nearly doubled from the late 1960s through the 1980s.[5]

Today, many scientists and policymakers continue to celebrate the Green Revolution's success. So why do we conclude that this approach to farming can't end hunger?

Industrial agriculture, some critics argue, cannot end hunger because its narrow focus on producing more for short-term profit—sometimes called "productivism"—can't incorporate the interests of hungry people.[6] Let's explore the question more fully.

A NARROW LENS

The essence of the problem of industrial agriculture is that it cannot end hunger because it disconnects farming from its rich context.

Within the industrial model, the sole challenge is to produce. Excluded from its calculus is the intricate web of food-connected relationships, both those among people and those involving the natural world. Yet, it is these relationships—how fair and balanced they are and how well they align with nature—that are decisive. They determine how food is grown, what is grown, who eats, who doesn't eat, and, ultimately, whether the Earth is enriched or degraded.

The model's worldview also disconnects farming from knowledge acquired over millennia. By denigrating practices deemed "not modern," it filters the world through a narrow lens of superiority that makes it more difficult to appreciate agroecological practices, explored in the next chapter, which build on time-tested, experiential knowledge.

To sum up, the industrial model is rooted in a worldview in which life is made up of independent and disconnected spheres. Within this model, the goal of agriculture is to produce food and other agricultural products. That's it. Determining who has access to food and avoiding environmental and societal downsides of producing it are matters outside the purview of the model.

India's Green Revolution offers us powerful lessons about what happens once people are trapped inside industrial agriculture's narrow view. During the two decades most associated with its takeoff—the 1960s and '70s—grain output in India increased dramatically, but supply *for each Indian* barely improved.[7]

One reason? The industrial model's narrow focus did not alter harmful power imbalances in the countryside, such as the disempowerment of women, including their inferior access to farm credit, education, and means of determining their own fertility.[8] As noted in Myth 1, these are among the deprivations associated with elevated birthrates; so with more people, of course more production alone can't be expected to result in more food for each person. Also recall that later on, when per capita production increased by almost a third between 1990 and 2012, the number of undernourished Indians declined much more slowly—by just 10 percent.[9]

And poverty?

With little attention devoted to making relationships in the countryside more equitable, we should not be surprised that in India "there is no statistical correlation between the GR [Green Revolution] and poverty-reduction," as Raju J. Das concludes in the journal *Geoforum*.[10] A mid-1990s review of hundreds of studies of the Green Revolution's impact found that most reported an increase in inequality.[11]

A RELATIONAL LENS

By contrast, we strive in this book to look through a relational lens, one that focuses on the *interactions of all elements*—what is often described as "systems thinking." It assumes that within any dimension of life, it's the organization of relationships within the entire system—or what we here call simply "the model"—that determines outcomes. Thus, in agriculture it's the model—for good or for bad—that shapes who has access to food, whether it is nourishing, and whether food-producing resources will continue to support life.

This shift from a disassociated to a relational way of thinking is arising across many fields, within both the physical and the social sciences.[12] It's also reflected in the ever-growing numbers of farmers and agricultural scientists worldwide who are rejecting the single-focus view of agriculture as they create highly effective alternatives featured throughout our book.

To understand why we face the deep and multifaceted trouble this chapter describes, it's critical to acknowledge that all systems—human and beyond—operate within rules. The industrial model of agriculture, we argue, organizes economic relationships within a market driven almost exclusively by a single rule—what brings the highest, quickest return to existing wealth. In Myth 6, we lay out the hunger-making logic of this rule. Note how different the premise of a "one-rule-driven market" is from the commonly held notion of a "free market" that flows from unfettered, fair competition among individual self-interests.[13]

Before we continue, we want to be perfectly clear: Our critique is not of farmers. Most are struggling to do the best they can, and many farmers throughout the world are leading efforts to transform a deeply flawed model. Our critique is of the industrial model of farming itself.

LESSONS FROM THE HOME OF "BIG AG"

Don't be surprised that we include here evidence of the industrial model's impact not just in the Global South, where most hungry people live, but also in the United States. Our rationale? If the model is failing where it is most fully developed, it makes no sense to promote it as a solution elsewhere.

Of course, many people hardly view U.S. agriculture as failing; instead it's commonly seen as hands-down proof of industrial agriculture's success—even though, as we noted in Myth 2, U.S. agriculture feeds fewer people per acre than does Indian or Chinese agriculture.

To those celebrating its success, we ask, Success *for whom*?

As we noted in Myth 1, one in six Americans is "food insecure," meaning not always knowing where the next meal is coming from.[14] For households with children, make that one in five Americans.[15]

And how are most U.S. farmers faring?

For every ten farms in the United States in 1950, only four are left today.[16] Half of all those remaining, as well as half of those most associated with family farming—i.e., midsize operations where farming is the principal occupation of the operator—lost money from farming in 2012, a typical year.[17] That's true despite the $23 billion in tax-funded subsidies that goes to farmers each year—more than half of it going to the top 10 percent.[18]

While many Americans likely imagine our farmland as family-owned and -run, in reality nonfarmers own almost 30 percent of it, and tenant farmers operate 40 percent.[19]

All of these numbers suggest a lot of insecurity for those who grow our food. So the extreme psychological distress showing up in India's farmers who take their own lives, discussed below, has parallels here. Since the 1980s, suicide among male farmers in the United States has been almost twice the rate of the general population.[20] And a landmark twenty-year study of male farmers found a significant correlation between commonly used pesticides and depression.[21]

Now, as we jump in more deeply to examine the industrial model of food production, be prepared! This chapter might feel like an unrelenting string of bad news, for it turns out that much of industrial agriculture's apparent success disappears as we register its inequities, losses, and hazards. But stay with us—the rewards are great. We ourselves find that only by appreciating the model's pitfalls is it possible to grasp why the promising news in the next chapter, and elsewhere in our book, is so important.

TUNNEL VISION AND ITS TRAPS

Along with the industrial model of agriculture have come a multitude of changes in our relationships with each other and with the Earth. Many are directly undercutting human well-being and nature's capacity to support life more broadly—in the here and now as well as for future generations. Let's start with the human side—with a glimpse of the changes wrought in the nature of our relationships with one another.

Deepening Dependency, Heightening Vulnerability

While the seeds that birthed the Green Revolution are often called "high yield," more accurately, they are "high response." Typically, they respond well to commercial fertilizers and pesticides that farmers need to purchase. Plus, compared with the seeds they replace, the new seeds often require more water for specific or extended periods, thus making irrigation a necessity in many areas.[22] So, dryland farmers are out of luck.

A key to understanding the impact on farmers is that these seeds are hybrid varieties, produced by crossing two parents with different desired traits. So farmers cannot plant seeds saved from a hybrid crop and expect the same traits to show up in their next harvest. Thus, once using hybrids, farmers who had previously used seeds they'd saved, and even shared with neighbors, now must make considerable up-front investment year after year not only in their fertilizer and pesticides, but also in new seeds.

The speed of deepening dependence on chemical inputs is striking: Worldwide pesticide sales increased, after adjustment for inflation, over 5.5-fold between 1960 and 2005.[23] In that market,

> Worldwide pesticide sales increased, after adjustment for inflation, over 5.5-fold between 1960 and 2005.

three corporations control about half of the sales.[24] (In the United States, agriculture accounts for about 80 percent of pesticide use; and in the Global South it's likely an even greater share.)[25]

Moreover, the use of manufactured—or "synthetic"—fertilizer, by weight, has grown worldwide at roughly the same pace.[26]

And what have these changes in farming meant for poorer farmers?

Farmers' cash expenditures now play a huge role. In Andhra Pradesh, India, for example, purchases of seeds, pesticides, and fertilizer together account for almost half of the agricultural costs for small farmers.[27] To buy these inputs, poor farmers have to take out loans. The problem is that roughly three-fourths of Indian farmers—many lacking collateral—have no formal credit source like a bank.[28] So these poorer farmers must turn to moneylenders, charging monthly interest that amounts to an annual rate of 80 to more than 200 percent. What a handicap for poor farmers, especially when compared with those better-off, who can obtain a bank loan with an annual interest rate of only 12 to 15 percent.[29]

With farmers' financial dependency comes vulnerability.

Once a small-scale farmer goes into debt to pay for chemical supplies and commercial seeds, one bad harvest or family emergency can be ruinous. Among poor Indian farmers, desperation can set in, resulting in despair and even widespread suicide, reports the *New York Times*. Farmers are committing suicide "because they have to take money at huge interest rates from the moneylenders," lamented India's prime minister, Narendra Modi, in 2014.[30] Because suicide in India carries a stigma and families reporting it risk losing benefits, it's likely to be underreported. Some even accuse the government of deliberate undercounting.[31]

Moreover, in a speculative, volatile global market, dependency on imported farm inputs adds to the risk. Imagine the blow to small-scale farmers dependent on synthetic fertilizers when the price of a key fertilizer jumped more than threefold in just two years, 2006 to 2008.[32]

Dependence on public subsidies is another aspect of farmers' heightened vulnerability. From the beginning of the industrial model's takeoff in India, government subsidies have helped to keep down the price of chemical inputs for farmers. But subsidy fluctuations mean great uncertainty for farmers. Moreover, the yearly cost to the Indian government of fertilizer subsidies—in part benefiting the global agrichemical corporations—now equals about a third of India's military budget.[33] This huge, but rarely noted, cost to the public is yet another reason this model of farming is unsustainable.

A parallel in East Africa is striking. In 2006 Malawi's government began heavily subsidizing synthetic fertilizer and, to a lesser extent, hybrid seeds, and ended up with a result far from what officials, and farmers, had hoped for. Malawi saw synthetic fertilizer use climb to "shockingly high levels" and farmers trapped in a "cycle of debt"—all while bringing only "minor yield increases," concludes a 2014 review by the African Centre for Biosafety.[34] The study also found "net transfers away from farming households to [foreign] agribusinesses" such as Yara International. And it all came about at "great financial costs to [farmers] themselves and the public purse," notes the study's lead researcher, Dr. Stephen Greenberg.[35]

Hidden Harms

Now, let's turn to the impact of industrial agriculture on the Earth itself, on whose health we all depend.

The Loss of Precious Soil

First, soil.

Both the quantity and the quality of food the Earth produces depend on dirt—yes, on healthy soils—as we explore in the following chapter. Yet, worldwide, the soil of more than half of all farmland is now at least moderately degraded, and farming is responsible for about three-quarters of the global soil-erosion crisis.[36]

> **Staggering Soil Losses**
> ~ Fertile soil washed or blown away worldwide could fill 4 pickup trucks with soil for every person on Earth each year.
> ~ Topsoil is eroding 13 to 40 times faster than nature can replenish it.
> —calculated from Pimentel et al., *Frontiers*, 1997, and UN Convention to Combat Desertification, 2012

Farming practices in the industrial model typically make soil more vulnerable to erosion in part because they result in less soil organic matter—the reservoir of nutrients available for plants and a key to healthy soil structure.[37]

Each year, seventy-five billion tons of fertile soil—what a UN agency calls "the most significant, nonrenewable geo-resource"—is washed or blown away.[38] Think of it: That's four pickup trucks full of fertile soil gone each year for every person on Earth.[39]

The result? Worldwide, we're losing topsoil ten to forty times faster than nature can replenish it, so that in just four decades almost a third of the world's arable land has been degraded or rendered unproductive.[40]

Actually, topsoil *is* renewable, but given the typical treatment of soil within the industrial model, renewal is slow. Very slow. Generating just one inch of topsoil in current conditions takes a minimum of two hundred years.[41] (In the following chapter, we learn how it's possible, using very different farming practices, to radically shorten this time.)

While soil loss is a huge problem, so is the depletion of soil fertility.

So now we explore the industrial model's attempt to enhance soil fertility, and thus plant growth, by adding manufactured fertilizers. It's a revealing story of the consequences of seeing life as operating in independent and disconnected spheres, not as continuously interacting elements. Within this worldview, developers of industrial agriculture apparently did not foresee that synthetic fertilizers, used to enhance soil fertility, can actually do the opposite.

Nitrogen Negatives

Take nitrogen, which is essential to plant growth and makes up 60 percent of the synthetic fertilizer used in the United States.[42]

Synthetic nitrogen fertilizers can result in a net loss of soil nitrogen—so concludes a study carried out over half a century at the University of Illinois and reinforced by numerous other field studies worldwide.[43]

Carbon, too, is key to fertility, yet continuous use of nitrogen fertilizer has also been linked to a "net decrease in soil organic carbon," in another University of Illinois long-term study. After five decades, the loss came to almost five tons per acre and rose with a "higher nitrogen rate," noted the study's coauthor, Saeed Khan. Moreover, synthetic nitrogen is linked to the depletion of soil organic matter, which is key in maintaining healthy soil structure, air and water availability, and the supply of usable nutrients.[44]

Then consider microbes, whose job it is make nutrients available to plants, as we describe more fully in the following chapter. There are a lot of them! Within a single teaspoon of healthy soil live more microbes than there are human beings on Earth.[45]

But what happens to these multitudes when a farmer applies synthetic nitrogen?

The plants feast on this synthetic abundance, and the microbes lose their jobs. With plenty of nitrogen to go around, microbes aren't stimulated to do their critical work of generating plant nutrients themselves. Later in the growing season, though, the synthetic nitrogen is all used up by the plants—or gone elsewhere to create the havoc we describe below—and the microbes are needed to feed them. So the plants try to jump-start the microbes by feeding them carbon compounds they exude from their roots. But that creates another problem. Plants are then diverting precious water and carbon to the microbes that could otherwise have gone to the crop, explains soil scientist Kristine Nichols.

So what's a farmer to do?

Likely apply even more nitrogen the next year to meet yield goals, but this only perpetuates the cycle. Attempting to address the problem by applying fertilizer at intervals throughout the growing season "only serves to keep this system decoupled from soil microbiology by keeping the microbes dormant," Nichols reports.[46]

In other words, nitrogen fertilizers disrupt the symbiotic relationship of plants and the community of soil microbes on which they—and we—depend.

A related problem is that synthetic fertilizers, especially the common ammonium-based nitrogen fertilizers, tend to make soil more acidic.[47] Soil acidity can limit plants' access to nutrients and to water, in part because acidity contributes to soil compaction.[48] And some plants simply can't survive in acidic soils, so synthetic fertilizers shrink biodiversity as well. Another consequence of acidic soils is that more acidic compounds and toxic elements leak into drainage water. Soil minerals also suffer structural damage in acidic soils.[49]

But this loss is only the beginning of the havoc wrought by overuse of nitrogen fertilizers.

As you read of this fertilizer's harm, keep in mind that 50 to 60 percent of the nitrogen applied to fields worldwide isn't even taken up by plants.[50] Thus, within the industrial model of farming, nitrogen excess leaching into waterways is unavoidable. Even agroecological farms that don't apply synthetic nitrogen experience nitrogen loss, though less than industrial farms; and they are working to reduce it further.[51]

And there's more: All that nitrogen unused by plants remains active and, with phosphorus, leaches into water to create "blooms" of what's popularly called blue-green algae and phytoplankton whose decomposition soaks up oxygen and suffocates marine life. Now we know what we're dealing with: the innocuous-sounding algae are actually a type of bacteria.

And the consequences?

More than four hundred aquatic "dead zones," created when the bacteria deplete water of oxygen, thus devastating marine life worldwide. Added together, this assault covers an area the size of the United Kingdom.[52] Humans are also harmed directly, as these bacteria are associated with liver and neurological disorders in humans.[53] (Keep this in mind when you read below of the incident in 2014 when people dependent on Lake Erie for water were told not to bathe in, much less drink, it.)

Nitrogen fertilizer runoff also contributes to nitrate contamination of drinking water in many parts of the world.[54] Water in one of fourteen household wells in agricultural areas across the United States has nitrate levels exceeding Environmental Protection Agency (EPA) safety standards, with an "elevated concentration" of nitrate near intensively farmed land such as the Midwest corn belt and California Central Valley.[55] Infants who drink water from the wells face heightened risk of "blue baby syndrome"—decreased oxygen-carrying capacity of the baby's blood—as well as certain cancers.[56]

> Fertilizer runoff into waterways has contributed to four hundred aquatic "dead zones" decimating marine life worldwide, now covering an area the size of the United Kingdom.
>
> —drawn from Diaz and Rosenberg, *Science*, 2008

Moreover, some of the vast amount of nitrogen not taken up in plants becomes nitrous oxide that is released into the atmosphere as a long-lived greenhouse gas three hundred times more potent than carbon dioxide.[57] Nitrous oxide also destroys the protective ozone layer in the stratosphere.[58] Additionally, it speeds climate warming, because it weakens the capacity of plant communities to remove heat-trapping carbon dioxide from the air and store it.[59]

The Radical Remaking of Nitrogen's Place on Our Planet

But let's back up to see how we got here.

Nitrogen—a key to plant growth—is everywhere. Since it's 78 percent of the air we breathe, it's hard to imagine that nitrogen might cause harm. But this atmospheric nitrogen—almost all of an unreactive sort—undergoes huge changes by organisms within the Earth's soil and water, and increasingly by human hands as well.

We've seen how soil microbes turn nitrogen into a form that plants—and then we eaters of plants—can use. In our bodies, it's essential, for example, to protein and to DNA and RNA.

All good. But from the early twentieth century onward, we've revolutionized the place of this vital element in the Earth's biological and chemical systems. The revolution started when two German chemists, Fritz Haber and Carl Bosch, developed the first practical way to convert atmospheric nitrogen to ammonia, which could then become nitrogen fertilizer.

At the same time, farming in the United States was moving toward single crop varieties over large areas, and thus away from diversity. Simultaneously, it was also shifting from a system that integrated livestock into farming—a system through which manure supplied nitrogen—toward one concentrating manure in hazardous "holding ponds," as we see below. Both moves—and related changes in farming practices arising to some extent in the rest of the industrial world as well—reduced plants' access to nitrogen.

Soon, we were pulling atmospheric nitrogen from the air and turning it into fertilizer at a remarkable pace. Since 1961, synthetic nitrogen fertilizer applied each year worldwide has grown much faster than any other fertilizer. It had jumped ninefold, to more than a hundred million tons, by 2012.[60] That's almost thirty-five pounds yearly for every human being on Earth.[61] By one estimate, half of the nitrogen in our bodies, mainly in protein, is made up of atoms from synthetic fertilizer.[62]

While it's now widely appreciated that extracting vast stores of fossil carbon in coal and oil from inside the Earth and then releasing it into the atmosphere as carbon dioxide ends up devastating life on our planet, many fewer of us realize that with nitrogen we're doing the reverse. We're pulling a fairly inert form of nitrogen out of the air, then remaking it into what's called "reactive nitrogen," although it's actually active—very active. We then move this form into Earth's soil, from which it goes into soil organisms, plants, water, and air.

In all this, we've disrupted the nitrogen cycle more radically than even the carbon cycle—also with dire consequences.[63]

We not only have moved it from air to Earth, but have created more—much more—reactive nitrogen, the type that feeds plants in the soil and algae in the water, harming marine life and polluting air.[64]

Each year, humans add three times more of the active form of nitrogen to our earthly environment than what natural terrestrial processes produce.

Globally, humans add each year three times more of this active form of nitrogen to our earthly environment than natural terrestrial processes, mainly bacteria, produce.[65]

In the United States, make that almost four and a half times.[66] Some comes from industry and transportation, but about two-thirds comes from synthetic fertilizers along with the vast planting of nitrogen-fixing legume crops such as beans and lentils.[67] Some studies show that this microbe-generated reactive nitrogen, which feeds leguminous crops, leaches less into waterways than does synthetic fertilizer.[68] So for some it wouldn't register as problematic.

But we've made it into a problem.

Industrial-model farming has so increased legume production that it, too, has become a big contributor to nitrogen overload. Embedded in a market driven by the highest return to existing wealth, the industrial model moves resources into meat production for the better-off, as we've seen. One result? Globally, since 1961, a tenfold increase in the production of the nitrogen-fixing legume, soybeans, primarily destined for livestock feed; a radical change reflecting "market demand" by those who can afford soy-fed meat.[69] Here again, we are struck by the negative interplay of forces not registered in the narrow production fixation of the industrial model.

Given such a radical remaking of the cycle of nitrogen, an element essential to life, it's no wonder that the consequences noted above for human health, soil fertility, and aquatic life are grave.

Looking beyond nitrogen, the next most ubiquitous fertilizer is the element phosphorus—the backbone of DNA and cellular membranes, and essential to plant and animal growth. Worldwide, its use has grown more than fourfold since 1961.[70]

Phosphorus Folly?

Since phosphorus arises in the natural weathering of rocks, and more than two-thirds of the world's cropland has "surplus" phosphorus, why are humans adding so much more via manufactured fertilizers?[71]

Part of the answer is that naturally occurring soil phosphorus—all but about one-tenth of 1 percent—can't be absorbed by plants without help. Bacteria and fungi must do their part, dissolving phosphorus so it's available for plant uptake.[72]

But here's the hard truth.

Industrial agriculture has not focused on building healthy soils that are full of microbes to do the critical work of making phosphorus usable. At the same time, we've been moving to a meat-heavy diet that requires nearly five times as many acres to feed each of us compared with a plant-centric diet—and each of those many more acres requires phosphorus and other nutrients.[73]

Another explanation for the vast increase in synthetic fertilizers is that in the modern era we humans have broken the cycle of nature by failing to return nutrients, including phosphorus, back to the soils. So we keep pouring more on.

And from where do we get phosphorus?

Not from the air, as with nitrogen, but from phosphate rock that's accumulated over tens of millions of years, and thus, in human time, is finite.[74] Today there are enough known reserves of phosphate rock to last about three hundred years at the current rate of use, reports David Vaccari of the Stevens Institute of Technology.[75] But Dana Cordell and Stuart White of the Global Phosphate Research Initiative in Sydney, Australia, stress that the crunch point is likely to occur far sooner than when the last ton of phosphate is mined. They cite six studies suggesting that within ten to seventy years we will hit "peak phosphorus"—the point at which extraction costs are so high, and prices are out of reach for so many, that global phosphorus production begins to decline.[76]

And then what?

Even if we don't literally "run out" of phosphate rock for a few hundred years, imagine the vulnerability in these facts. Almost 75 percent of world phosphate rock reserves are in a single area, Morocco and Western Sahara.[77] That degree of concentration alone would be unsettling, but there's more: The stability of the region—and thus the world's phosphorus reserves—is "threatened by extremist, terrorist, and criminal elements," reports UN envoy Christopher Ross.[78]

> Just one of the vulnerabilities built into industrial agriculture:
> Almost 75% of world phosphate rock reserves, essential for synthetic fertilizers, are concentrated in one small area in North Africa—politically unstable Morocco and Western Sahara.
> —drawn from Cordell and White, *Annual Review Environment and Resources*, 2014

And the U.S. share of phosphate reserves? Under 2 percent.[79]

Additional vulnerability lies in phosphorus's wild price swings. At the peak of the 2008 food-price crisis, the phosphate-rock price jumped almost ninefold within a single year.[80]

Phosphorus Turns Toxic

And if this news were not enough to persuade us to take another path, note that—as with nitrogen pollution— runoff of phosphorus in fertilizer can also contribute to bacterial "algal blooms" hazardous both to marine life and to humans.[81] In mid-2014 in Ohio, the highly toxic microcystin was discovered in phosphorus-contaminated Lake Erie, on which almost half a million people depend. Because the toxin can seriously damage the liver, residents were warned over several days not to bathe in, much less drink from, the public water supply.[82]

Little did most Americans appreciate that this crisis was not a freak accident, but instead a rather predictable outcome of the industrial model of agriculture.

So in a nutshell, here's what humanity has done: In just seventy years, we've extracted an essential-to-life "fossil" resource, processed it using climate-harming fossil fuels, spread four times more of it on the soil than occurred naturally, and failed to recycle it.[83]

It's ended up in our water, doing great harm—ultimately settling in ocean sediment, where it remains inaccessible to humans for tens of millions of years.

> In just seventy years, we've spread four times more phosphorus on the soil globally than occurred naturally, all mined from rapidly depleting reserves.
> —drawn from Cordell and White, *Annual Review Environment and Resources*, 2014

In Myth 4, we suggest solutions to these worsening crises.

Still, some scientists argue, synthetic fertilizer is a net gain for limiting climate change. Without it, they warn, food production would drop, so more land, including carbon-storing forests, would have to be cleared and cropped to feed us.[84] But, as the next chapter explores, this argument ignores the proven successes—and even greater potential—of modern-day ecological agriculture using little or no synthetic fertilizer. It is already achieving impressive productivity while simultaneously contributing to climate solutions.

The Pesticide and Antibiotic Treadmill

In addition to the harms and hazards of synthetic fertilizer, problems created by pesticides are also inevitable within the industrial model.

Pushing the overuse of pesticides is the evolutionary process itself—especially within farming systems where uniform crop varieties are grown over large areas. Over generations, insects and pathogens, like all organisms, develop resistance to threats. In Asia, the "brown plant hopper"—which threatened rice in the early years of the Green Revolution—has returned in recent years in "outbreaks in record numbers," notes a study in the *Journal of Asia-Pacific Entomology*. Insecticides are largely to blame, say the authors, because they harm the "natural enemies" of the plant hopper. Also implicated is the "heavy use of nitrogen fertilizer."[85]

In the United States, scientists have recorded five hundred cases of insects developing resistance since the introduction of insecticides.[86]

Weeds, of course, also develop resistance. Given the extensive use of glyphosate—the active ingredient in the herbicide in Roundup, promoted by Monsanto together with its "Roundup Ready" GMO corn and soy—it is not surprising that by 2013 almost half of U.S. farmers studied faced glyphosate-resistant "superweeds."[87]

Confronted with the widespread outbreak of weeds resistant to glyphosate, Dow Chemical in 2014 won regulatory approval of corn and soy seeds that can handle its new pesticide—unfortunately, one blending glyphosate and the infamous 2,4-D, which was used by the U.S. military in the defoliant Agent Orange in Vietnam.[88] As a result, the use of the controversial herbicide 2,4-D will increase twofold to sixfold by 2020, the USDA predicts.[89]

Whether the problem is weeds or insects, what's a farmer to do?

Many see no alternatives but to apply an ever-greater amount of a pesticide at an ever-greater cost or to turn to a new one. On this chemical treadmill, we're losing ground: U.S. insecticide use multiplied tenfold from the mid-'40s to the late '80s, yet crop losses to insects almost doubled.[90] Since then, public data shows a leveling-off of insecticide use.[91]

In the 1990s, agrochemical companies began coating seeds with pesticides, especially with a class of insecticides called neonicotinoids. Sales of such pesticide-treated seed have tripled globally in the last decade and the seeds are now common for row crops.[92] Neonicotinoids are strongly implicated in what's called "colony collapse disorder." Because cultivation of so many foods depends on bee pollination, the disorder potentially threatens a third of the global diet.[93]

So consider this startling comparison: The magnitude of the threat to food production posed by the collapse of bee colonies—possibly linked to insecticides—is on par with that of climate change.

Health Hazards

And what about the immediate consequences of pesticides for humans?

Roughly 355,000 people die each year worldwide from unintentional agrochemical poisoning—two-thirds of them in the Global South, reports the World Health Organization (WHO), using data from the 1990s. It notes that three million poisonings are reported annually, both accidental and deliberate.[94] Unfortunately, WHO offers no current estimates, but arguably the problem has worsened.

Consumers as well as workers are exposed. People tested in the United States carry in their bodies on average residues of thirteen pesticides, based on 2003 data from U.S. Centers for Disease Control.[95] The exposure, of course, is worse for farmworker families. Their children, who are exposed to pesticides through their mothers' milk, are particularly vulnerable.[96] Two dozen studies of the effects of prenatal exposure to a widely used class of pesticides—of which glyphosate is one—show measurable harm to children's cognitive, behavioral, and motor development.[97]

And what is the true extent of this harm?

Unfortunately, given that it is a vital question, we know surprisingly little. What we do know is that among EPA-registered pesticides marketed today, forty chemicals in them are classified as "known, probable, or possible" cancer-causing agents, according to the U.S. President's Cancer Panel.[98] And a number of studies find a correlation between pesticide exposure and increased risk of a wide range of cancers.[99]

Fifty years ago Rachel Carson's *Silent Spring* awakened the nation to pesticides' harm to wildlife and to human life,[100] and within ten years, DDT had been banned.[101] Yet, the threat from pesticide overuse has intensified. Echoing Carson, a twenty-five-year EPA veteran, E. G. Vallianatos, exposed what he experienced as industry's domination of the agency in his 2014 book, *Poison Spring*.

"With few exceptions . . . chemicals [are] dumped on the market without ever having been tested," he writes. Moreover, to justify approval, "EPA staffers routinely cut and paste studies by the very industries they are supposed to be regulating."

At the same time, "a market-good/government-bad" ideology has taken hold in our culture, influencing Congress to cut funding for testing pesticides, even as their numbers have increased. So testing labs serving the EPA's pesticide divisions dropped from a dozen in 1971 to just two by 2004.[102]

Also disconcerting is that in real life we're exposed to several pesticides at the same time, but that's not how they are tested. Regulatory agencies do not commonly test chemical combinations, contrary to the President's Cancer Panel recommendation. In one study that did, mice exposed to a mixture of aldicarb, atrazine, and nitrate in concentrations found in U.S. groundwater showed signs of altered immune, endocrine, and nervous system function.[103] Moreover, warns the President's Cancer Panel, pesticides contain many ingredients that their manufacturers call "inert" but that in fact are toxic; yet their testing is not required.[104]

One particularly worrisome pesticide is glyphosate.

As mentioned above, it's the active compound in Monsanto's Roundup, the most widely used herbicide in the United States: Its use doubled just in the first six years in this century.[105] Peer-reviewed studies link glyphosate exposure—for both consumers and farmworkers—to a range of ailments including gastrointestinal disorders, heart disease, infertility, and depression.[106] Even more troublingly, in the United States the maximum glyphosate residue allowed on a range of foodstuffs is typically higher, sometimes two times or more, than what is permitted in Europe, and higher than what is recommended by the FAO and the WHO.[107]

None of this has moved U.S. authorities to act. Then in 2015, the cancer research arm of the WHO declared glyphosate a probable carcinogen.[108] After four decades of this pesticide's harm, with citizen pressure, this judgment could be a turning point.

Atrazine, the second most widely used herbicide in America, though long banned in Europe, has been found in 90 percent of drinking water samples and more than a quarter of groundwater samples randomly tested by the USDA in 2012.[109]

A brief story raises worrying questions about how little we understand the risks to which such pesticides expose us.

In the 1990s, biology professor Tyrone Hayes of the University of California, Berkeley, began extensive research on atrazine's possible effect on amphibians, a project originally funded by the herbicide's maker, the world's second-largest agrochemical company, Syngenta. Hayes discovered disruption of sexual development in frogs, even in some exposed to atrazine at a level thirty times below EPA's legal limit.[110] Then, a 2011 study by other scientists revealed a correlation between atrazine and women's disrupted menstrual cycles, along with reduced reproductive hormone levels, even when the herbicide was present in tap water at levels below the allowed concentration.[111]

As you read about Hayes's findings, keep in mind another troubling reality: Twenty-three midwestern cities and towns sued Switzerland-based Syngenta for contaminating public drinking water, accusing it of "concealing atrazine's true dangerous nature." When Syngenta settled the suit in 2012, documents were released revealing the company's extensive and deceptive campaign to discredit its former consultant Hayes and even his wife.[112] Hayes defended his research and has retained his University of California professorship.

Finally, antibiotics.

In the United States, about 80 percent of antibiotics are used not to cure people but to speed growth in livestock.[113] Antibiotic overuse in animals is helping to spread antibiotic-resistant strains of bacteria to humans, and the toll is heavy. "More than two million people are sickened every year with antibiotic-resistant infections," and at least 23,000 people die as a direct result, reports the Centers for Disease Control.[114]

While part of the problem of antibiotic-resistant pathogens comes from overuse and other improper use of antibiotics in humans, overuse in livestock is strongly implicated. Be aware: Examining public data, Food & Water Watch researchers found widespread antibiotic-resistant germs in a range of meats. In 11 percent of ground-turkey samples the researchers discovered resistant salmonella, and in half of chicken samples they found resistant E. coli.[115]

In 2006, the European Union's ban on antibiotics to promote growth in livestock took effect, while the United States is just beginning to grapple with the crisis. In 2014, the FDA advised pharmaceutical companies to voluntarily alter product labeling in order to reduce the use of "medically important antimicrobials" in livestock, except in treating disease. If the labels are changed, it would inhibit the use of these medicines simply to enhance livestock growth.[116]

"Fossil Water" Dwindling

Beyond these multiple harms, now let's explore how the industrial model of farming also squanders precious resources. One is water.

By "fossil water" we mean underground water stores called aquifers, which make up 30 percent of the world's freshwater.[117] These aquifers—some of which have accumulated over millennia—provide much of the water agriculture draws on. In fact, about 40 percent of the world's food depends on irrigation, which draws largely from groundwater.[118]

Unfortunately, it's being rapidly depleted.[119]

In the United States, the Ogallala Aquifer—one of the world's largest underground bodies of water—spans eight states in the High Plains. It supports our nation's farm belt from South Dakota all the way to Texas and supplies almost a third of the groundwater used for irrigation in the whole country. Scientists warn that within the next 30 years over one-third of the southern High Plains region will be unable to support irrigation.[120] In Kansas the picture is bleaker: If existing trends continue, scientists estimate that about 70 percent of the Ogallala groundwater there will be depleted by the year 2060.[121]

Clearly, a model of agriculture that is depleting one of our food supply's most essential resources will not sustain us.

Shrinking Diversity, Shrinking Nutrition

With its narrow focus on producing crops with the highest return, the industrial model of agriculture also leads to what's called "monoculture"—growing a single crop over vast acreage, rapidly diminishing genetic richness. Over millennia peasant farmers bred almost two million plant varieties, but during the last century—as we've moved toward monoculture—roughly three-fourths of plant genetic diversity has been lost.[122]

As discussed above, with the introduction of hybrid seeds, farmers became dependent on purchasing new seeds each year. Then came the patenting of seeds and control over genetic material by a fast-concentrating seed industry, as we highlight in Myth 6. Once in this system, farmers must purchase patented seeds each planting season and cannot share them or use them to breed new seeds themselves, as farmers have done since the beginning of agriculture.

All these changes contribute to genetic impoverishment.[123] Fortunately, while this system is spreading, it is not yet the norm in the Global South, where most small farmers still use seeds they select and save from season to season.[124] The direction globally, however, is clear: toward monocultures and loss of plant varieties, narrowing our genetic options and increasing vulnerability to weather and disease—a risk warned against in the old adage *not* to put all one's eggs in one basket.

The genetic uniformity of the Irish potato monoculture, recall from Myth 2, contributed to famine killing a million people. In 1970, just one pathogen, known as Southern Corn Leaf Blight, spread rapidly across the United States, in some regions cutting yields in half. The following year, just when demand for different varieties of corn was high but availability was low, the blight struck again.[125]

Imagine the huge handicap today as farmers worldwide must quickly adapt to climate change, breeding new varieties for new conditions—all while working with rapidly diminishing genetic variety.

Moreover, within the industrial model of agriculture—with its assumption of disassociated parts—there's no built-in connection between food production and nutrition. So the model has led not only to the extreme narrowing of our diets but also to the wider degradation of food quality identified in Myth 1.

In the process, the plant foods that supply most of our calories have shrunk to a handful.[126] Today, just three plant species—rice, maize, and wheat—provide nearly 60 percent of the calories and proteins that we humans get from plants.[127] This narrowing of our diets contributes to nutritional deprivation discussed in Myth 1, showing up today in the alarming estimate that two billion of us suffer from deficiency in at least one nutrient essential to health.[128] Over 30 percent of the world's people consume insufficient iron, for example, with especially deleterious effects for women.[129]

Recall the physician in rural India we quoted in that chapter, who told us that most of his patients had adequate calories yet were ill with diabetes and heart ailments. They largely eat polished white rice, rather than the variety of grains, lentils, and vegetables common in the diets of many in earlier generations. So, while we think of poor Indians and American children as worlds apart, here they are in a similar fix: In the U.S. system—also disconnecting production and nutrition—about 40 percent of the calories children consume are now nutritionally empty.[130]

Animal Factories—Harm We Can't See and Don't Count

The industrial model is based, as we've said, on viewing all elements of a system as separate—neither interacting with, nor affecting, other elements. In agriculture, this way of seeing shows up in perhaps its starkest form in the U.S. livestock industry.

As recently as five decades ago, most animals that Americans ate were raised on farms, but today concentrated animal feeding operations, or CAFOs—about 5 percent of all livestock producers—account for more than half of U.S. food animals.[131]

For a lot of people these operations seem more like factories than farms, so CAFOs also get called "factory farms." Whatever one calls them, they entail large-scale and high-density confinement of animals. CAFOs are spreading fast. Globally, 80 percent of growth in the livestock sector is in animal feeding operations not integrated into farms, reports the FAO.[132] They provide roughly two-thirds of the world's poultry and egg production, and about half of the world's pork.[133] In the United States, 80 percent of chickens and pigs live confined—broiler chickens typically in a space no bigger than a sheet of paper, and gestating pigs commonly in 2-by-7-foot crates.[134]

Animals' natural behavior is of course suppressed, creating physical and mental stress, which commonly shows up in self-mutilation and excessive aggression.[135] A European Commission report noted abnormal repetitive behaviors such as bar biting and "sham chewing" (chewing on nothing).[136] To limit the damage, chickens' beaks are seared off so they cannot peck, piglets' tails are docked so other pigs can't bite them, and cattle's horns are either sawed off or chemically shortened.[137]

Chickens endure feather loss, blisters, tumors, deformities, and lung and heart disease, as well as "leg collapse" because breeding has made their bodies too heavy for their limbs to support.[138] Dairy cows often develop bacterial infections, while veal calves suffer from digestive disorders and anemia.[139] Some animal infections potentially threaten humans, too.[140]

Farmer and poet Wendell Berry powerfully captures the cruelty of the CAFO system in his *Stupidity in Concentration:*

The designers of animal factories appear to have had in mind the example of concentration camps or prisons, the aim of which is to house and feed the greatest numbers in the smallest space at the least expense of money, labor, and attention. To subject innocent creatures to such treatment has long been recognized as heartless.[141]

Berry's words resonate with us. But we want to stress that in the case of factory farms, unlike concentration camps, no one intentionally sets out to make them heartless. The real crisis is that even the most compassionate among us live within a system in which, increasingly, heartlessness is built in.

And this industrial system brings additional serious threats.

Focused on the highest yield per animal, a producer tries to speed up a creature's weight gain. One result is that CAFO operators treat animals with antibiotics, as well as arsenic, hormones, feed additives, and vaccines, both to accelerate growth and to keep the animals from getting sick in crowded conditions conducive to the spread of disease. Above we noted the implications for human health of the use of low doses of antibiotics in factory farms. Hormones used in livestock have been linked to disrupted child development as well as reproductive complications and some cancers.[142] Each year, food-borne illnesses sicken more than 9 million and kill well over 1,400 Americans—of which illnesses more than 40 percent were attributed to land animal foods.[143]

Disease-promoting conditions also lead to food waste. In 2010, for example, two Iowa egg companies recalled more than half a billion eggs for fear of deadly salmonella contamination.[144] That's nearly two eggs tossed for every American.

Ill health for humans also flows from CAFOs' concentration of livestock excrement. Decomposing animal waste produces not only foul odors but noxious gases—over 160, including hydrogen sulfide, ammonia, carbon dioxide, monoxide, and methane. Together with organic dusts from facilities, they contribute to respiratory and neurological problems in workers, as well as those living nearby.[145] The closer children live to CAFOs, the higher their risk of asthma, found a study supported by the Centers for Disease Control.[146]

Concentration of manure creates environmental havoc, too. Some factory farms generate more raw waste annually than do entire cities.[147] While large quantities of manure can be used on nearby cropland, it is often sprayed in excess, leading to runoff and subsequent surface-water pollution. Manure lagoons and holding ponds can also cause leaching of nutrients into groundwater, with all the negatives described above.[148]

All this, yet "no federal agency collects consistent, reliable data on CAFOs," laments the U.S. Government Accountability Office.[149]

We include CAFOs in a book about world hunger for at least three reasons.

As livestock production moves toward the industrial model in the Global South, many small producers' livelihoods are being destroyed, putting them at risk of hunger.[150] Second, as we've noted earlier, grain-fed cattle—a big part of CAFOs—return to humans only 3 percent of the calories that the animals eat in feed; so CAFOs help shrink the food supply and thus contribute to the illusion of scarcity.[151] Third, unawareness among many well-meaning Americans of the cruelty built into the factory-farm system seems to mirror a lack of awareness of the cruelty of hunger itself. Progress is made as more and more of us become conscious of all three dimensions and come to understand that all are needless.

Before going on from this review of the harm generated by the industrial model of agriculture, let's return to a country that many still see as its big success story—India.

AND HOW HAS THE GREEN REVOLUTION'S TESTING GROUND FARED?

The industrial model of farming in India took off in the 1960s when the government chose the large northern state of Punjab—the nation's "breadbasket"—to be its testing ground for Green Revolution hybrid seeds and chemical inputs.

About forty years into the experiment, in 2007, the Punjab State Council for Science and Technology released a 270-page report on the environmental and social consequences of this choice.[152] Upon the report's release, *The Economist*—long a proponent of chemical-and-capital-intensive farming—summarized its findings under the jarring headline: "Chemical Generation: Punjabis Are Poisoning Themselves."[153]

"In an effort to produce more grain," the report concludes, the state "has been overexploiting its land and water resources by changing traditional cropping patterns and resorting to high-input agriculture (instead of low-input, ecologically friendly farming practices)." Among problems arising from the "overintensification of agriculture," the report cites: groundwater depletion, reduced soil fertility and micronutrient deficiency, pesticide residues, reduced biodiversity, soil erosion, air and water pollution, and "overall degradation of the rather fragile agro ecosystem of the state."[154] In 2013, Punjab's minister of health reinforced the report's findings, stating, "A large number of cancer cases are due to excessive use of pesticides in the fields."[155]

The report also laments the fraying of the "socio-economic fabric of the state," divided between those who can afford "high expenditure life styles" and farmers struggling because of the "decrease in farm income." It describes the "country-wide phenomenon" of farmers being "crushed by heavy load of debt" and "committing suicide to escape from the vicious circle of indebtedness."

Also striking is that increasing production—the whole point of this radical change in the countryside—could not be sustained. A majority of Punjabi farmers, notes the report, have seen stagnant or declining yields of basic food crops.

The Punjab State Council for Science and Technology's report ends with a call for the Punjab to move to "sustainable low-input agriculture," including "organic agriculture."[156]

BUT DON'T GENETICALLY MODIFIED SEEDS SOLVE THESE PROBLEMS?

As all of these limits and hazards of the industrial model have become increasingly evident, its defenders suggest that new plant breeding technology can address them.

Can it?

In the 1970s, scientists began developing methods to create "genetically modified organisms" (GMOs) by inserting specific genes from one organism into the DNA of another. Proponents see the technique as a faster, targeted means to the same ends as traditional crossbreeding.[157] First introduced in the 1990s, GM seeds are often promoted as the solution to hunger, as in this ad by GMO manufacturer Monsanto: "Worrying about starving future generations won't feed them. Food biotechnology will."[158] And, today, this one GMO manufacturer controls the genetically engineered traits in seeds planted on more than 80 percent of U.S. corn and over 90 percent of U.S. soy acreage.[159]

So let us take in turn five common claims about GMOs and explore how they hold up to peer-reviewed evidence and field experience.

Five GMO Claims

1. GMOs have been thoroughly tested.
2. There is a scientific consensus that GMOs are safe.
3. GMOs produce higher crop yields, often with fewer pesticides.
4. Farmers and consumers worldwide are choosing GMOs.
5. GMOs are needed to feed a hungry world.

Claim One: GMOs have been thoroughly tested.

GMOs took off in the 1990s, and the United States has become the world's number-one GMO producer as well as consumer—all with no prior independent testing or broad public debate about their possible risks, or even about whether they were needed.[160]

Many Americans believe that GMOs are carefully regulated, but "contrary to popular belief, the FDA [Food and Drug Administration] has not formally approved a single GE [genetically engineered] crop as safe for human consumption," conclude William Freese and David Schubert in *Biotechnology and Genetic Engineering Reviews*.[161] In 2014, the *New York Times* reported that "the review process for new G.M.O. plant foods is voluntary. Producers are asked only to consult with the F.D.A. The agency 'does not conduct a comprehensive scientific review of data generated by the developer,' according to F.D.A. documents. Officials rely on producers to do their own safety and nutritional assessments, and they review summaries of those assessments."[162] Moreover, the FDA reviews but does not approve industry assessments.[163]

In the United States, some GM seeds are produced by inserting a gene from the Bt bacterium (*Bacillus thuringiensis*) that produces an endotoxin deadly to certain groups of insects. These seeds, actually carrying their own insecticide, must be approved by the EPA rather than the FDA. However, many fewer tests are required than for chemical pesticides—presumably because these crops are considered less potentially harmful.[164] (Recall the disturbing revelations above from a twenty-five-year EPA veteran about the dominance of the pesticide industry in the agency's testing practices.) In any case, our government's standards for approving pesticides are markedly weaker than those of many European countries.[165]

Long-term studies are not required for approval, and studies such as the Caen professor's that we discuss below are still a rare exception. Even in the EU, which officially adheres to the "precautionary principle," it wasn't until late 2013 that its food safety authority required a ninety-day feeding study.[166]

This pattern of failure of oversight of GMOs began more than two decades ago. The record shows that in 1992 the FDA acceded to the industry's requests and declared GMOs "substantially equivalent" to non-GM-bred crops. That one decision—ignoring the strong doubts of some of its own scientists—meant biotech companies could avoid independent, long-term testing and monitoring prior to release.[167] Also lacking is post-release monitoring of GMOs' environmental and health impacts.[168]

Claim Two: There is a scientific consensus that GMOs are safe.

Those claiming a scientific consensus on GMOs argue that genetic engineering is no different from traditional plant breeding. In fact, in genetic engineering, "DNA is isolated from its normal cell regulation, unlike conventional breeding that works with the whole cell system, developed by evolution," notes Christoph Then, executive director of a German biotech assessment organization.

Thus, one foundational concern among many scientists is that genetic engineering could be inherently more unpredictable than traditional plant breeding. Inserting a gene into another organism can result in multiple, unintended DNA changes and mutations, note scientists in *Biotechnology and Genetic Engineering Reviews*: "Unintended effects," they report, "are common in all cases where GE [genetic engineering] techniques are used."[169] An example was the inadvertent transfer of a Brazil nut allergen to soybeans in 1996.[170] All plant breeding risks unintended consequences, but genetic engineering can introduce a range of genes and traits—some never before present in our food supply or in plant ecosystems.

Evidence of harm has been found in peer-reviewed studies.[171] In one, pigs on a GMO diet were 2.6 times more likely to get severe stomach inflammation than were control pigs.[172] GMO proponents dismiss studies indicating harm, claiming they used inappropriate lab animals or feeding methods.[173]

In 2013, for example, University of Caen molecular biologist Gilles-Eric Séralini published a peer-reviewed study in *Food and Chemical Toxicology* that showed evidence of kidney and liver damage, hormone disruption, and more and earlier tumors in rats fed a GM diet. It was one of the few studies of the effects of GMOs over the life of a rat, typically two years.[174]

Immediately, it came under heavy attack, in part linked to the biotech industry.[175] Then, almost three hundred scientists and scholars signed a statement emphasizing the lack of scientific consensus on GMOs and calling for long-term, independent studies. Scientists also noted Séralini's research protocols are in line with those used in the biotech industry's own studies.[176]

The journal publishing Séralini's study then retracted it, saying its findings were "inconclusive," even though this reason is not among the journal's published standards for retraction.[177] In 2014, the study was republished in *Environmental Sciences Europe*. With the new release, Séralini made the study's supporting data publicly available—a level of transparency not found in industry studies.[178]

Two difficulties in coming to sound public judgment on GMOs are the dearth of long-term studies and the powerful influence of the industry itself. Consider this telling contrast: A review of ninety-four published studies on the effects of GM food or feed products found that of the studies in which an author is affiliated with the biotech industry, none revealed either health-related risks or lower nutrient values associated with consuming GM food or feed. By contrast, almost a quarter of the studies with no author affiliation with the biotech industry did find problems associated with consumption of GMO products.[179]

Thus, there is no scientific consensus on GMO safety.

Even in GMO studies where significant harm is not indicated, scientists have expressed concern, noting that "much more scientific effort and investigation is necessary" before they can be "satisfied" that eating food containing GM material is not likely to cause any health problems. Underscoring the lack of consensus, in early 2015 the UN Special Rapporteur on the Right to Food, legal scholar Hilal Elver, went on record opposing GMOs, noting that many unanswered questions about them remain, including concerns about long-term health effects.[180] Other scientists say, simply, that the technology is not worth the risks.[181]

Finally, note that as a class GMOs cannot be proved safe because each new GMO presents a new risk-benefit profile that needs independent testing.

Claim Three: GMOs produce higher crop yields, often with fewer pesticides.

In 2014, the U.S. Department of Agriculture stated flatly that after fifteen years genetically modified seeds "have not been shown to increase yield potentials . . . and [the yields] may be occasionally lower."[182] Manufacturers also like to claim that often GM crops require less pesticide than was previously used.[183] But official government estimates show that since GMOs arrived on the scene, pesticide use has held steady while an independent review found that overall U.S. use increased slightly.[184]

Claim Four: Farmers and consumers worldwide are choosing GMOs—proving that GMOs are meeting real needs.

Actually, just four countries account for more than 80 percent of GM crops, and many farmers are using GM seeds because they have no other seed option.[185] Among consumers in the United States few can be said to be choosing GMOs since most Americans believe they have never consumed any. Yet 70 percent of U.S. processed foods are derived from GM crops.[186]

One reason so few Americans are aware of GMOs?

There's no national requirement that GMOs be identified on product labels—despite a *New York Times* poll indicating that 93 percent of Americans favor mandatory GMO labeling, which already is the policy in sixty-four countries.[187]

At the state level, laws requiring labels on all foods made from GMOs are under way. In 2013 and 2014, three states passed labeling laws. Two of them are Connecticut and Maine, where the laws will take effect when northeastern states totaling twenty million residents pass labeling laws. Vermont's law requires GMO labeling to begin in 2016.[188] So far, the biotech industry's advertising campaigns have helped thwart labeling laws in two big states, California and Washington.[189]

Claim Five: GMOs are needed to feed a hungry world—both to provide enough calories and to add more nutrients to the diets of poor people.

We underscored in Myths 1 and 2 that not only is the global food supply abundant, but we have vast unused potential for increasing available food without genetic engineering. And, as noted, GMOs have not brought the promised yield gains.[190]

Given all this, promoting GMOs as a solution to hunger seems indefensible.

But, some still ask, what about biotechnology as a cure for specific and widespread nutritional deficiencies in poor countries?

GMO manufacturers promise nutritional boosts: overcoming, for example, vitamin A deficiency that causes blindness in as many as half a million children each year. For more than twenty years, the widely discussed genetically engineered, vitamin A–rich "Golden Rice" has been in development. But while debate over it rages, traditional plant breeding, without GMOs' risks, has produced vitamin A–enhanced corn and sweet potato varieties—already being enjoyed in Africa—and in less time with less money.[191]

But let's pause.

We fear debate over the best way to fortify specific foods sends the message that the crisis of nutritional deprivation is caused by deficiencies in these foods. But we can't blame the plants! Focusing on fortification can divert attention from the real reasons two billion of us suffer vitamin and mineral deficiencies: the combined impacts of poverty—including lack of access to proper sanitation and too little income to buy a range of healthy food—and the spread of nutritionally useless, sugar-laden processed foods heavily advertised by a handful of dominant corporations.

No scientist would be bothering to breed nutrients such as vitamin A into cereals and root crops if people everywhere had enough land to grow, or income to buy, fruits and vegetables already rich in these nutrients.

In responding to these five specific claims, let us not allow the hot debate over "the seed" to blind us to the problem of "the system": GMO seeds reflect and reinforce the industrial model of agriculture, with its inherent dependencies and dead ends.

During the century from the 1860s until the 1970s, U.S. farmers benefited from seeds developed through publicly funded agricultural research. Now corporate funding for plant breeding climbs while public funding shrinks.[192] So farmers here and around the world increasingly depend on seeds and other farm inputs developed and controlled by an ever-smaller band of global corporations.

Seed biotechnology exacerbates the problem. Monsanto's GM herbicide-tolerant seeds, for example, are genetically engineered to work only with certain herbicides, and farmers must therefore purchase the entire corporate package. Farmers must sign legal agreements that they may not save the seeds from their harvests, even for their own use. "In the United States, corn and soybean farmers no longer even actually buy seed—they merely lease a one-time use of patented, transgenic seed from a handful of dominant companies," observes University of Wisconsin professor Jack Kloppenburg, author of *First the Seed*.[193]

By increasing dependency on purchases from a highly concentrated industry, seed biotechnology ends up furthering the concentration of power at the root of hunger itself.

INDUSTRIAL AGRICULTURE & OUR DEEPENING DEMOCRACY DEFICIT

So far, we've stayed close to the ground. Now let's pull back to see the big-picture impact of industrial agriculture, asking: Does it undermine democracy?

By democracy, we mean here not a particular political structure but core democratic values—those of the wide sharing of power and its accountability—making possible transparency and self-determination. Only as we realize these values can we end hunger.

Industrial agriculture undermines democracy by concentrating economic power so that it can block transparency as it sways public policy. Recall how in the 1990s corporations developing GMOs were able to introduce a virtually untested technology without independent assessment and without public debate; as well as Syngenta's campaign to discredit a scientist whose research brought to light the risks of its product atrazine.

Evidence of GMO manufacturers' willingness to use their power to stifle scientific inquiry came early. In 1998, the first critical look at GMOs addressed to the broad public—cowritten by toxicologist Marc Lappé (Frances's late husband)—was almost in print when Monsanto threatened the publisher with legal action. The frightened publisher canceled the book, and only after many months did the authors find a more courageous publishing house.[194]

Recall how GMO manufacturers use their deep pockets for lobbying and manipulative advertising to defeat GMO labeling in the United States, despite overwhelming public support for it.

Democracy is compromised as well when the biotech industry uses U.S. government influence for its ends. In 2013, over 900 secret cables—made public via WikiLeaks—exposed the extensive U.S. diplomatic campaign to "pursue an active biotech agenda," breaking down resistance to GM products in other countries, including Nigeria, Kenya, South Africa, Nicaragua, and Vietnam.[195]

And in Myth 8 we will see that in Africa our government uses the "carrot" of economic aid to entice governments to "reform" laws and regulations not to the liking of foreign biotechnology companies.

WHY CONTINUE WITH A FAILING MODEL?

In this chapter, we've offered evidence that the industrial model of agriculture contributes to concentrating economic and political power that not only deepens inequalities at the root of hunger but also entails massive loss and risk. We've seen examples of concentrated private power exerting undue influence within public agencies, as well as its actively blocking transparency about hazards and seeking to blind us to alternatives.

With this understanding, we hope it is now easier to see why hunger ensues no matter how much food is produced.

Having reviewed the multiple negative consequences of this path, we must ask the obvious: *Why have so many continued along it?*

One easy answer is that the model's deeply entrenched way of seeing the world is backed by long-standing structures of power—making it inherently difficult to challenge and to dislodge.

An additional challenge is that the industrial model is continually being packaged and repackaged as pro-poor and environment-friendly. A recent example is "sustainable intensification"—defined by the Royal Society in the U.K. as increasing production "without adverse environmental impact" and without cultivating more land. It welcomes all tools, from genetically engineered seeds to organic techniques. Another new term is "climate smart" agriculture touted by agribusiness interests.[196] Both sound great.

But neither can work. Each ignores relationships of power that ultimately determine the impact of any technology. Where those relationships are highly skewed, even a technology that seems benign can further disempower poor people.

For example, "sustainable intensification" is promoted by Feed the Future, a U.S. government global initiative discussed in Myth 8. It calls for small farmers' participation in setting research priorities. It has failed, however, to follow through on that vision, even though it's clear to most people that whoever sits at a decision-making table will determine whose interests are served. Feed the Future does not challenge the industrial farming model that reflects a worldview of disassociated parts. It doesn't acknowledge power as embedded in relationships that determine who gains. Feed the Future therefore promotes integrating small farmers into the global market as a positive step, without addressing the rules governing trading and marketing, described in Myth 7, that are set against the interests of these farmers.[197]

The lesson?

Knowing the power of language, we can take care to look behind the label to analyze the impact of any proposed direction on the many interacting elements of a system. Only in this way can we know whether it will help or hurt the world's hungry people.

Finally, and going deeper still: One should never underestimate the power of fear in keeping us on this failing path. Those wearing blinders to the negative consequences we raise in this chapter make frightening declarations: "If everyone switched to organic farming, we couldn't support the earth's current population—maybe half," declared Nina Fedoroff, former president ot the American Association for the Advancement of Science, in 2008.[198]

If we believe claims that there are no alternative ways to produce enough food for all, then, of course, it's hard to acknowledge the losses and dangers of the dominant approach, much less envision other paths.

Fortunately, however, we need not simply envision, for other richly positive paths are already emerging. Let's now examine why they are working while the industrial model has failed.

myth 4

Organic & Ecological Farming Can't Feed a Hungry World

MYTH: Despite the many downsides laid out in the previous chapter, industrial agriculture and GMOs are essential. If all of us are to eat, we can and must learn how to make industrial agriculture more efficient and sustainable. If instead we were to attempt widespread food production without synthetic fertilizer and pesticides—organic farming—we'd be headed back to subsistence farming. Farmers wouldn't make a decent living, and there wouldn't be enough food for all of the world's people—much less the even larger population coming. Many would starve.

OUR RESPONSE: It's easy to absorb the notion that any alternative to industrial agriculture is romantic, a lovely vision that simply can't work. In truth, however, it's the industrial model that's now proved not to work, failing both to end hunger and to be a feasible path forward. Fortunately, mounting evidence points to other paths that are not only productive but also realigning human actions with the Earth's natural systems—as well as aligning human societies with human needs so that all of us can partake in the bounty.

Before jumping into this rich universe, first note that some studies show industrial agriculture producing higher yields than these alternative paths.[1] Taking a wider view, however, we'll see in this chapter and throughout our book that many small farmers in the Global South who reject industrial agriculture and adopt ecological practices have the opposite experience: They are enjoying yield increases, some quite dramatic. In the Global North differences in yields are often small and diminishing, while the benefits of shifting away from industrial farming are huge.[2]

Plus, in a world of abundant food—rife with the waste and inefficiencies tallied in previous chapters—wouldn't it be a shame to let fear of the possibility that yields might be even modestly lower block us from seizing solutions at hand?

It is precisely this shift—from a narrow focus to a wider view—that defines the very different lens we invite you now to try on.

If the downfall of industrial agriculture is its view of life as distinct, disconnected spheres, the key to the power of emergent, alternative approaches is a systemic view—seeing life's multiple dimensions in relation to one another, all connected and interacting. This systems approach encompasses the quality of our relationships both with other human beings and with the Earth. It is called "agroecology."

Thus, agroecology is more than a different way of growing food. Unlike the industrial model, it's not power-concentrating, but rather an evolving practice of growing food within communities that is power-*dispersing* and power-*creating*—enhancing the dignity, the knowledge, and thus the capacities of all involved. Agroecology thus helps to address the powerlessness at the root of hunger. Applying a systems view to farming, agroecology unites ecological science with time-tested traditional wisdom and farmers' ongoing experience. It is also a social movement, growing from and rooted in distinct cultures worldwide, as we take up in Myth 10.

As such, agroecology is not a formula. Rather it is a range of integrated practices, which we'll soon describe, adapted and developed in response to a farm's specific ecological niche. Agroecology weaves together traditional knowledge and ongoing scientific breakthroughs based on the integrative science of ecology.[3] By progressively eliminating all or most chemical fertilizers and pesticides, agroecological farmers free themselves—and therefore all of us—from reliance on climate-disrupting, finite fossil fuels, as well as on other purchased inputs posing environmental and health hazards.

One dimension of agroecology is organic farming, commonly understood to be farming with no synthetic pesticides and fertilizers.

"Organic" is also a legally defined certification standard for farming. Set by governments, it specifies what inputs can and cannot be used. Certified organic agriculture is one aspect of the much wider emergence of agroecological practices.

While organic farming can lay claim to a very small share—about 4 percent—of U.S food sales, it is growing rapidly. With fruits and vegetables, it's 10 percent. Even with minuscule public support compared with the billions in U.S. subsidies supporting industrial agriculture, organic farmland acreage in the United States has jumped 2.5-fold in a decade.[4]

Globally, organically farmed land more than doubled in the decade before 2011; in India, make that increase almost eightfold.[5] Today at least 164 countries are home to organic production that is certified under government-set standards and covers almost a hundred million acres.[6] Two million farmers—most of them small farmers in the Global South—are now certified organic, while many more use organic practices.[7]

> Globally, organically farmed land more than doubled in the decade before 2011; in India, make that increase almost eightfold.

The broader practices of agroecology may be spreading even faster.

Several Indian states, as we describe in Myth 10, are making significant investments in multiplying agroecological practices. In West Africa, a number of governments cooperated with the FAO to establish 3,500 "Farmer Field Schools" that train farmers in agroecological practices, already reaching 150,000 farmers.[8]

The agroecological innovation of interspersing crops and native trees, whose roots and leaves provide crops with nitrogen, is also taking hold rapidly in West Africa, as highlighted in Myth 2.

Moreover, producers of rice—which provides a fifth of the world's calories—are beginning to shift to an ecological process called "System of Rice Intensification" (SRI), mentioned in Myth 2. Breaking with tradition, in SRI rice paddies are kept moist but not saturated, plants are spaced more widely so all leaves receive sunlight, and seedlings are transplanted sooner. The happy consequence is that fewer seeds and less water and chemicals are used, while yields often increase significantly. Though the system gained attention only in the 1990s, its principles are now being applied to other crops, and farmers are beginning to try the approach in more than fifty countries.[9]

SO HOW DOES AGROECOLOGY WORK?

First, let's drop the idea that organic farming and agroecology are mainly about the absence of something—chemical inputs. They are about the addition of *a lot*. What's added are farmers' continually expanding knowledge and scientifically validated practices, all supporting the generative power of nature.[10] Agroecology is not about buying better products to coax more food out of your land; it's about using better practices to enable your land to give more—not only today but into the future. On this path, farmers learn continuously about the specific potential of their farm's own ecological niche.

"Accentuate the positive, eliminate the negative" was a catchy, if cheesy, 1940s song that even Paul McCartney latched on to. It pretty well sums up this model, one that takes advantage of the positive synergies within nature while cutting or eliminating the losses and the costs built into industrial farming.

A striking difference between the industrial model and agroecology can be captured in one word: "diversity." The opposite—uniform crop varieties and pesticides applied over vast acreage in the U.S. Midwest—typifies the industrial model. Agroecology thrives on the interactions not only among diverse crops but among insects, birds, animals, trees, and flowers, as well as among worms and soil microbes.

Now let's jump from this big picture to the ground level—in fact, right into the dirt.

Support the Invisible Helpers

And here, the first question is, What *is* dirt?

In industrial farming, soil is thought of largely as an inert substrate on which to work. In agroecology, soil is experienced as a living community whose health determines almost everything about what grows and how it grows.

Suspending the big debates of the previous chapter, we aim here to convey to our readers a feel for the Earth. So we begin with this question: Who are the members of this largely invisible "soil society" beneath our feet?

Microbes.

The U.S. Department of Agriculture (USDA) explains that microbes are "yeasts, algae, protozoa, bacteria, nematodes, and fungi that process soil into dark brown, spongy stuff called 'humus,' known for its pleasant, earthy smell."[11] Microbes "decompose the tough plant and animal residues in and on the soil and bring nitrogen from the air into the soil to feed plants."[12]

These hearty soil microorganisms do all this work, even though they "make up only one-half of one percent of the total soil mass."[13]

For healthy soil, an ecologically attuned farmer fosters conditions that enable soil organisms to transform "organic matter"—defined as "anything . . . that once lived"—into forms that nourish plants.[14] Organic matter—in weight only 2 to 10 percent of soil—is the secret ingredient enabling soil to support healthy plants.[15] Its carbon is "the main source of energy for the all-important soil microbes," explains the USDA, and "also the key for making nutrients available to plants."[16] Soil organic matter is a key source of minerals that plants need to grow; and as some of it decomposes very slowly, organic matter provides the substrate on which nutrients are transported to plant roots.

Beyond seeing soil as a living community, some ecological scientists and farmers are experiencing another mind shift as well.

They are now learning that *aboveground* synergies also matter in plant growth. Practitioners of SRI that's revolutionizing rice cultivation in dozens of countries in Asia and Africa—and now extending to crops beyond rice—stress that its success depends not only on microorganisms within the soil but on plants aboveground, too. "Bacteria and fungi . . . in, on, and around plants (and animals) provide the substrate for vast and intricate soil-plant 'food webs,'" write SRI specialists. These critters range from invisible microbes to creatures we can see, all feeding on one another and improving the environment for all.[17]

Beyond this "food web"—i.e., who eats whom to keep a system going—are many wider, beneficial interactions that scientists call the "interaction web": behavioral, chemical, and other interactions, such as when species produce resources for each other.[18]

Since broad-spectrum chemical applications can disrupt these interactive webs essential to life, it's easy to see how their use can be harmful to growing plants.

Going Plowless

Organic matter is key to this rich life within and above the soil; yet it has dropped by half in what the USDA calls "modern cultivated soils."[19] So today, a primary task of farmers is to reverse this loss. And that means refraining from both plowing and leaving soil barren. *Why?* Both turn raindrops into soil catastrophes "as devastating to soil microbes as a combination of an earthquake, hurricane, tornado, and forest fire would be to humans," reports the USDA.[20]

Perhaps counterintuitively, plowing "reduces water infiltration, increases runoff . . . damages the structure of the soil, and makes soil more susceptible to erosion."[21] Some might assume that because it breaks up the soil, plowing allows rain to penetrate, which is good. But it turns out that water infiltration doesn't depend on our machines. Earthworms do a great job: Their "burrowing creates continuous pores linking surface to subsurface soil layers," the USDA emphasizes.[22]

Common plowing practices in industrial agriculture also stir oxygen into the soil, stimulating too-rapid decomposition of essential organic matter.[23] Plus, such plowing can bury "the organic matter too deeply for the aerobic (air-breathing) microbes to work on it, and also seals off those wonderful worm tunnels so air and water can't penetrate as easily," notes coauthor of *The Soul of Soil* Grace Gershuny.[24]

In industrial agriculture, after plowing and harvesting, soils are often left uncovered—as mentioned above, exposed to harm from wind and rain, sometimes for months.

But there are other ways to work the land.

One approach, called "no-till" farming—a tool among a suite of practices collectively known as "conservation tillage"—was popularized by American agronomist Edward Faulkner's *Plowman's Folly*, published in the 1940s in America following the Dust Bowl's devastation. With no-till, farmers leave crop residues and weeds on their fields after harvest, rather than turning them into the soil. To place seeds in the ground, farmers use specially designed planters that guide seeds into the soil beneath the plant residues on the surface.[25]

No-till practices help the soil to retain organic matter and water and to sequester carbon. Taken up in earnest in the 1970s, primarily in North and South America, no-till acreage has spread threefold globally since 1999 to roughly three hundred million acres, an area almost the size of Venezuela.[26]

Critics are concerned about the increased herbicide use that's been observed in some no-till systems. Others say this isn't the fault of no-till. Rather, herbicide use flows from unsustainable practices, such as monocropping and leaving large tracts of soil exposed.[27] So to be beneficial, no-till or low-till must be integrated into ecological farming, not practiced as an isolated quick fix.[28]

Feed the "Soil Society": Keep It Growing, Keep It Covered

Of course, the best way to keep land covered is to grow perennial crops. So why not perennial grains, which don't require planting each year and thus avoid disturbing the soil?

In the 1970s that question turned into a life-changing "aha moment" for plant geneticist Wes Jackson. What if, he asked, we could feed ourselves by mimicking the prairie? So in 1976, Jackson founded the Land Institute in Salina, Kansas, to breed perennial grains. Jackson's mantra is "nature as model."[29] By 2014, his institute was showing modest progress toward developing perennial wheat, sorghum, and sunflowers.

It's hard to imagine a more revolutionary change—one vastly reducing humans' disruption of the Earth—than feeding ourselves with perennial crops. Yet, only "a handful of plant breeders" are pursuing this path, notes an overview in *BioScience*. It concludes that if "large programs to breed perennial grains [had] been initiated alongside the Green Revolution programs a half-century ago, farmers might well have seed of perennial varieties in their hands today."[30]

Nurturing healthy soil, including pathways for water to seep into, also means keeping roots alive.

"Living roots provide the easiest source of food for soil microbes . . . the foundational species of the soil food web," notes the USDA. How to keep roots alive at all times? Instead of growing a single annual crop, as in an industrial system, grow perennial crops or long-season cover crops, the department advises.[31] Fertility enhancers, cover crops are the grasses (such as ryegrass) and legumes (such as clover) grown in the period between plantings of the primary crop. They're also often called "green manure."

Cover crops, especially grasses with deep root systems, help prevent leaching of nutrients into groundwater.[32] And this is no small matter: Recall the four hundred aquatic "dead zones" from nutrient leaching we lamented earlier? For those applying synthetic nitrogen fertilizer, nonlegume cover crops can reduce nitrogen leaching by as much as 70 percent while maintaining crop yields.[33]

Mix It Up

Along with crop rotations, and often the combining of plant and animal production, agroecology also mixes multiple crops in the same field—a practice called intercropping, or polyculture. Indeed, American colonists in New England might never have survived if the Wampanoag people had not taught them an example of this neat trick, called "Three Sisters." In one field, three crops—beans, corn, and squash—help each other out. The *beans* fix nitrogen in the soil while climbing the stalks of *corn*, and the sprawling foliage of *squash* planted at their base deprives weeds of sunlight.

Common through parts of Mexico and Central America, Three Sisters demonstrates a big benefit of agroecology. The yield of each plant is typically greater than if it were the sole crop. The result is called "overyielding"—which must sound great to a farmer.[34]

But three crops are hardly the limit. In India, we've walked through agroecological fields with twenty crops—from millet to sunflowers to lentils to greens—growing in plots no larger than an acre. The crops benefit by mixing it up, and our health benefits, too. These intercropped fields in India, for example, combine protein-rich legumes with oilseeds and vegetables to meet the full range of human nutritional needs.

Rotating crops from season to season also enhances soil nutrients for plant health: Planting cereals one season and legumes the next fertilizes fields without adding synthetic nitrogen. It works because legumes—the family of peas, beans, lentils, chickpeas, clovers, and alfalfa—have bacteria in their roots that transform atmospheric nitrogen into forms that help plants grow, as described in the previous chapter. It's called "fixing" nitrogen. Then, when plants die, that nitrogen is released in the soil for other plants to use. These practices also help reverse the global soil-loss crisis because they store more carbon as soil organic matter, improving the soil's structure and its resistance to erosion.

Agroecological farming's enhanced attention to soil communities is one basis for confidence that humanity can restore the Earth's food-growing soils, essential to ensuring the end of hunger for all generations to come.

Pest Solutions Without Pesticides

In agroecology, protecting plants from pathogens, insects, and weeds isn't an "add-on"—or more precisely, it's not a "spray-on." Protection is in the farming system itself. In *Nature's Matrix*, three agroecology pioneers at the University of Michigan and California State University put it this way:

> The agronomist asks, "What are the problems the farmer faces and how can I help solve them?" The agro-ecologist asks, "How can we manage the agro-system to prevent the problems from arising in the first place?"[35]

So in agroecology, protection starts with a good defense. Since healthy plants can protect themselves from pests, "pest control" begins with soil structure rich in organic matter. Plus, agroecology's greater on-farm plant and animal diversity, as well as its crop rotations and cover crops, all enhance resistance to weeds, disease, and pests.[36] In fact, the value to farmers of intercropping—planting multiple crops in the same field—is as much in pest control as in soil and nutrition enhancement.

But, of course, even healthy plants sometimes need help. So farmers over the millennia have prevented pest damage by taking advantage of natural biological interactions among plants and insects. Today plant scientists are devising new schemes. One celebrated example is called "push-pull."

In many parts of Africa, push-pull protects corn plants from the stem borer. The flowering, low-growing desmodium is planted between corn rows because its odor repels this pest. That's the "push." Napier grass, planted at the border of the field, attracts the hungry stem borers away from the maize. That's the "pull." Plus, desmodium fixes atmospheric nitrogen, with the fertilizing advantage noted above, and napier grass also makes good fodder for livestock.[37]

It all works together.

Some argue that there's no need for absolutism when it comes to synthetic pesticides. In the last twenty years what's known as Integrated Pest Management (IPM) has emerged. It replaces scheduled pesticide applications with careful monitoring of pest pressures and minimal, only-as-needed use of the least toxic synthetic pesticide. Other experts, including agroecology trailblazer Miguel Altieri at the University of California and Colombian agronomist Clara Ines Nicholls, disagree. They see IPM as simply the continuing fixation on a single solution, not a systems solution.[38]

Andre Leu's farming experience tells him that eliminating pesticides is possible. Leu leads the International Federation of Organic Agriculture Movements, and told us that "in over forty years of experience in organic farming systems in more than a hundred countries on every arable continent, I've never met a pest, disease, or weed that farmers haven't been able to effectively manage through using ecological solutions."

> "In over forty years of experience in organic farming systems in more than a hundred countries on every arable continent, I've never met a pest, disease, or weed that farmers haven't been able to effectively manage through using ecological solutions."
> —Andre Leu, President, IFOAM

Leu noted his intentional use of the word "manage" to signal that ecological farmers appreciate that even what humans often call pests have important roles in the grand scheme, so it is best not to try to utterly wipe them out.

With this approach, agroecology reduces and, at its best, eliminates the poisoning of those who work the land. It lessens disease risk among both producers and consumers, who are no longer exposed to toxic agricultural chemicals.

All this while typically increasing food production on smallholder farms, as well as providing greater security for farm families, in part because agroecology's multiple crops offer in-the-ground "insurance." If one or two crops fail, all is not lost.

Agroecology is hardly a fringe movement. Note that in explaining how it works we've chosen to rely heavily on the USDA's often colorfully worded and accessible instructional materials. The very fact that the U.S. government's agricultural arm produces this great material speaks to the fast-growing appreciation of agroecological practices.

Cycling & Other Certainties

Synergistic interplay among plants and other organisms is one key to agroecology's power. Just as basic is what's called "nutrient cycling."

With carbon dioxide overload disrupting the planet, humanity is belatedly learning one certainty: that there is no "away" to which we can send stuff. So, too, with farming. Excess nitrogen and phosphorus that we're now adding to the soil in synthetic fertilizers must go *somewhere*. If not cycled back into the soil in regenerative ways, they destroy; witness the increasing harm to sea life and to humans described in the previous chapter.

So as much as possible agroecology seeks to avoid inputs such as synthetic fertilizer and to "close the nutrient loop" to eliminate loss and thereby maintain ongoing fertility, reducing agriculture's climate impact. Using cover crops, leaving residues on fields, returning compost, and, where possible, judiciously applying manure, as well as human waste—all these practices can return nutrients to the soil in quantities that feed new plants.

Fortunately, a farm's nutrient loop does not have to close completely, because in healthy soil ongoing biological and chemical processes involving the interaction of microorganisms, worms, and plant roots—often deep in the soil layers—continuously make nutrients available. As we've seen, for example, they "fix" atmospheric nitrogen so plants can absorb it.

Nonetheless, those who question the potential of ecological farming argue that soil nutrient generation and nutrient cycling can't supply enough nitrogen. "Synthetic nitrogenous fertilizers now provide just over half of the nutrients received by the world's crops," and without them, argues Canadian nitrogen expert Vaclav Smil, "we could not secure enough food for the prevailing diets of nearly 45 percent of the world's population, or roughly three billion people."[39]

This argument is countered by a 2007 University of Michigan study finding that "green manures"—cover crops that fix nitrogen and are plowed back into the soil—could provide roughly 70 percent more nitrogen than synthetic fertilizers currently provide.[40]

True, in our urbanized world, soil nutrients are exiting farms in significant quantities and not returning; and the science of agroecology is still evolving the best ways to ensure on-farm nutrient generation and cycling.

> "Green manures"—cover crops that fix nitrogen and are plowed back into the soil—could provide roughly 70 percent more nitrogen than synthetic fertilizers currently provide.

But the direction toward solutions for both nitrogen and phosphorus is clear:

To enhance fertility, we move away from synthetic fertilizer toward ecological practices, including, as much as possible, the cycling of nutrients back to the soil both within the farm system itself and beyond the farm.

In all this, we've emphasized nitrogen, but phosphorus is its own challenge. Phosphorus is the eleventh most prevalent element on Earth, but, as we stressed in the previous chapter, phosphate fertilizer draws on deposits that are finite, geographically highly concentrated, and likely to "peak" in the lifetimes of many alive today—making our dependence quite risky.

Solutions?

Reducing on-farm application of mined phosphorus takes first place.[41] To help, environmental scientist David Vaccari of the Stevens Institute of Technology advocates recycling animal waste. It is the "low-hanging fruit" of phosphorus conservation, because it is one of the most economically and technically feasible solutions, he argues.[42]

Spreading ecological farming practices to prevent erosion and thus phosphorus loss is also critical, stresses Vaccari. Since the U.S. cropland still loses almost two billion tons of topsoil a year, reducing erosion is a high priority on many counts.[43]

Moreover, we can each contribute to solving both the nitrogen problem and the phosphorus problem by improving our diets. Nitrogen experts Allison Leach and James Galloway, at the University of New Hampshire and the University of Virginia, respectively, have with colleagues devised a nitrogen footprint model, or "N-Print," similar in function to the "carbon footprint." Their goal is help us evaluate how to reduce our nitrogen footprint so that less escapes into the environment to cause multiple harms.[44] They emphasize reducing our intake of protein—which contains nitrogen—to recommended levels. Simultaneously, we'd cut our phosphorus consumption. (Recall that an average American eats about 70 percent more protein than recommended.)

We frame our positive choice slightly differently. Since eating animal-based foods accounts for almost three-fourths of an average person's "phosphorus footprint" worldwide, we can choose more plant foods.[45] And in so doing we help to reverse the shift to grain-fed, meat-centered diets. As Myth 2 underscores, this shift is unhealthy and uses massive quantities of grain and soy, generating nitrogen and phosphorous excess and waste.

So we can shrink our footprint of both essential elements by choosing plant-centered diets. Cutting our food waste reduces their loss as well.

Beyond our role as eaters, as citizens we can advocate for public policies to bring the nitrogen and phosphorus cycles back to health. They include policies to encourage the reduction of waste at every stage of the cycle—because four-fifths of the phosphorus we mine to produce food never even reaches our mouths.[46] Other crucial policy changes? Halting the use of cropland to grow crops for fuel, which uses nitrogen and phosphorus but feeds no one. And developing policies to encourage ecological farming and the recycling of urban waste.

And it's beginning.

In 2012, the EU called on members to reuse virtually 100 percent of phosphorus by 2020.[47] Sweden already requires 40 percent of phosphorus in sewage to be recycled back into the soil.[48]

In the United States, recycling nutrients through what are called "biosolids"—fertilizers produced from treated solid waste separated from municipal sewage—has big critics. As it turns out, removing heavy metals and other contaminants from sewage sludge, which includes waste from industrial sources, and safely applying it to fields is really difficult, although many are working toward solutions.[49]

Fortunately, a very different approach is catching on: It's called "nutrient recovery."

Instead of attempting to push out all the bad (toxic) stuff from sewage sludge and return what's left to the soil, there's a different approach. Since most of the nitrogen and phosphorus is in the wastewater, it focuses there, pulling out *only* the good stuff—the nutrients. It avoids the toxics altogether.

In a process called "struvite precipitation," phosphorus crystallizes with other elements, and is withdrawn from the wastewater to become a different kind of fertilizer. In this crystal form, the phosphorus is virtually insoluble in water, so it doesn't leach into waterways with all the harms we've noted.[50] Growing plants activate the phosphorus, but only when they need it, so the approach helps to reduce the vast waste of phosphorus we've noted.[51]

Scientists working in sewage and industrial facilities are taking the approach to scale.[52] One pioneer, the Canadian firm Ostara Nutrient Recovery Technologies, has plants in Europe and the United States, including one in Madison, Wisconsin. From wastewater, the company generates struvite pellets it calls Crystal Green. Unlike common powdery forms of phosphate that easily blow away, the pellets stay put in farm fields because of their larger size, CEO Phillip Abrary explained to us.[53] They safeguard against environmental contamination.

Earlier, we hailed microbes as the invisible enablers of life, and they turn out to play a key role in safe nutrient cycling. The phosphorus nutrient recovery approach is possible because some municipal waste treatment plants—so far, 10 to 20 percent of them in North America, according to Abrary—forgo the use of chemicals in retrieving phosphorus from waste and rely instead on hungry microbes that eat and exude phosphate that then allows for processes like Ostara's.

Unbeknownst to most of us, richest in both nitrogen and phosphorus is another part of the waste cycle: urine. Also little known is the fact that it is essentially sterile. For centuries humans have found simple ways to return urine, with these key nutrients, to the soil, notes the World Health Organization. Today, just 1 percent of our wastewater is urine, but in it are most of the nitrogen and phosphorus.[54] That's a lot of plant nutrition going to waste.

So turning urine into fertilizer is catching on in, for example, Finland and the Netherlands. In 2014, *National Geographic* reported that Amsterdam's public water utility invited male residents to use specially designated urinals whose purpose is collecting urine to fertilize rooftop gardens.[55] To help people get over the "ick" factor and encourage us to accept *all* of our nature, the Dutch playfully call it "peecycling"!

In West Africa, seven hundred families in eight Niger villages are cycling back to their fields all the nutrients in their own waste using waterless toilets and simple urinals—low-energy and low-cost—and enjoying yields equal to or better than what they got using chemical fertilizer.[56]

Clearly, a big part of making nutrient cycling practical is to reduce the distance between growing food and eating it. Note that three-quarters of food is still eaten in the country in which it's grown; and because most countries aren't as vast as the United States, this reconnection may be less daunting than it might seem.[57]

Moving in this direction, at least twelve governments in Latin America and the Caribbean have specific policies promoting urban and near-urban farming, reports the FAO. In Cuba, 40 percent of households grow some of their own food; and in Guatemala and Saint Lucia, it's 20 percent. Cuba helps move soil-nutrient sources, such as compost, to where they are needed. Its urban farming is agroecological and has helped to lift the country's average caloric availability from less than two thousand calories per capita per day in the 1990s to over three thousand by 2005.[58]

Of course, in much of the world, food is still quite local. In eastern Nigeria, for example, home gardens on just 2 percent of a family's farm generate half of its total production. In Indonesia, home gardens are estimated to supply more than a fifth of household income and 40 percent of the country's "domestic food supplies," notes the FAO.[59]

One home gardening breakthrough that's keeping food local is also changing lives in Africa's largest slum, Kibera, home to half a million people in Nairobi, the capital of Kenya. Since 2008, "sack farming" (sometimes called "vertical farming") has taken off. Women and children, primarily, grow vegetables, herbs, and other food crops in large nylon mesh bags of topsoil. In only in a few years, sack farming has quickly become a source of livelihood for thousands.[60] Costs are low, as used sacks are inexpensive, and soil is either scavenged from public areas or purchased cheaply from a vendor. "The size of the space isn't important," one sack farmer told researchers. Another added, "Sack gardening helps a lot. I now never go to sleep hungry; even your child can never sleep hungry."[61] With nearly one billion people living in slums, the innovation has "great potential . . . for livelihoods of slum dwellers," a researcher observes.[62] And it relates to our emphases here—because composting is easy with sack gardens, it has great potential for closing the nutrient cycle as it addresses hunger.

Now, let's delve deeper into how farmers in East Africa are using the principles explored in this chapter to better their lives.

FERTILITY WITHOUT SYNTHETICS: A STORY

The extremely cash-poor region of Ethiopia named Tigray is home to almost five million people, and had long been known for its degraded soils and poor crop yields. But in 1996 national and regional agencies, working with the Institute for Sustainable Development (ISD), based in Ethiopia's capital, Addis Ababa, launched a transformational strategy, with the goal of restoring soil fertility as well as community-environmental governance.[63] They called it, simply, the Tigray Project.

In part because of the region's low rainfall, the poorest farmers historically managed to grow only enough to store reserves that lasted four to five months. So the hunger season typically lasted more than half the year. Farming families "had to go through to harvest on almost empty stomachs," ISD's founder, Sue Edwards, explained.[64] Climate change only made their lives harder.

Starting in just four farming communities in 1996, the project worked with farmers to infuse a few basic agroecological practices. One is composting—returning organic matter from manure and plant residue to the soil. Unlike chemical fertilizer, which requires application every year, good compost can increase and maintain soil fertility for up to four years, according to Sue.

Thanks to healthier soil, farmers began getting higher yields, with fewer challenging weeds, and their plants also became more resistant to disease and pests.[65] By stopping the uncontrolled grazing of livestock, farmers in the project allowed degraded lands to revegetate, including steep slopes and gullies not suitable for agricultural production. This previously "useless" land now provides biomass that villagers harvest to feed to livestock and to make compost, returning nutrients to the soil.

Plus, the extra biomass not fed to livestock or used for compost becomes "feed" for biogas digesters that now supply all the fuel needed for cooking and lighting. Villagers are delighted by the money they save.[66]

So, in just five years, 2000 to 2005, farmers doubled their yields of cereals grown on compost-treated soil.[67] These farmers' experience confirms what the World Bank documents about corn yields: that for every additional ton of carbon in a soil's root zone, annual corn yields can rise by as much as 360 pounds per acre.[68] That's an extra pound of food for every day of the year.

The project also incorporated three other farmer innovations: the creation of small trenches along the bunds (low earthen ridges) between the fields to catch rain and soil runoff; tree planting and the nurturing of tree regrowth; and improvements in livestock grazing management to protect the soil.[69]

Throughout our book, we've stressed that the quality of relationships among people and with the Earth are inextricably linked, and Tigray's experience powerfully illustrates the point. From the project's beginning, villagers have assumed leadership via local associations, with elected

The Tigray Project's success is due in part to village-level associations, in which farmers make very specific public pledges, such as how many acres of land they commit to plant with trees, and hold one another accountable.

representatives who meet at least monthly. The associations create and enforce community bylaws, and, through them, villagers make a series of public commitments about, for example, conserving water and composting. Some of the public pledges are very specific, such as how many acres of land a person commits to plant with trees and the number of days of service she or he will contribute each year toward improving the environment of the community.[70]

Sue Edwards describes the result as a "wonderful pattern of farming and foods that all local people have developed, and that fits with their local culture and ecosystems."[71]

Tigray farmers using these practices now produce enough food to maintain a full year's reserve, and their farms' greater crop diversity enhances resiliency. By 2008, 86 percent of all the nearly seven hundred thousand farmers in the region were using natural fertilizer, on nearly half a million acres. Plus chemical fertilizer use fell 40 percent by weight between 1998 and 2005, while grain production climbed more than 80 percent.[72]

Some farmers even produce a surplus that they sell, raising their incomes more than tenfold, to roughly $700 a year. "These farmers—they're getting out of grinding poverty," Sue noted.

Tigray's story powerfully illustrates the heart of agroecology's success: Together, farmers pinpoint what's limiting plant growth in their area—in Tigray, it was water and soil fertility. Villagers then resolve those weak links and build from there.

And the ripples continue.

The gains were so obvious to everyone that within nine years these practices had been taken up by forty-two communities.[73] Continuing to combat poverty and land degradation, the Ethiopian government has since been spreading many of the Tigray-tested ecological practices—so far reaching about a quarter of the country's rural districts.[74]

COMBATING CLIMATE CHANGE, CONSERVING WATER

As underscored in Myth 2, agroecology makes a positive contribution to both addressing and adapting to climate change.

For one thing, soil holds much more carbon than the atmosphere does, and agroecology helps keep the carbon *there*, not in the air.[75] Minimal soil disturbance and cover crops—hallmarks of agroecology—enhance the absorption and storage of atmospheric carbon while they reduce the release of carbon dioxide. In part owing to the absence of greenhouse-gas-emitting

In part owing to the absence of greenhouse-gas-emitting fertilizers, organic farming methods generate *one-half to as little as one-third* the carbon dioxide emissions per acre that industrial agriculture generates.

fertilizers, organic farming methods generate *one-half to as little as one-third* the carbon dioxide emissions per acre that industrial agriculture generates.[76]

Water is a big concern on a warming Earth, and worldwide about 40 percent of irrigation draws on underground water, some held deep below the surface for millennia.[77] Groundwater depletion, stressed in the previous chapter, is one serious downside of the industrial farming model.

Given these realities, agroecological practices are especially helpful, as they enable plants to weather drought. Over more than two decades, the Rodale Institute in Kutztown, Pennsylvania, conducted field trials comparing organic and industrial corn-and-soy farming systems. In years of normal rainfall, yields were comparable, but in drought years, organic systems produced on average 30 percent more corn per acre.[78] Another long-term study published in 2003 found yields on organic farms superior in both drought and heavy-rain years.[79]

These studies provide more evidence that agroecology isn't a romantic notion, but rather a practical approach to ending hunger in the face of climate change.

Another powerful example is rice cultivation. Rice has traditionally been a very thirsty crop—think of flooded rice paddies—but in the last two decades the new type of agroecology mentioned as we opened this chapter, called SRI, has shaken up centuries-long rice-growing traditions.[80] The new method requires significantly less water, plus fewer chemicals and seeds—all while producing stronger plants and healthier soil. Not surprisingly, SRI lowers production costs, increasing the farmers' profit. In the Timbuktu region of Mali, SRI yields were 66 percent higher than those produced in conventional rice cultivation.[81]

In Africa, farmers are reviving ancient practices that help the Earth retain water. In the West African nation Burkina Faso, farmers create shallow circular indentations in the earth, called *zai*, as well as the just-mentioned earthen barriers called bunds. Both prevent water runoff. In Tigray, hand-built earthen dams and water conservation methods have in just a decade recharged aquifers, allowing springs to reappear.[82]

Another ecologically sound practice, agroforestry—growing trees and crops in the same area—which we highlighted in Myth 2, holds particular promise, especially in dry regions. It, too, reduces rainwater runoff, so that more is available to plants. In this case, it's the trees' roots that help the soil absorb rainwater. And recall the story of agroforestry in Niger, where it has revitalized a degraded area the size of Costa Rica and fortified food security for 2.5 million people.[83]

As evidence in Myth 2 suggests, if such agroecology practices, and agroforestry in particular, were taken up as widely as possible, they could play a significant role in righting Earth's carbon balance, and at the same time increase yields and build farms' resilience in the face of more erratic weather.

THE SEED OF IT ALL

We've focused on healthy soil as the key to the success of agroecological practices, but, of course, the seed is critical as well. In the previous chapter, and in Myth 6, we note a dramatic change in just a few decades: Within the industrial model, the "seed—formerly a free, renewable resource—has become a costly, non-renewable farm input," observes the Center for Food Safety.[84]

Half of commercial seeds are now controlled by three seed giants—Monsanto, DuPont, and Syngenta—limiting the capacity of farmers and plant breeders to create new, sustainable crop varieties.[85] Intellectual property rights on seeds, and thus on life itself, are diminishing biodiversity and creating vulnerability for farmers, and for all of us.

In response, a global movement of farmers and their allies is creating what Jane Goodall calls "arks for plants," to ensure that essential genetic diversity is not lost. Worldwide, such "arks" include roughly 1,400 seed banks, collaboratives, and exchanges, and 2,700 botanic gardens. The Millennium Seed Bank in West Sussex, U.K, is the world's largest, where almost two billion seeds from over 34,000 plant varieties are tucked away.[86]

At the opposite end of community seed saving is the controversial Svalbard Global Seed Vault in a man-made tunnel on the frozen island of Spitsbergen in Norway, eight hundred miles from the North Pole. Because Svalbard is supported by large corporations, the Center for Food Safety, among others, worries that it could become part of the problem, not the solution. The center is concerned that "corporations see in Svalbard an opportunity to gain further control of the world's plant genetics" and that the seed vault could become a "resource for germplasm . . . used for creating patentable hybrid or genetically engineered seed varieties," deepening farmers' dependence on a few global corporations.[87]

Thankfully, most of the world's 1,400 seed banks are community-based initiatives emphasizing seed sharing. One of the largest is Navdanya, founded in 1987 in India, which we take up again in Myth 10. For Gandhi, the British salt monopoly became the symbol of loss of self-determination. For Shiva, it is the seed, she told us. Regaining control over her culture's seed heritage is at the heart of Navdanya, which has created 111 community seed banks across India and trained over half a million farmers in seed saving, sharing, and ecological agriculture.[88]

For Navdanya, these efforts are part of "Seed Freedom"—freedom from dependence on corporate seed patents.

While seed banks alone don't have the capacity to deploy seeds to restore large areas, restoration projects are popping up that build on the seed bank movement. Southwestern Australia is home to the Gondwana Link project, "the world's largest integrated wild species restoration program in a biodiversity hotspot." The goal of the project is "to repatriate with local indigenous species many thousands of hectares of former farmland to create a biological corridor" spanning a distance greater than that from San Francisco to the Mexican border.[89]

Thriving in diversity, agroecology depends on the flowering of what some call the "open source seeds" movement. Think of it as "a genetic easement—or a national park for seeds," says University of Wisconsin professor of horticulture Irwin Goldman, a founder of the Open Source Seed Initiative (OSSI). In creating this "national park for seeds," the group attaches a "free seed" pledge to its seed packets, stating that genetic resources therein cannot be patented or in any other way legally protected, and that they must be shared. That way, they're "available in perpetuity in a protected commons," say the founders.[90]

In mandating seed sharing, the movement follows the example of free and open source software, with a clause in its pledge that the terms apply to any progeny of the OSSI seed. "This creates a pool of plant germplasm that is freely available to those who will share, but inaccessible to those who will not share," emphasizes cofounder and environmental-sociology professor Jack Kloppenburg, also of the University of Wisconsin.[91]

REUNITING FARMING AND EATING

Forgive us, but this is the perfect spot to call on a tired cliché.

If a Martian landed in a U.S. cornfield and was told that the plants sustain the human race, the visitor would assume—if logic holds on Mars—that there would be a one-to-one relationship between the thirty-three essential nutrients human bodies need to thrive and what grows in that field. But no. Increasingly, the food we grow and process into what we eat has become not only de-linked from what sustains healthy humans, but a threat our health.

Agroecology helps to reverse this deadly trend.

By its emphasis on plant diversity, and its insistence on healthy soils that generate healthy plants, agroecology creates healthier food. A meta-analysis of 343 peer-reviewed studies examining the nutritional value of food produced organically reveals that organics have significantly higher concentrations of antioxidants, including polyphenolics, flavonols, anthocyanins, and many others, which are linked to a reduced risk of chronic disease and cancer.[92]

Plus, agroecology's significant reduction in, or rejection of, chemical inputs means that it helps to protect eaters from the hazards noted in the previous chapter. Because pesticides are almost ubiquitous in the United States, even organic food is somewhat contaminated. However, the same meta-analysis found that in organic crops the frequency of occurence of pesticide residues was four times lower than their industrial counterparts, with significantly lower levels of the toxic metal cadmium.[93]

These findings confirm reports from farm families we've encountered in our research for this book. Their suffering from the symptoms of farm-chemical exposure triggered their move toward agroecological practices in order to regain their health.

AGROECOLOGY'S COMMUNITY-BUILDING RIPPLES

As in the Tigray story, let's again reconnect agroecology with the goal of ending hunger.

First, note the startling estimate that 70 percent of the world's hungry people live in rural areas and depend on farming.[94] So the progress of small farmers in the Global South is inherently an advance against hunger.

How do ecological practices help? For one, they improve farmers' income. With much lower production costs, the practices directly benefit the farm family's net income, even if there's no yield gain, which there often is, as we've seen. Lower costs and solid yields using agroecological methods in southern India, for example, can increase a farmer's net income by over 50 percent, notes a World Bank report.[95] More broadly, the World Bank finds that many agroecological practices increase both the farmer's profit *and* the soil's sequestration of carbon—with intercropping and a type of agroforestry dubbed "alley cropping" shown to be the standout practices on both counts.[96]

Another boon, in contrast to monocropping, is that an ecological farmstead's multiple crops mature at different times. So a family's income and its workload are spread out over a longer period of time. And because the health of farm families improves in the absence of chemical exposure, they can keep more of their income that may have otherwise been spent on medical care.[97]

While agroecology typically increases farmwork, on occasion agroecology can reduce it. When Campesino a Campesino, which you'll meet in Myth 10, spread velvet bean intercropped with corn in Central America and much of the Caribbean, not only did the practice increase the nitrogen available to plants; it also eliminated the need for weeding. Farmers were freed for other important tasks or for enjoying the fruits of their labor.[98]

In another positive social ripple, agroecology especially benefits women. In many areas, particularly in Africa, nearly half or more of farmers are women, but too often they lack access to credit.[99] So, agroecology—with no need for credit to buy synthetic inputs—can be a godsend for them.

Plus, agroecology enhances local economies as profits on farmers' purchases no longer seep away to corporate centers elsewhere. One example comes from the women in Andhra Pradesh, India, to whom we return in Myth 10. Once they switched to agroecological practices and stopped buying chemical inputs, the women began making and selling natural pesticides—mixtures of neem, chili, and garlic. Local farmers purchase the women's homemade potions and keep the money circulating within their community, benefiting all.[100] (If one wonders whether pests really are put off by what might read like a salsa recipe, be assured that the U.S. Environmental Protection Agency affirms that such approaches, called biopesticides, are indeed effective—and without hurting people.)[101]

Beyond these quantifiable gains, farmers' confidence and dignity are enhanced via agroecology. Its practices rely on farmers' judgments, which are based on their expanding knowledge of their land and its potential. Success depends on farmers' solving problems, not on following instructions from commercial fertilizer, pesticide, and seed sellers.

Developing better farming methods via continual learning, farmers also discover the value of collaborative working relationships. Freed from dependency on purchased inputs, they are more apt to turn to neighbors—sharing seed varieties and experiences of what works and what doesn't work in, say, composting or natural pest control. These relationships can build confidence in further experimentation for ongoing improvement. Sometimes they foster collaboration beyond the fields—such as in launching marketing and processing cooperatives that keep more of the return from farming in the hands of farmers.[102]

Beyond such localized collaboration, agroecological farmers are also building a global movement for "food sovereignty," which they define as the:

> right of peoples to healthy and culturally appropriate food produced through ecologically sound and sustainable methods, and their right to define their own food and agriculture systems. It puts those who produce, distribute and consume food at the heart of food systems and policies rather than the demands of markets and corporations. It defends the interests and inclusion of the next generation.

To achieve food sovereignty, members support each other and advocate for pro-peasant, pro-agroecology national and international policies through La Via Campesina. Founded in 1993, it now joins together two hundred million peasant farmers in 164 local and national organizations across seventy-three countries in Africa, Asia, Europe, and the Americas.

In Myth 10, we share illustrative stories of this fast-growing movement. Today, the International Federation of Organic Agriculture Movements, founded in 1972, brings together organic farmers in eight hundred affiliate organizations in 120 countries.

We hope it's now clear that agroecology is best understood not as a fixed set of steps—plow, plant, harvest. Instead, it is a never-ending process in which farmers work with multiple time horizons to ensure the soil is covered, plants are diverse, and plant growth is continuous. Its time-honored practices encourage synergies among plants, soil organisms, insects, and animals. Agroecology integrates ecology, culture, economics, and traditional knowledge in ways that create healthy, plentiful food, enrich soil, combat climate change, and conserve water and other vital resources, all at the same time.

And in human relationships, agroecology fosters mutuality, rather than dependency, whether within the family, village, or wider society.[103]

BUT CAN AGROECOLOGY ACTUALLY MEET THE WORLD'S NEEDS?

Having reviewed its many strengths, we end with what is, for some, the still urgent question of whether agroecology is up to the task of feeding us all.

Several overview studies offer a resounding "yes."

In 2006, a seminal study affirmed ecological farming's potential in the Global South. It compared yields in 198 projects in 55 countries and found that ecologically attuned farming increased crop yields by a mean relative average of almost 80 percent.[104]

A 2007 University of Michigan global study concluded that organic farming—in this case, including sustainable and ecological practices—could support the current human population, and expected increases, without expanding farmed land.

Then, a 2007 University of Michigan global study concluded that organic farming—in this case, including sustainable and ecological practices—could support the current human population, and expected increases, without expanding farmed land. This study confirms that it is in the Global South—where it matters most—that ecological farms typically enjoy higher yields, often dramatically higher.[105] Recall, for example, the yield gains achieved with agroforestry, featured here and in Myth 2. In

Zambia, corn yields produced without synthetic fertilizers, but within the canopy of the nitrogen-fixing native *Faidherbia* tree, produced yields three times higher than crops grown without the benefit of these trees.[106]

Then, in 2009, came a striking endorsement of ecological farming by fifty-nine governments and agencies, including the World Bank—"Agriculture at a Crossroads: International Assessment of Agricultural Knowledge, Science and Technology for Development" (IAASTD)—painstakingly prepared over four years by four hundred scientists. It urges support for "biological substitutes for industrial chemicals or fossil fuels."[107]

After taking in this promising picture, you may now be puzzled by the studies referenced as we opened this chapter that found lower yields on organic compared with industrial farms.

So let's look deeper at these findings. The more recent study notes that in organic fields of multiple crops and those using crop rotations—both common agroecological practices—the yield gap amounts to less than 10 percent. The authors also suggest that "investment in agro-ecological research" to improve practices further "could greatly reduce or eliminate the yield gap for some crops or regions."[108] Great point—and note that agroecology has achieved all the successes documented throughout our book with but a tiny fraction of the resources devoted to furthering industrial agriculture.

Unfortunately, in neither study do the authors mention that organic farmers can benefit significantly despite their somewhat lower yields, because their financial costs are lower. And note that these studies cover yields only on farms that are certified organic—that use zero synthetic inputs—and so they can't register the yield gains of the much broader universe of agroecological farmers who may use such inputs rarely, as a last resort.

Perhaps most important, studies comparing yields on certified organic farms and industrial farms cannot capture what is arguably making the greatest difference in the lives of the rural poor, who are the majority of the world's hungry: that is, the significant yield increases many are achieving, like the Zambians just mentioned, once they adopt agroeco-logical practices. That higher yield is in contrast to their previous yields, not to those of an industrial model, which they can't afford in any case.

Moreover, as we show throughout our book, higher yields found in industrial agriculture produced with chemical inputs not only entail disastrous health and environmental costs, but are *also unnecessary for all to eat well.*

Nonetheless, objections to this pathway continue . . .

- *Too few desire rural life.*

The argument goes like this: Agroecology can never be taken to scale worldwide because, compared with industrial systems, it requires more people working the land, and wouldn't most rural people, if given half a chance, head straight for the nearest city?

To weigh this proposition, first consider that today 2.6 billion people, well over a third of all of us, live in the countryside and depend directly on agriculture.[109] Imagine the intensified suffering if they continue filling urban slums because rural life has become unbearable or they have been forced off their land.

The great news is that considerable outmigration from rural areas, which many take as inevitable, may not be: People want to stay where they have roots— *if* they can also enjoy the rewards. In the central plateau of Burkina Faso in West Africa, for example, a study of twelve villages discovered that, with farmer-instituted water and soil-conservation practices, along with agroforestry integration of trees and crops, outmigration stopped. One village, which had lost a quarter of its population in the ten years before the new practices began, did not lose a single family once ecological methods increased crop yields and led to other advances.[110]

In other parts of West Africa as well, villagers who'd given up are coming back. You may recall the Niger village chief Moussa Sambo, quoted in the previous chapter, in whose village farmers are practicing ecological techniques similar to those studied in Burkina Faso. In late 2010, even as many in Niger were facing food shortages, young men in Sambo's village were returning. Apparently the ecological practices were working.[111]

Evidence abounds that many rural people so treasure their way of life that they will take great risks to protect it. In Myth 6, for example, you'll read of rural people resisting "land grabs" by powerful foreign interests. Forest communities from India to Central America are

> Brazil has achieved some form of protection for almost half of the forest of the Brazilian Amazon, with the majority of it now managed by rural people themselves.

also challenging the seizure and destruction of their forest homes. Brazil has achieved some form of protection for almost half of the forest of the Brazilian Amazon, with the majority of it now managed by rural people themselves.[112]

Moreover, many rural people highly value their way of life despite the profound inequities in income and essential services they suffer. In India's Maharashtra state, to pick just one example, even a relatively well-off farmer with fifteen acres—many times more than a poor farmer—earns about the minimum wage, working every day of the year.[113] Imagine the draw of the countryside if the rewards for effort were fairer.

In sum, growing numbers of rural people seem determined not only to stay put but to contribute to the solutions to the global climate-environment crisis.

- *Too much hard labor.*

Ecological farming often requires more labor for weed and pest control.[114] So readers might wonder, Are many people really willing to do the arduous hand labor that agroecology seems to require, if there's an alternative? It's an important question, especially considering that today roughly a billion farmers rely solely on hand tools.[115]

So we want to be clear: Agroecology does not eschew mechanical tools that lighten work. Whether machines contribute positively to our lives depends on who controls them, as well as on their scale, cost, and ecological impact. For example, some female farmer groups of the Deccan Development Society in Andhra Pradesh, India, lighten their workload using millet-processing equipment they own cooperatively. Elsewhere in that state small farmers are adopting the "Conoweeder"—a $16 device that from afar resembles a lawn mower with an extra-long handle—to radically reduce time spent weeding their rice paddies.[116]

- *Too elitist.*

Hardly. First, organic and ecological farming is spreading fast among the world's poor farming communities. And even in the United States, urban farming movements in low-income inner cities, from Boston to Chicago to Oakland, are beginning to make organic food available. Second, those most hurt by chemical agriculture are farm laborers, who are a huge proportion of the world's hungry people. So anyone purchasing food that hasn't been chemically treated benefits workers by helping to draw more acres toward safer farming practices. Third, in buying organic food, we help to build the market, potentially bringing prices down for everybody.

In the United States, for example, for those whose incomes give them choice, buying organic to meet recommended daily servings of fruits and vegetables would add about $10 a week to a food budget.[117] And, since more than a fifth of American households earn over $100,000 a year, there is vast room for the organic-food demand to grow, with all the benefits just noted for our whole society.[118]

All this and more make ecological farming and eating the opposite of "elitist."

- *Can't be scaled up.*
  Ah, this is a big one.

Despite all the positive news about agroecology's production successes, critics argue that agroecology can't rise to the global food challenge for another reason: Minimizing or forgoing synthetic fertilizers and pesticides, agroecology depends on returning to the soil most of the nutrients harvested—i.e., the "nutrient cycling" discussed earlier in this chapter. But, the critics' argument goes, nutrient cycling is possible only on a small scale—and surely not when food is shipped far and wide.

Our response begins with two observations.

First, the industrial model has already proved that it can't scale.

After half a century of expansion, it still leaves roughly one-quarter of the world's people suffering nutritional deprivation, as we explain in Myth 1. Then, add the growing number of those suffering from food-related diseases—including diabetes and a large share of heart conditions—and we see that the dominant food system is failing to support the health of a huge portion of humanity.

This failure, rooted in a perception of life as evolving in disconnected parts—not interacting elements within whole systems—has led to the harms and dead ends documented in the previous chapter.

Second, agroecology's core principle of aligning with nature's regenerative processes is the direction all human systems must take if humanity is to thrive, or even to survive. As we then shift global resources toward sustainable, regenerative practices, we will no doubt learn ever better ways of aligning farming with nature, and may go further in creating the healthy cycling of nutrients than anyone now imagines. Recall the promising "nutrient recovery" technology described earlier and the European countries that are prioritizing nutrient cycling. And note also, thankfully, we don't have to create perfectly "closed-loop" farming systems to achieve sustainability.

Surely, we're convinced, if humanity can master space travel and decode the genome, we can grasp the laws of biology and tackle the logistical challenge to achieve something even more basic—nutrient cycling at the level necessary to enable us all to eat. Worldwide, less than 1 percent of agricultural research focuses on advancing the knowledge and practice of organic farming.[119] Imagine the potential captured in that one statistic that would be released if we shifted course.

So, may we—scientists, engineers, public officials, farmers, and *all of us*—get on with this challenge, reweaving these broken stands in the web of life.

- *It's just too late.*

A final objection to scaling agroecological farming could be that our globalized, corporate-dominated food system is so entrenched that a more localized and farmer-empowered system is a pipe dream—or, indeed, is already irrelevant. Well, not so fast. As just noted, more than three-quarters of all food grown does not cross borders.[120] In the Global South, the number of small farms is growing, and small farmers—on only a quarter of the world's farmland—produce 80 percent of what's consumed in Asia and sub-Saharan Africa.[121] In these numbers is much strength on which to build.

A Choice?

In the previous chapter we laid out the consequences of the dependency-generating, environmentally damaging industrial model of agriculture. Here we've highlighted very different approaches. Instead of concentrating power and reinforcing dependency, these pathways create relationships of mutuality that generate new, distributed power—problem-solving power.

The choice between these very different directions is ours.

As we opened Myth 3, we noted that any model of agriculture reflects and shapes its wider context. So, readers may wonder, how can agroecology truly break through the current context of concentrated power?

Well, sometimes it takes a "moment of dissonance," that sudden, unnerving awareness of a conflict between one's long-held assumptions and one's experience.

We hope, therefore, that the evidence presented throughout our book will generate profound shock and honest hope, both strong enough to crack the assumptions that have kept us on a failing track. Then we can see that agroecology is not merely a particular set of agricultural practices. It is one dimension of a more democratic social order, in which growing nutritious food and transforming power relationships are inextricably linked—an agricultural model that sustains life in all its forms.

In Myth 10, we explore the positive eruption of broad, global citizen movements that are bringing this model to life, democratizing production via agroecology and in some cases working hand in hand with responsive governments.

myth 5

Greater Fairness or More Production?
We Have to Choose

MYTH: No matter how much we believe in the goal of greater fairness, we face a dilemma. Redistributing land so that big producers have less and land-poor farmers and farmworkers have more would undermine "economies of scale" and therefore reduce output. So, even well-intended reforms aiming for greater fairness in control over food-producing resources end up hurting the hungry people they are supposed to help.

OUR RESPONSE: Justice and production are not competing goals; instead, they are complementary. Perhaps the discouraging notion of an inevitable trade-off persists because too many of us do not perceive the ways in which inefficiency is built into unfair food systems, how systems dominated by a few underuse and misuse food-producing resources. People will understandably fear change in the direction of greater fairness until it becomes clear how unfairness itself blocks human progress.

The notion of "economies of scale" may be the economics dictum that makes the most intuitive sense to people. After all, didn't Henry

Ford prove it? Early in the twentieth century, assembly-line operations convinced us that "big" is necessarily more "efficient"—meaning, in Ford's example, able to produce cars cheap enough for even workers to buy.

Reality, however, is a lot more complicated. If farming were like making cars, then the huge collective farms of the Soviet Union in the 1950s might have fared better. To get a grip on what makes farming productive or not, let's first look at incentives built into our current food system, rife with inequity.

Wealthy landowners hardly need to use every acre, and by all indications they don't. In Brazil barely a fifth of the country's agricultural land is cultivated, largely because large farms control most of it and a disproportionate share of their land is not planted.[1] Throughout Latin America, this pattern holds.[2]

With so much land left unused by the big operators, it is no surprise that small farms in the Global South are almost always more productive than larger ones—they get more out of each acre of land. But this finding also holds true even in comparing output only on actually cultivated land.[3] In 1962, Nobel economist Amartya Sen first made this case about the superior productivity of small farms in India, and since then dozens of studies beyond India have confirmed his observation.[4] In Turkey, for example, every 1 percent increase in farm size is associated with a drop in productivity that's about one and a quarter percent.[5]

Thus we should not be surprised that small family farms in Brazil, with access to less than a quarter of the agricultural land, produce 40 percent of the total value of Brazil's agricultural output—including 70 percent of beans and almost 60 percent of both cow's milk and pork.[6]

> Brazil's small family farms, on less than a quarter of the country's agricultural land, provide 40 percent of the total value of Brazil's agricultural production, including 70 percent of beans.

Agricultural economists call it the "inverse relationship between farm size and productivity" and continue to debate the underlying reasons why small-scale farms in the Global South are typically so much more productive.[7]

One reason is that farmers with smallholdings work their land more intensively. So in addition to the productivity benefit, their greater use of labor, often of those formerly unemployed, is a social good in a world of widespread joblessness.[8]

Not only do they apply more labor, but small farmers also make more efficient use of space. The smaller the farm, the more likely it is to integrate diverse crops—cultivating them together or staggering them over time, or both. Smaller farms also often integrate livestock or fish—commonly fed crop residues—and use their waste as fertilizer. As noted in the preceding chapter, such integrated systems are far more productive on any given unit of land than farms growing a single crop, common in large operations.[9]

Plus, such crop diversity typically produces more nutrition per acre.[10] In India, for example, we've visited farms of little more than an acre whose crops meet all of a family's nutritional needs, from protein in dal (lentils) and sunflower seeds to vitamins in the many greens and cruciferous vegetables.

Moreover, in rethinking this myth it's important to register the productivity that greater gender equity could unleash. In low-income countries, for example, women are responsible for growing 60 to 80 percent of the food, yet they typically don't own the land, and receive only about 5 percent of agricultural extension services worldwide.[11] Lack of a land title diminishes women's access to credit, because they have nothing to serve as collateral; in Africa, for example, women access only 1 percent of potential agricultural credit.[12]

The cost to productivity is huge: It's estimated that if women were to have the same access to productive resources as men do, their agricultural yields could increase by 20 to 30 percent. The total *additional* food produced could fulfill the needs of as many as 150 million people, more than the combined populations of the U.K. and France.[13]

> If women were to have the same access to productive resources as men do, their agricultural yields could increase by 20 to 30 percent. The total *additional* food produced could fulfill the needs of as many as 150 million people.

Add to this suppressed potential production of small-scale female farmers the proven superior performance on average of small farms—generally with less fertile land than that of large landholders—and one begins to appreciate how profoundly *counter*productive present male-privileged and elite-controlled farm economies are.[14]

THE ILLUSION OF EFFICIENCY

In addition to production per unit of land, an increasingly important measure of productive success is energy use: Just how much energy does it take to produce a given amount of food? Farming uses many forms of energy—including sunlight, human and animal labor, wind, petroleum, and synthetic fertilizers, as well as coal, natural gas, and nuclear fission for electricity.

By this measure, large capital-intensive farms hardly shine. In fact, they rank at the low end of the energy-efficiency scale. Capital-intensive U.S. farming, characterized by large farms, uses seven to ten units of primarily fossil energy to produce just one unit of food energy.[15] By contrast, traditional farm systems not dependent on synthetic inputs not only yield more food energy per unit of energy used in its production, but typically rely on renewable energy sources—human labor, animals for hauling and plowing, and manure—whereas capital-intensive agriculture depends heavily on climate-disrupting fossil fuels.[16]

Large, highly capitalized farms also tend to waste more fertilizer. The extent is suggested in the finding that just 10 percent of global cropland accounts for at least a third of all the wasted nitrogen and phosphorus—euphemistically called "surplus." The excessive nitrogen and phosphorus are applied to fields to nourish plants, but instead ends up polluting water and air. This hugely disproportionate share of harmful waste is concentrated in four areas—China, northern India, the United States, and Western Europe: precisely where farms are more industrialized.[17]

With all this evidence to the contrary, why do so many people believe that bigger is better—that large-scale operations are most efficient?

At first glance, it is easy to mistake the very size of a large-scale operation for proof of its superiority. The real reasons for the success of big producers—advantages resulting from wealth and political clout—are all but invisible. They often include preferential access to credit, and to irrigation, chemical fertilizers, pesticides, technical assistance, and marketing services. Note that in India, for example, as we mentioned in Myth 3, a large farmer can obtain a bank loan at 12 to 15 percent annual interest, whereas small farmers relying on moneylenders pay as much as 5 to 10 percent monthly—so they are saddled with annual interest rates of 80 to over 200 percent.[18] The distribution of government farm subsidies is often skewed in favor of the bigger operators. In the United States in 2012, the top 1 percent of recipients of federal commodity subsidies captured on average $84,000 each, while those in the bottom 80 percent received on average only $1,500.[19]

Considering all this, small producers' ability to outperform big growers becomes all the more impressive.

Motivation for Good Farming Undermined

Where a few large landowners control most of the land, often they rent out much of it. Typically, hired hands do much of the work, and sometimes renters pay with a share of the crop. Because these nonowners working the land have less of a stake in its long-term productivity than do owners, such arrangements undermine the best, most careful use of the land.

Without secure rights to their land, how could one expect the millions of poor tenant farmers in the Global South to invest in long-term land improvements, to judiciously rotate crops, or to leave land fallow for the sake of long-term soil fertility? Plus, in parts of India where long-term tenants can assert legal right to the land, owners often prefer to evict tenants regularly. Given the threat of eviction, why would tenants make long-term investments to improve the farm?[20]

Short leases with no assurance of renewal create the "preconditions for neglect of conservation," observes renowned development economist Michael Lipton. Consider the disincentive built into the sharecropping system in Bangladesh. Because there is no guarantee of being able to farm the same plot in subsequent years, the sharecropper lacks incentive to invest in long-term land improvements. And, even if motivated, a sharecropper has little with which to invest: Typically, although sharecroppers do all the work and cover production costs, a third of the harvest goes to the landowner.[21]

Before moving on, let's consider an instructive parallel.

Big collective farms of the Soviet era were also a form of large-scale operations in which workers often did not have a direct stake in output, and their productivity was notoriously low. Similarly, in Cuba huge state farms underperformed small-farm productivity for decades, but the country got by anyway because of massive foreign aid from the Soviet Union. Then, 1989 brought the collapse of the Soviet Union and an abrupt cutoff of aid, forcing Cuba to rethink its farming. In 1993 it began the conversion of state farms into small, semiprivate farms belonging to former farmworkers and cooperatives. Several hundred thousand urban dwellers were allotted plots for growing vegetables and fruits. Between 1994 and 2000, vegetable production significantly increased every year.[22] By 2005, calorie supply per Cuban had exceeded the world average.

Cooperation Thwarted

Monopoly control over food-producing resources also thwarts their full utilization by undermining the motivation for community cooperation, which historically has played a crucial role in agricultural development. For example, "repair and maintenance of rural water courses, reservoirs, tanks, etc., have always been the responsibility of the community," notes a study by the UN Food and Agriculture Organization (FAO).[23]

In Bangladesh, for example, cooperation in digging and maintaining ponds for irrigation and fish cultivation was common before 1793, the year the British instituted individual land ownership. By the mid-1990s, the bottom 40 percent of households owned less than 2 percent of all the land, and landlessness continues to worsen.[24] Village cooperation in many areas is a thing of the past. Traveling in rural Bangladesh in the 1980s, an agrarian development expert noted that most ponds, once a village asset, were silted up and useless.[25]

Why should the land-poor majority pitch in if the benefits go mainly to the village's few better-off landowners? For their part, the landowners find that, thanks to aid programs, they can irrigate with a pump, dispensing with the need for a pond altogether.

The Misuse of Food Resources

So far we've explored how farming systems controlled by a few thwart the fullest, most careful use of resources by those who work the fields. But antidemocratic systems also drive the landowner or absentee investor to misuse the land.

Seeking the greatest profits in the shortest time, many big growers often overuse the soil, water, and chemical inputs, thoughtlessly eroding the soil, depleting the groundwater, and poisoning the environment. Since they are likely to have other income-generating investments and can take over additional land if need be, why should they concern themselves about the long-term viability of a particular piece of land, to say nothing of the health of workers and the larger community exposed to toxic chemicals?

Plantations growing bananas, coffee, oilseeds, cocoa, and other crops for global corporations may offer the most extreme examples.

Most banana plantations, for instance, use huge quantities of agrochemicals.[26] Fungicide is widely used, and when it decomposes, a primary metabolite—linked in lab animals to cancer and birth defects—moves through the environment. In Mexico's banana-growing areas, scientists find high concentration of fungicide in surface water.[27] In Costa Rica's protected conservation area, the endangered spectacled caiman—a relative of the crocodile—has been found to carry in its blood nine pesticides used on banana plantations, seven of which have been banned as too toxic by an international convention.[28]

The Wealth Produced Leaves Town

Surely more bunches of bananas and more bushels of grain are not the only goals of farming. Agriculture must also generate wealth for bettering the lives of those in rural communities where worldwide 70 percent of all those who are food insecure live.[29] Income from farming is needed for better housing, health care, transport, and other improvements. But when a few control most of the land, what happens to the wealth produced? Much of it flows out of the community in pursuit of higher returns elsewhere.

We remember visiting northwest Mexico, an area with some of the country's most lucrative farm production, thanks in part to publicly funded irrigation. But stopping in neighboring towns, we saw only squalor. Wealth from the rich farmland, pocketed by a few, had gone into foreign bank accounts, fancy cars, and private planes for hopping over the border on shopping sprees. Virtually none of that wealth stayed around to enrich the community.

Our firsthand observations are confirmed by rural economists. The most extreme examples of wealth produced "leaving town" involve export crops.

Above we mentioned Costa Rica, where banana plantations occupy some of the most prized agricultural land and just three corporations—Del Monte, Dole, and Chiquita (recently merged with Fyffes)—control 84 percent of banana exports.[30] Out of every dollar a consumer pays for a banana, about ten or eleven cents, and sometimes as little as five cents, returns to the farm or plantation that produced it. Most of the rest benefits the corporations that control shipping, ripening, distributing, and retailing.[31]

And the workers who grow the fruit? They receive as little as $3.50 *a day* in wages.[32]

This suction force draining wealth from the Global South is intensifying. In the phenomenon now known as "land grabbing," which we describe in the following chapter, between 2000 and 2010 investors purchased or were negotiating for a land area slightly larger than Mexico.[33] Purchasers typically promised "modern" farming, new local jobs, and enhanced food security. But in Sierra Leone—which, sadly, is no exception—a Swiss company bought 154 square miles of land in 2008, promising two thousand new jobs; yet three years later, there were only fifty.[34] In the Global South two-thirds of such "investors" expect to produce for export, not for locals.[35] Their "investment" is not in the community, but in extracting exportable wealth.

THE FRUITS OF REFORM

As we came to grasp the many ways in which agrarian systems controlled by elites both underuse and misuse resources, we found ourselves asking if virtually any alternative could be worse, and whether greater production—and more sustainable production—wouldn't naturally follow if access to land, credit, and knowledge were more fairly shared.

Many studies have considered the likely impact on productivity of a more equitable distribution of land and concluded that land reform could bring significant production gains.[36] In a 2004 World Bank–funded pilot land reform project carried out by the Malawi government, almost four acres were distributed to each of more than fifteen thousand households. By 2009, their corn yields had increased measurably, and farm incomes had climbed an average of 40 percent. Surrounding communities experienced positive impacts as well.[37]

Unfortunately, testing predictions about the benefits of fairer access to land isn't easy. Big landowners use all of their considerable power to resist.

In the mid-1980s, for example, large landowners in Brazil were so frightened of land reform that they invested an estimated $5 million to buy arms and hire gunmen to protect their land against peasant occupations.[38] And as landless workers have continued pressing for land since then, over a thousand have been killed by landowners, their agents, and police, according to Brazil's large Landless Workers' Movement.[39] (We discuss the Landless Workers' Movement, known by its Portuguese acronym, MST, below and in Myth 10.)

In the face of reforms, and fearing that land and power will be taken from them, landowning elites have been known to sabotage production. Either they liquidate what they can of their resources—slaughtering livestock, for instance—in order to take their wealth out of the country, or they try to disrupt the economy in hopes of stirring up popular discontent against the government carrying out the reform. If support from foreign sources gives the privileged oligarchy hope that it can successfully resist the changes, it is all the more likely to attempt such sabotage.[40]

Although modern history offers relatively few examples of genuine agrarian reform, in the twentieth century several far-reaching reforms were carried out. Even though much has changed in the intervening decades, examining their impact can tell us a great deal about the concerns in this chapter.

Japan. Fearing social unrest in the aftermath of World War II, a conservative government carried out a major land reform with prodding and support from U.S. occupation forces. Transforming tenant farmers into owner-cultivators, the reform not only brought in greater equity but also may have removed constraints on the growth of Japanese agriculture.[41]

South Korea. Before the early 1950s, more than half of South Korea's agricultural households were landless, but after a sweeping land reform almost all of them owned land. Between 1945 and 1950 more than a fifth of the cropland was redistributed, with all beneficiaries ending up with roughly the same amount of land. Within a decade, yields far surpassed pre-reform levels.[42]

Taiwan. In the years 1949 to 1953, Chinese Nationalist forces (KMT) were driven out of China by the peasant-based Red Army, which championed land reform. The Nationalists then occupied Taiwan, where they themselves imposed reforms on the Taiwanese landed aristocracy— nearly doubling the proportion of farm families owning land, from 33 to 55 percent. Their reforms also cut the proportion of farmland worked by tenants from 39 to 21 percent and lowered rents, lessening insecurity on the remaining tenancies.[43] As a result, agricultural productivity rose, income distribution evened out, and rural and social stability improved.[44]

China. In the early 1950s, the Chinese undertook perhaps the most far-reaching land reform ever attempted. Half a billion rural people were affected. Looking back over almost fifty years of agrarian change in China, we can say that when authority over the use of the land was wrested from wealthy landlords and eventually passed into the hands of large administrative units, production increases were modest. But when responsibility for land was further devolved to individual families beginning in 1978, and rewards were made more commensurate with effort, production advanced well ahead of population growth.[45] And this was true even though the drop in fertility rates had slowed. In just six years, from 1978 through 1984, agricultural output per capita grew a phenomenal 33 percent.[46]

Such reform contributed to China's success in decreasing the number of hungry people by 138 million since 1990, contributing almost 70 percent of all progress in reducing the number of calorie-deficient people in the entire developing world.[47]

Kerala, India. The 1969 land reform in Kerala, a state in southern India, abolished tenancy, allowing for a massive redistribution of land rights. In effect, absentee land ownership—associated with high levels of exploitation and oppression—was abolished, and the foundation was laid for participatory democracy and protection of human rights.[48] Over two million acres were redistributed, with 1.5 million tenants becoming small owners.[49]

Unlike most land reforms, redistribution in Kerala hurt farm productivity, according to analysts at the London School of Economics.[50] However, for the vast majority of people affected in the countryside, food access and the general quality of life improved greatly, thanks to popular political demands that resulted in government investments in health and nutrition programs—more evidence that quality of life is not all about the quantity of production. In Myth 10 we feature this state's remarkable continuing progress.[51]

Latin America. Major land reforms were carried out in a number of Latin American countries in the last century. Mexico, Bolivia, and Chile provide especially useful lessons. The Chilean reforms were cut short in 1973 by the U.S.-supported military overthrow of the elected government, but the indirect effects were considerable. The Bolivian reform, as in Mexico, shook up the rigid class structure. It turns out, says land reform analyst Thomas Carroll, that equity of "citizenship"—a voice in the political process—was a more important outcome of Latin American reform than greater income equity, and eventually, a rural middle class emerged.

In addition to these specific examples, we have examined several overview studies of the impact of land reform. Those focusing on *genuine* land redistribution show net benefits in productivity, efficiency, and alleviation of poverty, as well as positive ripple effects through the larger economy.[52] Nevertheless, many without specific knowledge of land reforms tend to belittle their effectiveness. According to Michael Lipton, the development economist quoted earlier:

> Many otherwise well-informed people believe that, since the post-war reforms in Japan, Korea and Taiwan, there are few if any success stories of land reform. They believe that it has been legislated with so many loopholes that very little has happened; or that it has been evaded, or has failed to get land to the poor, or has been reversed; that it runs against modern farming, with (alleged) economies of scale and complex techniques and marketing; that, even where desirable, it is politically infeasible; in short, that the golden age of land reform, if it existed at all, was over by the mid-1960s. *All these statements are false, and are generally agreed to be so by subject specialists, whatever their analytical or political preference.* There is almost no area of anti-poverty policy where popular, even professional, opinion is so far removed from expert analysis and evidence as land reform.[53]

The naysayers typically confuse, intentionally or not, genuine and egalitarian efforts to redistribute quality land with inegalitarian colonization projects that send the landless to remote areas where the soil is poor, or with window-dressing reforms that fail to address underlying inequities. Examples of phony reforms abound—from the Philippines and Pakistan to India and other countries. Carried out by governments beholden to rural oligarchies, they inevitably leave intact extreme imbalances of power.[54] When the potential of reform is measured by the consequences of these fake land reforms, the conclusions are, of course, pretty discouraging.

Land Reform from Below

While many policymakers argue that land reform is no longer politically feasible, the dispossessed are proving them wrong.

In the 1994 Zapatista uprising in Chiapas, Mexico, landless peasants seized thousands of acres belonging to wealthy landlords.[55] Soon, thirty-two new municipalities—independent of the Mexican government—were created in Zapatista territory. The threat of loss of control over food production once Mexico signed the 1994 North American Free Trade Agreement (See Myth 7) reinforced Zapatista communities' drive for self-provisioning and local exchange, independent of the global market.[56] So, in the new communities, local people have organized "production societies," producing food, cattle, coffee, and artisanry for vibrant local economies.

In Brazil, with the end of the dictatorship in 1985, Brazilians could organize freely, and the Landless Workers' Movement took off. Known (as noted above) by its Portuguese acronym MST, the group seeks nationwide land reform—"Brazil without *latifundios*" (large estates). And in 1988, as we explain in Myth 10, Brazil's new constitution provided the movement legal grounds for its claim that agricultural land must serve a social purpose. Over the decades, the movement has led more than 2,500 occupations of idle land held by a rural elite monopolizing the best land; and it has successfully settled 370,000 families on almost twenty million acres of land to which they won the legal right.[57]

In Myth 10, we dig deeper into the lessons of this approach of justice *with* production.

Not Size Alone, but How Decisions Are Made

Shifting the power balance to favor the poor majority is the heart of genuine land reform. And we should not let the question of size confuse us on this point. Here, we are evaluating the consequences of different structures of power and accountability. Size is often a handy stand-in for these concepts. But to refer to "big landowners" is actually to refer to a particular authoritarian structure of power, in which a few make all the decisions over the use of a vital resource: farmland.

Just because an agricultural system is characterized by small farms, we should not assume it is necessarily equitable. If small farmers are at the mercy of those who control distribution of farm inputs and marketing of farm commodities, they are powerless even if large operators do not monopolize the land. Thus, more than land tenure must be addressed. In the more effective land reforms, smallholder farmers have been supported with marketing mechanisms, infrastructure, credit, and technical assistance.[58]

PRODUCTION FOR WHAT?

While historical experience shows us that fairer control over agricultural resources can bring greater, not lower, output, a more fundamental consideration ought never to be forgotten. Even if one could prove that an elite-dominated agricultural system were more productive, we would still ask, So what? Is not food production of value only if it fulfills the needs of all? If a society's agricultural system is strikingly productive but its citizens go hungry, of what use is it?

The problem of production must therefore never be posed in isolation. The question must not be what system can produce more but under what system—elite-controlled or democratically controlled—hunger is more likely to be alleviated.

If as a result of a redistribution of assets the poor gained buying power, the composition of crops would likely respond to their needs so production of luxury crops might slump. Families with their own land for the first time might consume the food they grow, rather than sell it and increase marketed production. Indeed, the market value of production might fall—but hunger would be falling, too. Our point is simply a reminder never to focus so narrowly on production that we forget why we care in the first place. Our real concern is how to end hunger.

Many people have been made to believe that we must choose between a fairer economic system and efficient production. This trade-off is an illusion. In fact, the most inefficient and destructive food systems are those controlled by a few in the interests of a few. Not only can greater fairness release untapped productive potential and make long-term sustainability possible; but it is also the *only* way that production will contribute to ending hunger.

143

myth 6

The Free Market Can End Hunger

MYTH: Governments are increasingly incompetent, corrupt, and dead-locked. So ending hunger requires freeing people up to innovate their own solutions. Without government regulations blocking efficient production and distribution, food would be more affordable for poor people. It's best if governments just get out of the way so the free market can solve the hunger problem.

OUR RESPONSE: Sadly, a free-market-is-all-we-need formula blinds us, preventing us from seeing that a well-functioning market is impossible without democratic government. Such dogma even ends up destroying the conditions necessary to realize the prized strengths of the kind of market we need to end hunger—one with openness, competition, and transparency. Worst of all, it keeps us from asking critical questions about the market—and about government—that we must ask to end hunger:

- What is the market good at, and what can the public sphere best accomplish?
- How does a society create the conditions that enable both the market's virtues to shine and government to succeed in its essential roles?

When we answer these questions, things get really interesting. But before we explore them, there's an even more foundational question:

What's the market *not* designed to do?

Certainly, most humans seem to agree that it's not designed to distribute life, at least not human life. Only Iran allows the selling of human organs.[1] No society allows the selling of babies—or, for that matter, grown-ups. (Unfortunately, this universal legal norm doesn't mean that slavery has yet disappeared.)[2]

Thus, it could follow logically that treating food—on which life depends—as *simply* a commodity violates the spirit of a widely shared human sensibility as well as common sense. "Food is not a commodity like others," President Bill Clinton pointed out in 2008. "It is crazy for us to think we can" treat "food like it was a color television set," he added, arguing for policies enabling "maximum agricultural self-sufficiency."[3]

> "Food is not a commodity like others. . . . It is crazy for us to think we can" treat "food like it was a color television set."
>
> —Bill Clinton, 2008

So food is certainly a commodity, but it is *more than a commodity*.

In fact, the right to adequate food has been accepted by most countries, beginning in 1948 with the Universal Declaration of Human Rights, which stated that "everyone has the right to a standard of living adequate for . . . health . . . including food."

In 1966 the right to food was reaffirmed with the adoption of the UN International Covenant on Economic, Social and Cultural Rights. A decade later it came into force, and by 2015 one hundred sixty-four countries had made themselves legally bound as state parties. The United States is not among them.[4] Additionally, the right to food is spelled out as an individual human right in the constitutions of more than two dozen nations, ranging from India to Ireland to Ethiopia.[5] Virtually all are market economies.

Declaring a right is easy enough. But, what does that mean for something that's also a commodity within a market economy?

Anything deemed to be a "human right" cannot logically be left entirely at the mercy of market exchange: A right to protection from bodily harm, for example, wouldn't mean much if you had to pay the policeman to catch an intruder. Similarly, without clear and enforced standards, a "right to food" can feel meaningless. The people of India have a right to food, but their country is home to more hungry people than any other.

Since food is both necessary to life itself *and* a market commodity, the only way this right can be realized is if democratic government ensures that every person has the means to secure enough healthy food through access to employment at a living wage; or, if blocked from paid employment by old age, ill health, or family responsibilities, one has access to public support. So we will close this chapter with a story about a major city in a country that's taken this approach to heart to make progress against hunger.

First, however, let's examine some commonly held notions concerning what is often called the "free market" about which we need clarity if we are to make real the right to eat.

SIX FREE-MARKET FICTIONS

Consider in turn each of these six thought barriers on the road to realizing food as a human right:

One: A "free market" works best to meet human needs.

If by "free market" we mean one unbounded by rules, it does not exist. All market economies are governed by rules. Ours covers everything from whether a company can keep its profits offshore to avoid taxes all the way to what livestock you can legally keep in the backyard.

> Six Free-Market Fictions
> 1. A "free market" works best to meet human needs.
> 2. Government necessarily impedes a vital market.
> 3. A free market serves individual freedom.
> 4. A free market gives us all the choices we need.
> 5. A free market maximizes a nation's efficiency.
> 6. The market is "value neutral"; it's merely a tool.

But behind most economies today one unspoken rule has become the dominant driver: *Do what brings highest return to existing wealth.* We mean simply that choices dominating most economies are guided by what will bring the corporations' executives and shareholders the greatest immediate gain.[6] By this rule, wealth accrues to wealth until we end up in the United States with inequality more extreme than in Turkey or India; and in a world with two-thirds of adults trying to survive on 3 percent of global wealth.[7] In recent decades, it's gotten worse: Between 1988 and 2008, the income of the world's top 1 percent increased 60 percent, while that of the bottom 5 percent didn't budge.[8]

Applying this unspoken rule to farming, what happens?

While in theory the market rewards hard work, in reality it requires hard work but largely rewards those who already have considerable resources—and in farming that means equity in land. Those who possess it have easier access to credit, often on better terms, and can therefore withstand market swings that would wipe out a small farm. Better-off farmers, moreover, can expand to make up in volume what all farmers lose in per-acre profits. (Note that farmers' costs often rise faster than the prices they receive, so periodic price-depressing gluts are common in a world where many can't afford to buy the food they need.) One consequence in the United States has been the wholesale folding of family farms over many decades. Of the 2.2 million farms remaining, only 9 percent support themselves from farm income alone.[9]

These consequences are not the fault of greedy people. They reflect the rules we allow to govern our economies. They are rules that we can transform, as examples throughout our book demonstrate.

Two: *Government necessarily impedes a vital market.*

In fact, a market economy cannot thrive without government. Think of the essentials of economic success that government provides, from legal structure to infrastructure. As for government being bad for business, this can hardly be true: Note that in economies ranking among the world's most successful, government spending contributes a big part of the GDP. Take three of the five countries deemed most economically "competitive" by the business-oriented World Economic Forum: Switzerland, Finland, and Germany. In each, government spending accounts for about a third to more than half of the country's GDP. And in the United States, which ranks fifth in global competitiveness, make that 40 percent of the GDP.[10]

This correlation—between a significant government role and a strong market economy—makes sense as soon as we register one key fact: that a lot of what government spends money on directly benefits an economy.

Education is a great example. In the United States, education's actual contribution to national wealth is estimated to be almost four times greater than the cost of public education as counted in the GDP. Or consider clean air standards: During the first two decades of the Clean Air Act, enforcement cost the U.S. government $500 billion. But the resulting health savings alone—for people directly and the economy more generally—came to $22 trillion. That's a benefit forty-three times greater than is registered in the GDP.[11]

Of course, government spending—in societies where policymakers often serve concentrated private power, as in the United States—can also promote competition-killing monopolies and oligopolies. Case in point: The U.S. government spends almost $92 billion annually to subsidize agribusiness and fossil fuel industries. That's 56 percent more than what goes into social benefit programs, such as housing assistance and food stamps, reports the conservative Cato Institute.[12]

Three: A free market serves individual freedom.

In the 1980s, in an auditorium on the UC Berkeley campus, I (Frances) had the opportunity to debate perhaps America's most celebrated "free market" champion, Milton Friedman, author of *Capitalism and Freedom* and coauthor of *Free to Choose*. Friedman claimed that the market serves freedom by enabling people to make choices based on their values.[13]

I then pointed to the obvious: If this is true, the market serves human freedom *only on one condition*: that people have purchasing power to express their values in the market. Thus freedom, using Friedman's own definition, actually expands as societies set rules ensuring that wealth is widely and fairly spread. By the same logic, a market operating without rules to keep wealth circulating widely and fairly denies individuals' "freedom to choose."

Four: A free market gives us all the choices we need.

Of course, we all enjoy the purchasing choices that a market economy offers. But before we celebrate too much, let's consider some choices our market economy denies, illustrated in a metaphor we've borrowed from political philosopher Benjamin Barber:

In our market economy, we get to join a giant cafeteria line with plentiful dishes where—if we have the money—we can grab whatever appears appetizing. Surely it feels like a lot of choice. But notice what we *don't* get to choose. We cannot enter the kitchen and select the menu.

In other words, we don't get to say, "No, it's not seemingly endless choices among processed foods that I want. I want more plentiful, and less expensive, fresh fruits and vegetables. And I choose to make that choice possible by shifting my tax dollar–paid subsidies now benefiting meat and processed food companies toward those helping family farms that grow diverse crops without toxic chemicals."

Or, "No, I don't want to have agribusiness marketers using the latest psychological tricks to lure my kids toward what's unhealthy. I choose to have my kids tempted in the media and in grocery stores by beautiful, healthy, whole foods."

True, our supermarkets typically carry thirty thousand items. That's a lot of choice! But unless I have "menu making" power and a voice in public decisions, my choices—including those protecting my family's' health as well as healthy soil and water—are extremely limited.

Five: A free market maximizes a nation's efficiency.

Many cherish the market because we've learned it's more efficient than the only alternative we've heard of: the widely discredited, government-in-charge Soviet model. Problems arise, however, when our horror at the failure of economics-by-fiat leads us to assume a market economy is *necessarily* efficient. It isn't.

But first, what do we mean by efficiency?

Let's define it as getting the most benefit from resources, human and natural, while ensuring their ongoing health. So defined, most modern-day economies appear "efficient" only because their measures of economic success do not deduct huge built-in losses such as those laid out in Myth 3. Uncounted in the price of food, for example, are the estimated almost two billion tons of topsoil washed or blown away on cropland in the United States each year.[14]

Few would call it efficient to put up new walls in one's home using stones plucked from the foundation. But in too many ways, our misunderstanding of what it takes to create well-functioning markets means that we're providing food today by destroying the essentials our progeny will need tomorrow. And this analogy doesn't even capture the deepest inefficiency of a power-concentrating market: the human potential wasted and destroyed by the devastation and indignity of hunger.

Six: The market is "value neutral."

We hope that, even in the little we've said so far, it's clear that how a society shapes its market's rules powerfully determines its ethical standing. Since no one willingly chooses to go hungry, the very existence of hunger belies the promise of democracy, itself grounded in the value of one person, one voice.

An economy is an expression of a society's values.

Ironically, other market champions claim the opposite: Rather than "value neutral," the market in the eyes of some is part of a divine plan. The idea took off, explains Princeton historian Kevin Kruse, after the 1930s had ushered in significant social protections, and business interests including various chambers of commerce fought back. One of their highly effective strategies was enlisting religious leaders to cloak the "free market" in religious sanctity. Perhaps their most famous enlistee was Reverend Billy Graham, known as the "the Big Business evangelist." Followers were encouraged to accept on faith what in this book we call a "one-rule" market inherently concentrating wealth.[15]

THE MARKET: MY, HOW YOU'VE CHANGED!

Having challenged some unfortunate but common assumptions about the market, let's turn to basics. What *is* "the market"? For sure, it isn't a fixed thing, for it has evolved over aeons and changed radically even within our lifetimes.

From early human societies, the market—meaning simply the exchange of things of value—has served us by taking advantage of our differences. (I can trade the extra peanuts I grow for the earthenware you make.) Exchange for mutual benefit was embedded in norms and rituals of family, community, and religious life. Expectations of reciprocity, conventions about what is sacred, loyalty to place and community—all characterized early markets' often unspoken rules.

In a historical blink of the eye, however, economic life has become detached from community ties, emerging as its own realm, and arguably, in recent years, rising above all others and driven by its own rules.[16]

Just over half a century ago, a team of distinguished economists observed that U.S. executives believed they had "four broad responsibilities: to consumers, to employees, to stockholders, and to the general public."[17] And in 1981 the Business Roundtable, representing CEOs of many of the biggest U.S. companies, stated that a corporation "must" consider its impact beyond shareholders to "the society at large."[18]

We do not want to romanticize U.S. economic history; we simply want to note that such sentiments suggest that the notion of economic life operating almost completely outside a framework of community responsibility is relatively new, at least in today's extreme form. Over time, the U.S. economy has been ripped from its social moorings—and this process has included the weakening of workers' voices, once ensured by trade unions representing as much as a third of the workforce.

The One-rule Economy Emerges

Increasingly, our economy is governed by "one rule," mentioned as we opened this chapter: Highest return goes to existing wealth. Economic enterprise is to serve the pecuniary interest of owners of assets—corporations' executives and shareholders. End of story. Therefore, value created by all of us flows inexorably to a few of us, in what can feel like a giant—and ultimately deadly—game of Monopoly.

When one company controls most of a market, it's called a monopoly. In 1890 the United States outlawed monopolies in the Sherman Antitrust Act, which the Supreme Court later described as a "charter of freedom."[19] Almost sixty years later, Supreme Court Justice William O. Douglas, dissenting in a steel-industry case, wrote:

> The Curse of Bigness shows how size . . . can be an industrial menace because it creates gross inequalities against existing or putative competitors. It can be a social menace—because of its control of prices. . . . [P]ower that controls the economy should be in the hands of elected representatives of the people, not in the hands of an industrial oligarchy. Industrial power should be decentralized . . . so that the fortunes of the people will not be dependent on the whim or caprice, the political prejudices, the emotional stability of a few self-appointed men.[20]

Unfortunately, in virtually all industries, including the business of food, we've not protected our freedom from "the curse of bigness." Oligopolies, almost as destructive as monopolies, exist when a handful of companies control a huge market share. But from grain trade to the store shelf, that's what we see. Four companies—Archer Daniels Midland (ADM), Bunge, Cargill, and Louis Dreyfus—control as much as 90 percent of the world's grain trade.[21] In the United States, just ten companies account for more than half of all food and drink sales, despite tens of thousands of items in a typical U.S. supermarket.[22]

Concentration of the buying power of consumers is just as harmful to freedom. Our one-rule market responds to the tastes of those who can pay, an increasingly small minority able to make what economists call "effective demand." And that minority is a shrinking slice of the whole: Since 2009, 95 percent of all income gains in the United States have gone to the top 1 percent.[23]

Inevitably, then, production shifts to items they desire, some fairly ridiculous. How about a $20,000 designer espresso machine, or a diamond-studded iPhone case or fishing lure?[24] Seriously.

All of this is hardly "value neutral," as the consequences for hunger are profound.

Tightening Concentration of Control

In one outcome of this radical redistribution to the top, three-quarters of all agricultural land worldwide, as we've seen, moves into the production of food from livestock; yet most of the world's people eat little meat.[25]

And prime land in the Global South increasingly grows export crops. In Africa, for example, since the 1980s, cocoa production has jumped about threefold.[26] In Kenya, where one in three young children is stunted by malnutrition, prime resources are used to grow flowers that end up on jets headed to Europe.[27] Perhaps most tellingly, luxury agricultural crops now include *nonfood*—agrofuel for gas tanks displacing basic food crops.

During the last two decades, this pattern of tightening control has entered a new stage. Industries producing actual goods—at least potentially useful to our society—have been eclipsed by dealers in money itself, who often provide no useful service. In 1985, the financial industry's share of corporate profits was about 13 percent. By 2012, its share had more than doubled to almost 30 percent.[28]

Economists dub this dramatic change the "financialization" of an economy.[29]

> In 1985, the financial industry's share of corporate profits was about 13 percent. By 2012, its share had more than doubled to almost 30 percent.

And in this new world, "financial institutions" aren't necessarily what we've always imagined: Among the eight top global food-retail corporations, their financial assets—businesses such as checking and loan services—made up almost a third of their total assets by 2007, up by half over the previous decade. And, no surprise, Wal-Mart is among them.[30]

Such private monopoly power—itself now dominated by the financial industry—is still called a "free market," but in reality it is the opposite, killing competition, openness to newcomers, and transparency, and thus generating hunger from plenty.

And worst of all, our "one-rule" economy creates a self-reinforcing cycle: Concentrated *private* power usurps and distorts *public* decision making. In Washington today, there are almost two dozen lobbyists, mainly serving corporate interests, for every one official elected to protect citizens' interests.[31] So government policies increasingly benefit the minority.

Many people observe how power at the top works to fuel this quickening spiral of concentration, and say, in effect, "See, I told you so. Government *is* bad. Let's weaken it further." They fail to grasp that the root cause of ineffective government is our failure as citizens to keep it accountable by removing the control of private wealth within our increasingly rigged democracy.

Interestingly, the idea for the game Monopoly originated with a concerned Quaker, Lizzie Magie, in the early twentieth century. She wanted to help us grasp what happens when we make rules that concentrate wealth. It might take a while, but in the end one player gets all the property and the game is over!

Apparently, her lesson was lost on us, but it's not too late to listen to Lizzie.[32]

"FREEING" THE MARKET PUT TO THE TEST

One smart way to learn is to reflect on experience, and in the 1980s, the international financial organizations—the World Bank and the International Monetary Fund—offered us all a case study in what happens when the government-is-bad-and-market-is-good philosophy meets real life.

These two organizations reset terms for their loans. Their message to nations needing their services was: "Yes, we'll help you. Well, not exactly . . . we'll help *if* you agree to our terms: First, you must substantially dismantle government supports, whether for schooling or farming or health care, and let the market reign. Sell off public assets and pull back government functions, services, and rules governing the market.

To earn foreign exchange, grow export crops; and then, if you need to, just import food from countries that grow it more efficiently."

Lenders were convinced that their prescription, known by the dry label "structural adjustment," would free the market. And poor countries dependent on international assistance felt they had no choice but to comply.

So in country after country, what were formerly governmental functions, such as helping remote farmers sell their crops, were turned over to private companies. Public funds for schools, health care, and roads to get crops to market were cut; and the number of agricultural "extension" agents—that is, public employees spreading knowledge of best farming practices—was slashed. In Malawi, for example, even though the population increased by two-thirds, the number of such hands-on farming educators is only a third of what it was in the 1990s.[33]

Plus, as we explore more fully in the next chapter, opening local markets to imports—a central tenet of "structural adjustment"—meant relaxing trade rules that protected local farmers from cheap, often subsidized, imports from industrial countries.

And what happened?

The experiment flopped. Over the 1980s and 1990s, these policies "contributed to the further impoverishment and marginalization of local populations, while increasing economic inequality." So concludes a 2002 study by a global network of civil society groups, based on a participatory review that included the World Bank itself.[34]

Over this period, the least developed countries experienced a stunning turnaround. They moved from being net exporters of agricultural products to being net importers. From the 1980s to 2011, imports of food by the UN-defined "least developed countries" leaped about sevenfold to $37 billion in 2011.[35]

The experiment failed in part because it undercut local farmers as well as the very public assets that create the conditions all markets require to work—from educated kids to roads to enterprising producers.

HOW A TILTED PLAYING FIELD GROWS HUNGER

We've seen that once economic power is tightly concentrated, aided by these flawed prescriptions for developing economies, a market's supposed virtues disappear. Openness to competitive newcomers, fair bargaining, and efficient use of resources are the first victims. And, for sure, hunger follows, even in the world's wealthiest country, the United States. As we've seen, one in six Americans struggles with hunger.[36]

Seeds—a Showcase for the Unfree Market

As global corporate control of commercial food and farming tightens, the promise of "freedom" in the market becomes freedom denied to farmers. They include the 1.5 billion people on small farms producing most of the world's food, and in Asia and sub-Saharan Africa, it's 80 percent.[37] Here is what we mean: Unless farmers use agroecological practices, as explored in Myth 4, every season they must purchase seeds, fertilizers, and pesticides—making their families' survival dependent on pricing decisions in faraway corporate boardrooms.

Take seeds.

As we learned in Myth 3, throughout the twentieth century most new seed varieties were developed with publicly funded research and breeding. In fact, in the United States, as late as 1980, public-sector-generated seed was used on 70 percent of land planted in soybeans, and accounted for 70 to 85 percent of common varieties of wheat.[38]

But over the last forty years, the commercial seed industry has replaced the public's role. The seed market has gone from being a competitive arena of small, family-owned firms to one in which just three companies—Monsanto, DuPont, and Syngenta—control over half of the global, proprietary (brand-name) seed market. Worldwide, over a period of just twelve years ending in 2008, a handful of corporations absorbed more than two hundred smaller independent companies.[39]

In the United States, by the late '90s just two companies had come to control half the seed market, the USDA reports.[40]

With genetically engineered seeds, the concentration has become even more intense. Eighty percent of U.S. corn acres and more than 90 percent of soybean acres use seeds with genetic traits controlled by a single company, Monsanto.[41]

Seed patenting, also discussed in Myth 3, is key to understanding the dependency created by concentrated power. Before hybrid seeds and seed patenting, most seeds were common property, enabling farmers to save and share their seeds. Under patent law, it is illegal—an infringement on company property rights—for a farmer to save seed from one season to the next.[42]

Monopoly power plus patents, as we were taught in "Econ 101," brings unjustifiably higher prices. In the United States, the combination has helped double the per-acre cost of soy and corn seed—almost all genetically modified—between 1996 and 2012.[43] For fertilizer, the global price jump has been

> In the United States, the per-acre cost of soy and corn seed—almost all genetically modified—doubled between 1996 and 2012.

even steeper, as we learned in Myth 3. Note that we've adjusted these price leaps to account for inflation.

In Ankeny, Iowa, eight hundred farmers crowded into a community college auditorium in 2010 for hearings convened by the U.S. Departments of Justice and Agriculture. The topic: corporate control and anti-competitive practices. One grain farmer told the crowd that he had no choice but to buy Monsanto seeds, yet "the prices eroded any gains he got from higher yields." Tufts University development expert Timothy Wise covered these hearings and describes Monsanto's grip as "a classic case of monopoly selling power."[44]

If we take all this in, perhaps we should not be shocked that in the United States median farm income was a *negative* $1,141 in 2013.[45] Far away from Iowa, in Andhra Pradesh, India, many farmers are caught in a similar bind, robbing them of financial return. There, as we noted in Myth 3, purchases by a typical nonorganic farmer of just three corporate inputs—pesticides, seeds, and fertilizer—now amount to almost half of the farm's total costs, reports the World Bank.[46] If crops fail, farmers' indebtedness for these purchases can lead to despair and even suicide.

The pain wrought by monopoly and oligopoly power sure seemed like one clear takeaway from the market theory we learned in school. But free-market dogma has apparently blinded many people to the danger that founder Thomas Jefferson identified in the "aristocracy of our monied corporations," and that Justice Douglas later called the "curse of bigness."[47]

"Free Market" Lessons Unlearned: A Story

Here's a revealing story about how what's called the "free market" actually works to ensure hunger amid plenty.

The year is 2007. Suddenly, food prices shoot up. It is a terrifying time, especially for poor households in poor countries—households that typically spend upwards of half their income on food, mainly grain.[48] Between 2006 and 2008, the average world price of oil, corn, wheat, and soybeans more than doubles. The price of rice, the staple food for more than 1.6 billion people, climbs threefold. And much of that jump happens in a few months' time![49]

Desperate people take to the streets in over two dozen countries. Thousands die.[50]

By 2010, prices fall somewhat, and people are breathing a sigh of relief, but prices soon jump again, by 2011 surpassing the first spike.[51] Experts on the Middle East, including *New York Times* columnist Thomas Friedman, identify food-price pressure as a factor in the popular eruption in 2011 known as the "Arab Spring" and in the violence in a number of countries that follows.[52]

Moreover, even after these radical spikes abate, higher prices remain: The UN's inflation-adjusted Food Price Index is still elevated—as of 2014 hovering around 50 percent higher than the norm during the two decades before the crisis.[53]

In the years following the acute crisis, experts have scrambled to explain what happened: Just why did prices shoot up?

Note first that there was never a shortage of food. As prices began to climb in 2007, world food production per person actually rose. In the worst crisis years—2008 and 2011—world food production per person hit historic highs.[54]

So What *Had* Changed?

Many point to a drought in Australia in 2006 and 2007 that cut the country's wheat from 12 percent to 6 percent of global exports. But no one seems to suggest this shortfall was key to the ensuing disaster. Some cite increasing price pressure from Chinese demand for grain-fed meat. But China's cereal imports, largely for feed, were at the time of the crisis well below what they had been at the beginning of the decade.[55]

While the debate continues, four forces stand out:

One: Speculation on food commodities exploded.

After the Clinton administration removed significant public oversight of the financial industry in 1999 and 2000, investor dollars flooded into new commodity index funds.[56] Individuals and institutions that buy into such funds are, in effect, betting on the future prices of commodities, including food crops.[57] Testifying before a U.S. congressional committee at the height of the initial 2008 crisis, portfolio manager Michael Masters noted that "assets allocated to commodity index trading"—including major food crops—increased twentyfold after 2003, reaching $260 billion in March 2008. Unlike "traditional speculators," Masters noted, "index speculators . . . buy futures and then roll their positions . . ." The result is that "demand . . . actually increases the more [that] prices increase," he testified.

In fact, commodity prices increased more in the aggregate over the five years before 2008 than at any other time in U.S. history, Masters estimates. In his view, trading strategies of "index speculators" amount to virtual hoarding via the commodities futures markets, and "provide no benefit" to the public.[58]

Speculation is blamed not just for upward pressure on prices but for food-price volatility. For wheat—a staple food in many parts of the world—the duration of what's called "excessive price volatility" more than doubled in the period following the price crisis.[59]

Two: An unexpectedly "large surge in diversion" of corn into agrofuel production helped push prices up, observe two agricultural economists.[60]

Worldwide, in the early 2000s acres producing agrofuel—not food—began growing rapidly, and by 2011 they amounted to an area the size of Sweden. (It's projected that by 2020, agrofuel acreage will almost double.)

Moving prime farmland out of food crops might seem odd in a world worried about its food supply.[61] So how did it happen?

Lobbyists for agrofuel in the United States and Europe succeeded in convincing governments to mandate its production—obviously contradicting the tenets of a "free market."[62] Oh, how easily "free market" ideology is discarded when it doesn't serve the interests of those who have most vigorously touted its wonders.

Three: The price of petroleum rose.

And indeed it did. From 2002 to the height of the crisis in 2008, oil prices leaped sevenfold, hitting $147 a barrel.[63] But blaming food-price spikes on what happened to the price of oil leads us to this question: Why has our food economy become so vulnerable to the fossil fuel industry? Obvious answers include the greater and greater distances our food travels and the petroleum used to manufacture synthetic fertilizers and pesticides.

Another layer of explanation lies in the way economic power translates into political power: The U.S. fossil fuel industry is among the largest contributors in political contests, spending $145 million in lobbying efforts in 2013 alone.[64] The industry's political power helps shape public policies. In 2013, for example, worldwide fossil fuel consumption subsidies were "over four times the value of subsidies to renewable energy and more than four times the amount invested globally in improving energy efficiency," reports the International Energy Agency.[65]

Fourth and finally: Some argue that world grain stocks were a factor. Relative to use, they dipped in 2007–08 in part because there is no world grain reserve system to fill in the gap.

"Stocks" are carryover grain stored both on farm and off, while "reserves" are grain officially held for release to avert price crises. Although there are no global reserves, some countries and at least one region have a reserve system.

Even as world grain production hit historic highs in the price-crisis years 2008 and 2011, grain stocks fell by a fifth in the 2007–08 season compared with two years earlier. At the time of the second price peak in 2010–11—when the FAO Food Price Index was about 70 percent higher than a decade earlier—the dip in stocks was 5 percent compared with the previous year.[66]

In both price crises, grain stocks available to quell the price takeoff were down a bit. But, clearly, stock levels hardly determine price levels: Today, stocks are back up to where they were before the crises, even as world food prices remain well above precrises trends.[67]

United States grain exports make up a huge share of total world exports, so stabilizing our supply, and thus prices, is critical for the world's food stability. During the Great Depression, President Roosevelt created a grain reserve to help keep prices high enough to save American farmers from collapse and to be on hand in case of a natural disaster.[68] Then, following the food and petroleum price crisis of the early 1970s, USDA in 1977 created a price-stabilizing program, the Farmer-Owned Grain Reserve, offering farmers low-interest loans plus reimbursement to cover the storage costs of the grain.[69]

But, soon, the ideology of "market freedom" denigrated government's role and led to the 1996 farm bill—nicknamed the Freedom to Farm Act. By 2008 it had abolished our national system of grain reserves.[70]

Today, the critical need for stocks and reserves remains one lesson unlearned. Encouraged to trust the "free market," poor countries remain dependent on food imports in a highly concentrated and volatile market.

It has not worked. The price crises brought unnecessary hardship and hunger for millions. And since then almost nothing has been done to prevent a repeat of this pain.

Hunger-making Ripples Continue

The price spikes of the 2000s brought even tighter consolidation of control—from grain trading to farmland ownership itself. Between 2007 and 2008, at the peak of food-price escalation, earnings of the giant grain trader Cargill went up over 50 percent, from $2.34 billion in 2007 to $3.64 billion in 2008.[71]

As food prices rose, suddenly agricultural land became an attractive target not only for private investors but for governments, too, mainly in the Middle East and China. They began to acquire farmland abroad out of fear that their citizens, if rattled again by food-price hikes, might rise up against them as they did in 2008. Private investors saw profit, too.

And where could they get land on the cheap?

Where rights are often governed more by custom than by formal deeds—Africa and parts of Asia. So it is not surprising that two-thirds of purchases of agricultural land by foreign investors during the 2000s were in poor countries with serious hunger.[72] Their purpose was not to meet unmet local needs but to profit by producing for export.

> Between 2000 and 2010, private corporations and governments in the Global North acquired enough land in the Global South to feed *a billion people*.

And the scale is vast: Between 2000 and 2010, private corporations and governments in the Global North acquired enough land in the Global South to feed *a billion people*.[73] One story captures this tragic development.

In 2014, southwest of Tanzania's capital, Dar es Salaam, Timothy Wise, quoted above, spoke with a task force representing eleven villages that had lost twenty thousand acres of land to the British-owned Sun Biofuels.

"Under Tanzanian law," Wise explains, the deal "had converted the title of the property from 'village land' to 'general land,' putting it under the control of the Tanzanian national government." The government then signed a ninety-nine-year lease with Sun Biofuels to produce biodiesel for export to Europe.

Villagers agreed. They were "promised compensation for it and, more importantly, more than a thousand jobs, a variety of community development projects—roads, wells, schools, health clinics—and agricultural investment in local farms," Wise reports.

But in just five years, the plantation failed. What's left is five thousand acres of *Jatropha* trees—a genus of flowering plants prized for its oil—"surrounded by guards hired to keep villagers off what used to be their land."

Sun Biofuels had sold its lease to the Tanzanian company Mtanga Farms. Initially, Mtanga showed no willingness to take responsibility for the commitments Sun Biofuels had made to the villagers, but as of this writing public pressure could be prompting a shift in position.

For now, thousands of farmers and herders, with no legal ground on which to defend themselves, are even worse off than before. To make way for its plantation, Sun Biofuels had cut down a forest integral to the lives of the villagers. "We used to fetch water—it was close," Salima Nasoro, a brightly clad woman from Muhaga village, told Wise. "We used clay for handicrafts. We cut poles for construction. We made timber. We got charcoal. We kept bees. We collected traditional medicines."[74]

All that is gone now.

And this story is being repeated throughout the Global South. Sixty million indigenous people worldwide may be at risk of losing their land "as a result of large-scale agro-fuel expansion," reports a UN representative.[75]

Here we've briefly covered a historic unfolding of destruction and displacement of lives, most of it happening in less than a decade. We hope its vast pain brings home to our readers the inevitable consequences of a market economy that is profoundly unfree. Before moving toward a resolution of this problem—that is, how to create a truly efficient and life-serving market—let's remind ourselves that our power-consolidating market not only destroys lives but also wipes out common wealth: those resources on which all life depends, including healthy soil, water, and diverse species, as explored in Myth 3. The impact of greenhouse gases on all aspects of life on our planet, explored in Myth 2, may be the greatest loss uncounted in our misnamed "free market."

Clearly, it is possible for some to celebrate our "efficient" food system only because they do not deduct these losses. In a truly efficient market economy, *all* would be incorporated into its logic. Prices would take into account the cost of preventing these losses, and societies would therefore make very different choices.

But how to achieve this efficient market economy?

Through democratic rules we create together, including protective standards that prevent harm before it happens.

RULES FOR LIFE

Once we appreciate government's essential role in a healthy market, the temptation is to debate *how much* government versus *how much* market is best. We can fail to see that it's not the size of government but *what it does* that matters most. Does a government responsive to its citizenry help to ensure the continuing dispersion of power and transparency, or does it do the opposite?

How these questions get answered depends on how a government is chosen, and thus to whom it feels accountable. So the removal of the power of private wealth in political contests is not a separate concern. It is essential to ending hunger.

In other words, ending hunger is not possible without a particular kind of government: what we call Living Democracy, engaging citizens and accountable to them. Only it can unlock the market from competition-killing monopoly and free up policymaking from domination by corporate elites. Citizens all over the world are waking up to the challenge. It's a goal that could unite those typically thought of as left and right, since many across the political spectrum understand that

Six "Rules for Life"

1. *Remove private power over public decision making.*
2. *Shift public support to farming practices that contribute to the public good.*
3. *Create publicly held food reserves to stabilize prices and meet emergencies.*
4. *Protect from harm.*
5. *Protect the right to know and ensure transparency in our food system.*
6. *Ensure that all people have the capacity to free themselves from hunger.*

concentrated power—whether public or private—undermines freedom and leads to the death of effective markets.

Among the most important roles for government is that of public rule setter, so the key to whether government can help end hunger is: *What kind of "rules" does it set*?

We call rules that support vital communities "rules for life." Here is only a taste of what we mean. Of course, none is the product of a perfect democracy. What they do reflect is policymakers' sense of accountability to the citizenry rather than to a moneyed minority.

They are rules, for example, that do the following.

One: Remove private power over public decision making.

Foundational to all others are rules that remove the power of private wealth in politics and effectively fend off monopoly power. In 1938 Franklin Delano Roosevelt warned a joint session of Congress: "[T]he liberty of a democracy is not safe if the people tolerate the growth of private power to a point where it becomes stronger than their democratic state itself. That, in its essence, is Fascism. . . ."[76] A very strong statement. And it speaks powerfully to the challenge posed to democracy by corporate interests today.

Many Americans believe, wrongly, that Supreme Court decisions unleashing unfettered—even secret—corporate and union spending in elections mean that only a constitutional amendment can remove money's grip on our public life.[77] Fortunately—because amending the Constitution is a very lengthy and risky process—there are legislative steps the American people can insist Congress take right now that could achieve much of the same end. We return to this vital point in Myth 10.

Two: Shift public support to farming practices that contribute to the public good.

In Western Europe, a number of governments provide subsidies for as much as five years to help farmers through the transition from chemical to organic farming.[78] In Myth 10, we describe state government actions in India supporting village-learning networks that spread ecological farming practices. The UN's agricultural arm, funded by world governments, sponsors what are called Farmer Field Schools, which have already helped two million Asian farmers transform their fields into learning labs, enhancing soil organisms and increasing yields—often dramatically: In Kenya, Farmer Field Schools have helped some farmers increase their yield by 80 percent.[79]

Governments can also create rules to foster a market for healthy food.

- The government of the Netherlands helps to shift farming toward safe practices by, for example, setting a goal of reducing the negative impact of pesticides. As a result, the Netherlands cut pesticide harm 85 percent between 1998 and 2010, in part by mandating safer spraying practices and banning the most polluting pesticides.[80] To meet its national goals, the government also asked all food services in public institutions, including schools, to procure only "sustainably" grown food. Food services in government ministries were required to make at least 40 percent of their food organic. (For some ministries, it was 75 percent.)
- Similarly, Italians passed a law in 1999 requiring their public schools and hospitals to use organic food.[81]

Just imagine the United States making a "subsidy shift"—moving support from big, industrial farms to instead encourage ecological farming and eating.

Three: Create publicly held food reserves to stabilize prices and meet emergencies.

After all, creatures from ants to bears stock up for tough times. And we know from the Bible that food storage was a big topic way back then. And, more than 2,500 years ago, a celebrated Chinese emperor developed a system of granaries that bought grain at above-market prices when it was plentiful and released it when prices rose, all to ensure a steady supply at a steady price.[82]

So, if we'd stayed true to our ancestors' wisdom, the horrific pain caused by the price crises of the last decade could never have happened.

Few Americans are aware that after a serious food-price spike in the early 1970s—triggered by a U.S. corn blight and weather-related crop failures elsewhere, as noted above—the USDA created a Farmer-Owned Reserve in 1977, mostly of corn and wheat. It appeared to reduce year-to-year price variability.[83] With the election of Ronald Reagan in 1980, however, the White House preached the "magic of the marketplace," a sentiment leading to the 1996 farm bill that, as mentioned above, effectively killed U.S. grain reserves.

Fortunately, many other countries have continued to see food reserves as essential to protecting lives. During the 2007–08 price spike, Bangladesh was one of three dozen countries that released public stocks to dampen the price rise and protect their people from hunger. Access to food in Bangladesh "would probably have been worse if there were no public stocks and public distribution system in place," concluded the FAO.[84] Based on a successful pilot starting in 2004, the East Asia Emergency Rice Reserve became permanent in 2013—committing thirteen countries to providing food assistance in disaster-caused emergencies as well as for poverty alleviation.[85]

Forty years ago President Gerald Ford, a Republican, advocated for a global grain reserve at the first World Food Conference.[86] Today shedding dogma and again embracing this commonsense hunger-fighting solution becomes ever more urgent as climate change brings more erratic weather.

Four: Protect from harm.

Rules for life also include standards and inspections for food safety. In Myth 3 we noted a government's essential role in ensuring independent, long-term testing of, for example, new gene technologies, food additives, and farm chemicals, and in enforcing safe-use standards.

Five: Protect the right to know and ensure transparency in our food system.

We can, for example, join the sixty-four countries that require GMO labeling. We can go further, demanding that our food also carry information (such as a Fair Trade label) on whether workers producing it are paid a living wage, and information on its climate impact. One transparency tool for consumers is Buycott.com.

Six: Ensure that all people have the capacity to free themselves from hunger.

This "rule for life" forms the bedrock of the widening global commitment to food as a human right, with which this chapter began. The right to food is a public good that only accountable government can achieve by ensuring the opportunity for every able-bodied community member to secure income needed to function and contribute, and by protecting those unable to work.

Thus one rule for life is a minimum wage high enough to support a family in dignity. Such a public good might look quite different in rural societies as contrasted with urbanized societies. In Rwanda, for example, where the 1994 genocide also killed 90 percent of cattle, a new policy ensures "one cow per poor family." Cows provide milk as well as a source of natural fertilizer. Each beneficiary family gives its first female calf to a neighbor, who is then obliged also to pass on its first female calf, and on it goes. By one account, this policy and similar measures have helped over one million Rwandans emerge from poverty.[87]

A RIGHT TO EAT

With a good set of rules, the market can work beautifully. So let's look at how a "right to food" takes shape in a country when many people work to give it teeth. The result is not doing away with the market but striving to ensure it works for all.

In Brazil, beginning in the 1980s, "social activism based on human rights principles" created

> The right to food is spelled out as an individual human right in the constitutions of more than two dozen nations, ranging from India to Ireland to Ethiopia.
> —FAO, 2014

seven thousand committees that worked on everything from income generation to urban gardens to support for agrarian reform. "This movement," according to the FAO, was an "essential trigger" for generating the "policy commitments to the right to food."

After the end of dictatorship in 1985, Brazil's social movements gained such strength that in 2002 they were key in electing the first working-class president, Luis Inácio "Lula" da Silva. He launched an ambitious, multifaceted national campaign: Zero Hunger.

Brazilians understand that the right to food is not about handouts, notes the UN, but about "fostering people's capacity to produce or purchase food."[88] In Myth 10, we return to the right to food in Brazil, including the "conditional cash transfer" it pays directly to poor women, an income support now being adapted in forty countries.[89]

The Right to Food in Belo Horizonte

Take, for example, the Brazilian city of Belo Horizonte, home to 2.5 million people.

In 1995, 20 percent of its young children were malnourished.[90] But things had begun to change in 1993. A newly elected administration declared food a right of citizenship. City officials effectively said: "If you are too poor to buy food in the market, you are no less a citizen. We are still accountable to you."

The new mayor, Patrus Ananias, created a city agency to act as convener as well as facilitator in setting new rules. He assembled a twenty-member council of citizen, labor, business, and church representatives to advise in the design and implementation of a new food system.

Among the dozens of Belo Horizonte's right-to-food initiatives is one providing local family farmers with small, prime spots of public space in the inner city to sell produce to urban consumers. By avoiding the "middleman" markups of the typical retailer—which often reach 100 percent—poor consumers gain access to fresh, healthy food, and farmers' profits grow, too.

In addition to these farmer-run produce stands, the city makes good food available by offering entrepreneurs the opportunity to bid on the right to use well-trafficked plots of city land for grocery stores called "ABC" markets, a Portuguese acronym meaning "food at low prices." The city determines a set price—about two-thirds of the market price—for roughly twenty healthy items. Everything else the proprietors, now at about three dozen locations, can sell at the market price.

"We're fighting the concept that the state is a terrible, incompetent administrator," said Adriana Aranha, who coordinated these efforts in the 2000s. "We're showing that the state doesn't have to provide everything—it can facilitate. It can create channels for people to find solutions themselves."[91]

The city partners with a local university to "keep the market honest, in part simply by providing information," Adriana told us. The prices of forty-five basic foods and household items at dozens of supermarkets are posted at bus stops, online, on television and radio, and in newspapers so people know where they can get the best deals.

The right to food also shows up in three attractive "people's restaurants" in Belo Horizonte, part of a hundred across the country. These restaurants serve nutritious meals—costing about 50 cents—to tens of thousands of citizens each day, and all with without "means testing" or any stigma. Local farmers supply the food.[92]

The result of these innovations?

Belo Horizonte cut its child death rate—widely used as evidence of hunger—by 72 percent in twelve years.[93] Today these initiatives benefit almost 40 percent of the city's 2.5 million population.

> Today, Belo Horizonte's initiatives benefit almost 40 percent of the city's 2.5 million population. And their cost? Less than 2 percent of the city budget, or about a nickel a day per Belo resident.

And their cost?

Less than 2 percent of the city budget, or about a nickel a day per Belo resident.[94]

Behind this dramatic, lifesaving change is what Aranha calls a "new social mentality"— the realization that "everyone in our city benefits if all of us have access to good food, so, like health care or education, quality food for all is a public good."

Belo Horizonte provides the world with many lessons. For us, most profound is the evidence of the extraordinary progress that can be accomplished quickly when people let go of dogma and roll up their sleeves together, enabling more and more people to participate in the market.

And Belo's lessons are igniting change as far away as Africa. In 2013 and 2015, the Hamburg-based World Future Council, on which coauthor Frances serves, organized two study tours to Belo for African municipal leaders. Moved and motivated by the effective strategies they saw in action there, several Namibian mayors are developing ongoing relationships with their counterparts in Belo and adapting what they've learned, such as urban agriculture, to Namibia's realities.

BEYOND DOGMA: DEFINING A FREE MARKET

We hope the stories throughout this chapter can help us all see the market for what it is—a tool, not an absolute.

As we've seen, taken as dogma, the market can be deadly. An early example is the Irish Potato Famine, which two new histories of the tragedy now tell us flowed in part from the zealotry of British officials. Enamored of Adam Smith, who, in the century before the famine, famously wrote of the power of the market's "invisible hand," the British forced the Irish peasants to abandon their use of barter in favor of market exchange. Their failed strategy was one early consequence of the market taken as gospel, in this case contributing to mass starvation.[95]

In the same vein, we hope the information in this chapter can help us all let go the notion of a tug-of-war between the market (good) and government (bad), a casting of the challenge that blocks us from creating solutions. An effective market and an accountable government are essential allies, not enemies. Since a "free market" operating without rules is a fiction, the key questions about hunger become, What *kind of market* works, on what terms, and for the benefit of whom?

Exploring these questions, perhaps our societies can move toward redefining the term "free market" itself, realizing that a market is free only when *all people are free to participate fairly in it to meet their needs.* In challenging the six "free-market fictions" with which we began, and probing the meaning of freedom within the market, we can shape democratic markets valuing life, and put an end to hunger.

myth 7

Free Trade Is the Answer

MYTH: Freed from government meddling, world trade could help reduce hunger. Every country could benefit from its "comparative advantage"—each exporting what it can produce most cheaply and importing what it cannot. So countries with severe hunger and poverty could increase exports of those commodities best suited to their geography. Then, with greater foreign exchange earnings, they could import food and other essentials to alleviate hunger and poverty.

OUR RESPONSE: The theory of comparative advantage sounds perfectly sensible. Didn't all of us learn in junior-high geography class how "natural" and fortunate it is that Juan Valdez in South America grows coffee for us while we in turn export industrial goods that his country needs? So, in a world of unhampered free trade we all win.

Such an appealing theory! It falls apart only when we apply it to the real world. When our country was getting on its feet, it certainly didn't adhere to the theory. Even though textiles and other manufactured goods could have been imported more cheaply from England than produced at home, our new nation put up high tariff walls to protect its own fledgling industries.[1]

Moreover, if increased exports led to the alleviation of poverty and hunger, how do we explain what's happened in so many countries in

All too often, export crops displace local food crops, as well as small-scale farmers, who make up the majority of hungry people worldwide.

the Global South where exports have boomed while hunger has remained untouched or has even worsened?

Understanding begins by appreciating that those profiting from exports typically are large growers, international trading companies, foreign investors, and others who have no incentive to use their profits to benefit hungry people. Plus, all too often export crops displace local food crops, as well as small-scale farmers, who make up the majority of hungry people worldwide.

First, we examine just a few examples from the Global South that illustrate how little, if at all, poor and hungry people benefit from booming exports. Then we challenge the very premise that such a thing as "free trade" can exist at all, and look at how trade rules today are tailored to benefit the richest countries, not the poorest.

So what has "free trade" brought to three countries known as big successes in boosting exports?

Paraguay. Soybean production started to take off in 1990, and with the country's excellent growing conditions, Paraguay soon captured fourth place among world soy exporters.[2] The soy boom fueled double-digit annual economic growth, often surpassed only by China. Today, soybeans—almost all grown on large-scale plantations—blanket 80 percent of the nation's cultivated land.[3] Yet, half of rural Paraguayans are impoverished.[4]

To cash in on the soy boom financed by the World Bank and big U.S. investment companies, the country's largest landowners—fewer than 2 percent of all landowners—have taken over more than three-fourths of the nation's arable land. In the process, over twenty years a hundred thousand small-scale farmers have been forced to relocate.[5] Since soy plantations are highly mechanized and most beans are not processed in Paraguay, rural jobs are scarce, forcing the newly landless people to try to survive in urban slums.[6]

Neither has the soy boom generated much in the way of tax revenue, which in theory could help the impoverished majority.[7] Thanks to tax exemptions benefiting the big soy operators and the absence of a tax on exports, the big winners from Paraguay's soy boom include the grain trading giants Cargill, Archer Daniels Midland (ADM), and Bunge.[8]

And if you hold out hope that at least Paraguay's massive soy exports are feeding hungry people in other countries, please note: Paraguay's soy exports are instead fattening livestock in China and elsewhere for those who can afford grain-fed meat. Or the soybean oil is turned into diesel, fueling cars in Argentina and Europe.[9]

Ivory Coast. It's likely that the chocolate in your favorite candy bar started out here, as this West African country is the world's top cocoa producer.[10] Over just two decades, its cocoa-bean exports grew 60 percent[11] while its cocoa-paste exports shot up more than eightfold.[12] Even beyond cocoa, the Ivory Coast exports more food overall than it imports.[13] Measured by exports, the Ivory Coast is a big success.

Sadly, however, children make up much of the cocoa industry's labor force, working under what the U.S. Department of Labor calls "forced conditions"—i.e., slavery—with many child workers trafficked into the Ivory Coast from Ghana, Mali, and Burkina Faso.[14] Such horrific conditions make the Ivory Coast "essentially the candy-coated equivalent of a narco-state, dominated by three multinational processor/traders," writes investigative journalist Tom Philpott, "two of them based in the United States: Cargill, Archer Daniels Midland, and the Switzerland-based Barry Callebaut."[15]

In the Ivory Coast, export success has not reduced hunger or poverty. The number of calorie-deficient doubled from just under 1.5 million in 1990 to three million in 2014, a growth rate much higher than that of the country's population. The poverty rate rose dramatically as well during the same time period.[16]

Guatemala. Industrial monoculture—especially in sugar and oil palm—has been rapidly expanding in Guatemala, one of the countries with the highest rate of poverty in the Western Hemisphere. Most of the palm oil is exported, mainly to Mexico, where it ends up in processed foods and used in agrofuel.[17]

Palm oil plantations have spread onto land where indigenous families once produced corn and beans for locals, a 2014 Oxfam field investigation found.[18] One strong-arm tactic the big planters have used to pressure farming communities to sell their land is to surround their plots with fences and armed security guards, blocking farmers' access to their own land.[19] Now displaced farmers, who once produced food for their families and local markets, must try to survive on seasonal and low-paying plantation jobs.[20]

Profits from exports go primarily to just six companies that dominate Guatemala's palm oil industry and control input supply, production, processing, marketing, and prices.[21] Some capital for one of the biggest companies came from the New York–based global investment bank Goldman Sachs.[22]

Given these realities, it's not at all surprising that, even with increasing agricultural exports, nearly half of Guatemala children are stunted.[23]

These far-flung examples tell a story. Where the majority of people have been deprived of the power to access and develop the agricultural resources of their countries, those in control will orient production to the most lucrative markets. And those markets are abroad, in a globalizing food market where wealth is becoming ever more concentrated. So even pets in Europe and North America, enjoying chow made with palm oil, now "outbid" the impoverished half of humanity.

Not surprisingly, those benefiting most from this system will also be the most ardent champions of "free trade."

"ADVANTAGE" FOR WHOM?

The logic of comparative advantage is supposed to lie in countries' differing natural endowments, but in reality differences in soils and climate often have little or nothing to do with a country's exports. Mexico exports tomatoes to the United States not because its climate is more advantageous than Florida's or California's (except for a few weeks in late winter, its climate is no better), but mainly because tomato pickers in the Mexican state of Sinaloa are paid a paltry $10 a day, even less than pickers in Florida.[24]

So, when the headline of a Philippines' Ministry of Trade ad in U.S. business magazines read "WE WANT YOU TO TAKE (COMPARATIVE) ADVANTAGE OF US," its message was right on target.[25] The real "advantages" of countries in the Global South are their low labor and land costs, along with the virtual absence of laws protecting labor and the environment.

And countries in the Global South learn that if their workers do demand and receive decent pay and benefits they can lose their "advantage."

Such is the story of pineapples: In the twentieth century, Hawaii became the world's leading pineapple exporter, a sector largely controlled by just three companies: Dole, Del Monte, and Fyffes/Chiquita.[26] Then, in the 1970s, fieldworkers in Hawaii organized and won livable wages and benefits. Moreover, land costs skyrocketed.

What did Del Monte and Dole do? They simply shifted production of virtually all canned pineapple to the Philippines, where, under the dictatorship of Ferdinand Marcos, labor organizing was effectively prohibited, and they could obtain land by dispossessing poor farmers.[27]

Continuing technological advances in transportation and electronic communications allow global corporations to search the world for the lowest wages, most lenient labor and environmental regulations, and cheapest resources. Working people and whole countries are, in effect, forced to answer: Who will work for less? Who will make do with seasonal or other forms of part-time work? Who will forgo health insurance and occupational safety regulations? Who will allow toxic dumping in their backyard?[28] The result is what some have aptly called the "global race to the bottom."

EQUAL RULES, UNEQUAL CONSEQUENCES

International food-trade volume more than doubled between 1986 and 2009, and today 23 percent of global food production enters international trade.[29]

What are the implications of this huge shift?

Historically, goods crossing borders have enabled governments to use import duties, export taxes, customs fees, and more to raise revenue to spend on public goods—everything from hospitals to schools. Using these tools, governments could potentially shape economies in the public interest by, for instance, influencing whether vital goods, like food, are best produced locally or imported. For a government accountable to its citizens, these tools are vital. But to global corporations they're anathema. They reduce profits on exports and imports, thus infringing on a corporation's bottom line.

As explained in the previous chapter, the World Bank and the IMF, starting in the early 1980s, often made new loans to debt-mired countries on the condition that they would adopt a set of economic policies known as "structural adjustment programs."

The loan conditions required cutting back government's role in the economy and instituting other "free market" policies. In the previous chapter, we stressed cutbacks in education and other social services. Loan conditions also included the elimination of government's role in buying, storing, and distributing food, and at the same time they mandated opening the economy to imports, regardless of the impact on local producers.

The effect of "structural adjustment" was to eliminate or substantially weaken tariffs and other trade controls, as well as the government's ability to fix exchange rates for national currencies, use its spending power as an economic stimulus tool, and enact regulations on foreign investment. Losing these tools, governments found their economies increasingly vulnerable to economic forces and actors beyond their borders.

In all, "structural adjustment" programs—though piecemeal and impermanent—dramatically weakened the ability of many governments in the Global South to shape their own economies.[30]

Agriculture, Separate and Unequal

In grasping the impact of trade, it helps to know the backstory. So what are the rules of international trade and how did they come to be?

In 1947, the General Agreement on Tariffs and Trade (GATT) was created as a result of the United Nations Conference on Trade and Employment. GATT was a global trade accord among twenty-three nations primarily meant to reduce trade barriers.

From its inception, GATT treated agriculture like any other part of the economy. But the United States and then, once it was established, what is now known as the European Union (EU) created agricultural policies protecting their own farmers that violated GATT rules. How could they get away with it?

These two powerful players threatened to leave GATT otherwise, giving fellow GATT members little choice but to go along with exempting their agricultural policies from the rules. This exemption was to be a "temporary" waiver, but in fact it remained in force for almost forty years, during which powerful Northern countries protected their own producers by restricting imports of sugar, peanuts, and dairy products.[31]

However, in 1986, the "Uruguay Round" of trade negotiations, within the framework of GATT, got under way. It lasted over seven years, with eventually 123 nations participating. By its conclusion in 1994, the Uruguay Round had created a separate agreement for agriculture and established a permanent international organization to oversee trade negotiations and arbitrate trade disputes. Thus was born the World Trade Organization (WTO).[32] And its Agreement on Agriculture (AoA) made permanent a number of "structural adjustment" policies, including a reduction of tariffs and an end to allowing the use of quotas to limit imports.

With the establishment of the WTO in 1995, and its AoA, pressure mounted to eliminate special treatment for the United States and EU.

The new agriculture agreement, after all, was supposed to be about expanding export opportunities for countries in the Global South. The hope was that their lower production costs would give them a positive leg up in global competition, and under the right conditions—outlined in the final section of this chapter—help ordinary farmers. Many expected a crackdown on the array of U.S. and EU government-financed programs that promoted their agricultural exports and shielded domestic producers from less expensive imports from the Global South.[33]

But no. These powerful governments figured out a way to continue protecting their own agricultural producers by simply changing their policy *methods*.[34] The WTO rules put pressure on the U.S. government to end its policy of setting a floor under farm prices using a guaranteed loan rate, so policy shifted instead to providing income support for farmers disconnected from how much farmers produce—a tack WTO rules do permit. This means U.S. "dumping"—exporting at prices below actual production costs—continued, and EU countries, which had made comparable changes in their programs, also continued dumping. At this writing, EU export subsidies are minimal, but if world agricultural commodity prices fall again, that could trigger new EU farm supports.

Thus the potential for dumping by the United States and Europe continues. In recent years, however, international prices, although volatile, have tended to be high, bringing some relief to commodity exporters North and South, and a little breathing room for producers. But several export crops that matter most to many countries in the Global South—sugar, rice, cotton—continue to be highly protected in the Global North, untouched by WTO reforms, despite disputes such as one over cotton that we will touch upon below.

Overall, dependence on food imports in the Global South has continued to deepen—an alarming trend, especially for the countries least able to pay for imports.[35] And the recent big increase in land leasing and purchasing in low-income countries by interests in the Global North, described in the previous chapter, is designed to meet, not local food needs, but agrofuel and food demand in countries with greater purchasing power.[36]

On the positive side, the AoA does give Global South countries just a little more say. Unlike the World Bank and IMF, where voting is weighted by the percentage share of capital stock each member holds, in the WTO power sharing is fairer. It is a one-country-one-vote, consensus-based system.[37] Yet critics argue that "actual decision-making" is made by "big trading powers impos[ing] a consensus arrived at among themselves."[38]

The Balance of Power Remains Dramatically Unequal

On the surface, trade rules that apply equally to all and are consensus-based seem fair. But when implementation brings radically different impacts, are they really?

For example, the WTO's primary enforcement mechanism is trade sanctions; and that leaves small countries, typically with few trading partners, vulnerable. By contrast, large countries have many trading partners, and thus are at much less risk. Imagine a trade dispute between the United States and Bangladesh: Should Bangladesh break the rules and the United States be allowed to impose sanctions, those sanctions would bite. How could Bangladeshi exporters expect to quickly replace this huge market? But suppose that the United States breaks the rules, and Bangladesh is invited to impose sanctions. Then what? The U.S. companies would hardly notice.

Additionally, small countries in the Global South simply do not have the budget to launch disputes at the WTO when other countries break the rules.

Moreover, when the EU and the United States are formally judged to have broken the rules, what happens? They often pay a fine rather than change a damaging policy. For example, the WTO's dispute resolution body has found that U.S. cotton programs subsidizing U.S. producers violate the rules and depress world prices, thus harming millions of small-scale farmers in the Sahel region of Africa. These poor African farmers grow high-quality cotton as a cash crop for income critical for family survival. Yet the United States refuses to reform its programs and instead pays tens of millions of dollars each year to Brazil, the chief plaintiff in the dispute.[39]

But because their countries did not have the financial resources to be plaintiffs, the impoverished African farmers, major cotton suppliers for world markets, get no relief from the dumping, and receive no compensation, either.[40]

Trade is vital to most countries' food security—both as a source of food and, partly via international sales, as a source of rural income and employment. To address hunger, governments need tools to stabilize food supplies, including measures to maintain public grain reserves and to regulate trade by, for example, keeping out "dumped" food when it undermines local, often poor, farmers.

The WTO's Agreement on Agriculture cries out for more fairness.

"FREE TRADE" AND MEXICO'S RURAL DEMISE

When President Bill Clinton signed the North American Free Trade Agreement (NAFTA) in 1993, he described it as creating "the world's largest trade zone," through which the "United States must seek nothing less than a new trading system that benefits all nations [and] lifts workers and the environment up without dragging people down."[41] The agreement's aim? To wipe out all trade and investment barriers among the United States, Canada, and Mexico. And it worked. Within one decade of its implementation, all United States–Mexico tariffs were eliminated except for those on a handful of U.S. agricultural exports to Mexico.[42]

But let's back up one step.

In 1985, upon joining GATT, the Mexican government undertook a sweeping "liberalization" of the economy. In agriculture, liberalization meant eliminating almost all government subsidies, credit, and other supports to the traditional small-farm sector that supplied most of the country's staple foods, including corn, beans, rice, and wheat.[43] Such policies drew labor, land, and capital away from the traditional sector and toward stepped-up output by larger-scale agribusinesses producing fruits and vegetables for export to the United States. Mexico's signing on to NAFTA in 1993 can be viewed as the culmination of that liberalization.

Changes wrought by NAFTA threw millions of Mexico's small-scale farm families into competition with agricultural imports from the United States. Mexican growers hardly had a fighting chance. The U.S. producers got big tax-paid subsidies that reduced the prices of their exports—subsidies that NAFTA allowed the more powerful United States even as it denied this type of support for farmers in Mexico.[44]

Not surprisingly, two decades after the launching of this far-reaching "liberalization" of trilateral North American trade, agriculture and rural livelihoods of the less powerful partner, Mexico, had been devastated. Between 1993 and 2008, roughly 2.3 million rural Mexicans—mostly farm families—left agriculture and searched for work in their cities, the United States, or elsewhere.[45] "This trajectory, from country to city to border crossing, is one that has been imposed on them," observes development expert Raj Patel.[46]

NAFTA also undercut Mexican food security by increasing the country's dependence on imported food, the nation's most traditional staple (think of tortillas), first domesticated by indigenous peoples in Mesoamerica thousands of years ago. Imported corn as a proportion of consumption more than tripled, from 10 percent to 32 percent, from 1994 to 2008.[47] At the same time, the price of imported corn was driven up by U.S. diversion of corn to ethanol, thus placing in greater jeopardy the food security of low-income households in Mexico.[48]

Increasing food imports and the displacement of traditional farming meant Mexican diets shifted dramatically toward processed food and away from the age-old staples, corn and beans. Consumption of wheat-based instant noodles is now higher than that of beans and rice![49] Grain-fed pork and poultry consumption also grew markedly, as U.S. pork exports to Mexico increased more than eightfold between the early 1990s, when NAFTA was signed, and 2008.[50]

Among the biggest winners from NAFTA is the world's largest pork producer, Smithfield Foods.[51] On the U.S. side of the border, Smithfield benefits from a steady supply of cheap feed for its extensive hog operations, thanks to the big taxpayer subsidies for corn and soy we just mentioned. Plus, NAFTA granted the company tariff-free exports of its pork to Mexico, a welcoming investment climate for expanding into Mexico, and offered tariff-free importation of cheap U.S. feed for its expanding Mexican livestock operations. Moreover, displaced small and medium-sized corn and hog producers, undercut by the flood of U.S. imports, provided Smithfield a steady supply of low-wage workers, not just for its Mexican facilities, but also for its U.S. meatpacking plants. In the United States a growing pool of undocumented Mexican workers helps Smithfield hold down wages and undermine any union-organizing activities.[52]

By 2010, foreign companies controlled an estimated 35 percent of Mexico's pork industry, with Smithfield likely accounting for half of that.[53]

In 2013, Shuanghui, China's largest pork producer, purchased Smithfield Foods for almost $5 billion,[54] the biggest Chinese takeover of a U.S. company to date.[55] As a result, a large non–North American corporation now benefits from NAFTA.

Mexico's campesinos—Spanish for small-scale farmers—have understandably protested NAFTA, which they see as the root cause of their painful dislocation and the loss of livelihood for those who remain. In one mass demonstration of anger, more than a hundred thousand campesinos hit the streets in Mexico City. "We are the people of corn, survivors who refuse to disappear," read a signboard.[56]

TRADE RULES WITH NO TRANSPARENCY

As we were writing this book, a new trade agreement reminiscent of NAFTA was being negotiated—the Trans-Pacific Partnership (TPP), involving a dozen Pacific Rim countries, with the prospect of others joining later. If completed, the TPP would legally bind all signatory countries to its new rules, which would govern everything from food safety and farm subsidies to financial institutions and the Internet.[57] Former secretary of labor Robert Reich has called the TPP "NAFTA on steroids."[58]

The implications for democracy are troubling: The new agreement, like all trade treaties, would require member nations to alter their laws to conform to the treaty or face the likelihood of trade sanctions against their exports.[59]

Although most of the participating countries call themselves democracies, their trade diplomats conduct TPP negotiations behind closed doors. Here, other than administration officials, the only people who enjoy access to the draft-treaty texts are some six hundred trade "advisers" dominated by representatives of big corporations, reports Public Citizen's Global Trade Watch. In 2013, its spokespeople noted: "So far, the executive branch has . . . denied requests from members [of Congress] to attend negotiations as observers—reversing past practice." Only in 2015 were legislators finally allowed limited access to even view the text.[60]

Negotiating in secrecy may reflect the Obama administration's awareness that, if allowed inside, those in Congress might convey their constituencies' concerns. A substantial majority of Americans, according to the Pew Research Center, are skeptical about increasing trade. Only 17 percent believe that more trade would lead to higher wages here, and 50 percent say that it would destroy U.S. jobs.[61]

> Only 17 percent of Americans believe that more trade would lead to higher wages here, and 50 percent say that it would destroy U.S. jobs.

So legislators are sidelined even though the agreement, as noted, would modify public policy and even laws. And we've learned from previous "free trade" negotiations that the interests of corporations usually end up trumping those of citizens. Behind the lovely term "harmonization" of laws—in this case, laws aligning regulations across borders—are corporations and powerful lobbies. They push to lower environmental and labor standards in all partner countries, generating that "global race to the bottom."[62]

Perhaps most damaging for democracy is the provision that the TPP would allow foreign corporations to seek financial restitution from governments, claiming their profits are diminished as a result of government regulation, according to former World Bank chief economist and Nobel laureate Joseph Stiglitz. He offers an alarming example of such corporate power over governments, which would be permitted in the TPP:[63] In 2010, tobacco giant Philip Morris, now headquartered in Switzerland, sued Uruguay over antismoking regulations. Alledgedly Uruguay's health standards violated a bilateral trade agreement between Switzerland and Uruguay. Uruguay eventually watered down its antismoking legislation, effectively bowing to Philip Morris.[64]

REFLECTIONS ON TRADE'S PITFALLS AND PROMISES

Some have taken the title of our earlier book *Food First* to mean that we advocate "autarky"—a policy of economic self-sufficiency—with all people eating from their own backyards, and thus that we oppose trade. But putting "food first" to meet local needs does not mean that export crop production is in itself the enemy of the hungry.

Our point in this chapter is to warn against the uncritical notion that trade necessarily helps alleviate hunger and poverty. In most societies of the Global South today, "free trade," encouraging export-oriented agriculture, hurts the poor for many reasons. It:

- Allows local economic elites to do business unperturbed by the poverty around them that curtails the buying power of local people. By exporting to well-heeled markets abroad, they can profit anyway.
- Provides both local and foreign elites incentives to increase their dominion over Global South economies and fuels their resistance to reforms that might shift production away from exports.
- Fosters low wages and miserable working conditions. Global South countries "compete" effectively in global markets only by crushing labor organizations and exploiting workers, especially women and children.
- Throws poor farmers in the Global South into unfair competition with foreign producers that dump surplus food in their economies. Driving local producers out of business, these producers heighten the vulnerability of ever more food-dependent nations to the capricious swings of global commodity markets.

But recognizing the positive potential of trade, we must then ask the same question we asked of the free market in the preceding chapter: *What conditions are required for trade to contribute to development that benefits the poor and hungry?*

- Citizens of the Global South achieving a more equitable voice in public decision-making bodies and in control of their nations' resources, including the use of foreign exchange generated by exports.
- Agricultural workers gaining freedom to organize and bargain collectively and to build solidarity with their counterparts across national borders.

- Governments in the Global South cooperating to challenge transnational trading corporations' control over markets, which puts low-income countries into self-defeating competition with each other to lower their standards.

These are some of the changes needed for trade to contribute to genuine, broad-based development. Finally, truly rewarding trade requires a key aspect of "food sovereignty," the capacity of people to meet the basics for survival from their own production. For how can any nation be independent and hold bargaining power on any issue if its very survival depends on food imports and it must desperately seek foreign exchange to stave off famine?

Defenders of the current order might call our preconditions for beneficial trade utopian and dismiss them out of hand. But what is more utopian than clinging to a textbook model of comparative advantage and stubbornly refusing to peek out at the real world?

Trade and export agriculture are not in themselves the enemy of the hungry; but in the real world of extreme power differentials, both reflect and fuel the forces generating needless hunger.

myth 8

U.S. Foreign Aid Is the Best Way to Help the Hungry

MYTH: Our country is so rich, and we are blessed with a surplus of food. Increasing our government's foreign aid is surely the most important way we can help the world's hundreds of millions of hungry people.

OUR RESPONSE: Despite the desire of so many Americans that their aid dollars alleviate hunger, U.S. foreign aid faces huge challenges in helping hungry people.

First, ending hunger requires profound changes allowing people who have been made powerless to gain a voice in their own futures. But much of our government's aid, unfortunately, goes to nations whose political and economic elites are likely to feel threatened by just such changes and would thus fight them with all their might. Resistance is likely even on the local or village level, where larger landowners, moneylenders, and others among the better-off could easily view advancement by the powerless as undermining them.

The second challenge starts at home. Whether aid helps to democratize or to further concentrate power depends on how a donor government defines its national interest; and, sadly, too much U.S. aid is driven by an understanding that is not aligned with the interests of the overwhelming majority of Americans. This dominant interpretation of our national interest also pits us against the well-being of the world's hungry people. So citizens must ask: Can we as a nation come to appreciate that our interests and those of hungry people are in many ways strongly aligned, thus enabling us to redefine the national interest?

We strive in the next chapter to contribute to such reevaluation by exploring the powerful parallels and interconnections between our lives and those of the hungry abroad. In all this, we try to keep in mind that more important than our nation's foreign aid—even at its best—are two especially powerful roles we can play in ending hunger. We return to them as we close this chapter and in the remaining chapters as well:

One is to work to *remove the obstacles* in the paths of hungry people, especially those created by policies of our government and activities of U.S.-based corporations.

The second is to *seize the power of positive example*. By transforming our own society into an exemplar of political and economic fairness, transparency, and citizen engagement, we can provide evidence for the world of what it takes—and prove that it is possible—to end hunger. What power!

Note that this chapter focuses on our government's direct country-to-country economic foreign aid. Evaluating nongovernmental aid and multilateral aid, such as that of United Nations agencies, is beyond this book's scope. But despite our limited focus, we believe that many of the questions we raise can be usefully asked of all aid.

Moreover, we focus on country-to-country economic aid because it is what U.S. citizens can most directly influence through their elected representatives.

U.S. FOREIGN AID—THE BIG PICTURE

Most Americans' hunches about the size of U.S. aid turn out to be way off. Year after year, Americans polled since 2009 on average have estimated that 28 percent of the federal budget goes to foreign aid.[1] If that were true, our foreign aid would constitute

> Americans polled on average believe that 28 percent of the U.S. federal budget goes to foreign aid. In fact, it's less than 1 percent.

a bigger slice of the budget than the Department of Defense![2] In fact, it's less than *1 percent*.[3]

WHAT DRIVES U.S. FOREIGN AID?

"Remember that foreign assistance is not charity or a favor we do for other nations. It is a strategic imperative for America," said Secretary of State John Kerry in 2013.[4] His words echo those of his predecessors from both major parties.

> "Remember that foreign assistance is not charity or a favor we do for other nations. It is a strategic imperative for America."
> —Secretary of State John Kerry, 2013

Arguably, every nation's foreign aid is a tool of its foreign policy. So here's what's key: Whether any nation's foreign aid can help end hunger is determined by the goals of those shaping that policy—*how* they define the national interest.

In the next chapter we make the case that the interests of the vast majority of Americans, including the need for security, are aligned with those of the world's poor and hungry people. A world of extreme and deepening inequality and unfairness is a world driven by humiliation, anger, and for some even a willingness to die to protect one's people and deepest values. Thus, addressing the roots of injustice and extreme deprivation is at the heart of our national interest. Yet U.S. policymakers appear to assess our national interest quite differently.

Over the past decades this interpretation has changed as the geopolitical context has evolved.

In the decades after World War II, during the "Cold War" between the Soviet Union and the West, a view of the world as divided into two rival "camps" shaped foreign aid. Officials saw aid as one of the tools they could use to defeat the other camp. They portrayed the popularity of alternative economic models in much of the Global South as threatening U.S.-style capitalism (not to mention the profits of U.S. corporations) and thus justifying the use of aid to arm and financially prop up authoritarian governments. So U.S. support went to even those regimes most notorious for repressing citizens working for the very reforms that could address hunger and poverty, in Iran, the Philippines, South Korea, South Vietnam, El Salvador, Indonesia, Nicaragua, Zaire, and many other countries. All were controlled by political elites the United States counted as loyal allies friendly to U.S.-style capitalism and U.S. corporations.

Robert White, U.S. ambassador to Paraguay (1977–80) and El Salvador (1980–81), had this to say about the consequences of this orientation:

> Because we feared revolution, we consistently opposed the forces of change while uncritically supporting dictatorships and small economic elites. We blinked at repression and participated in the perversion of democracy throughout the hemisphere.[5]

At the same time, U.S. policymakers sought to make us believe that "our" vital interests were threatened by any experiment that didn't mimic the U.S. economic model—that is, a system favoring unlimited private and corporate accumulation of productive assets. Any nation seeking to alter its economic ground rules in favor of the poor majority—through, for example, land reform—was labeled as having "gone over to the other camp," and thus was an enemy of the United States.

Such was Nicaragua's fate in the 1980s. When a popular uprising overthrew the U.S.-backed Somoza dictatorship, which had long monopolized the nation's agricultural and other economic resources, it was labeled "communist" by the Reagan administration, even though leaders of the uprising represented diverse philosophies and participants, including many from religious communities. The Reagan administration suspended aid and armed the new government's opponents, who sought to undermine the land reform and other support for small-scale farmers and rural workers—changes that offered the best hope for Nicaragua's hungry.[6]

Arguably, by heightening fear within Nicaragua, U.S. support of armed opponents not only diverted energy from pro-poor reforms but also helped to strengthen strong-arm elements within the new leadership. Consequences continue even to this day.

Driven by the fear that communism would defeat capitalism in the battle for the "hearts and minds" of impoverished populations, U.S. foreign aid to some countries did include some piecemeal poverty-alleviation programs, but they did not alter the lopsided control over land and other productive resources that generates needless poverty and hunger.[7]

Since the Cold War's end, U.S. aid has reflected even more clearly the goal of promoting worldwide corporate-friendly market and trade rules, as we discussed in Myths 6 and 7 and return to below.

MILITARY AID — A BIG PIECE OF U.S. FOREIGN AID

Before focusing on economic aid, note that our country's military aid is also significant. In fiscal year 2012, it was more than half the size of total U.S. economic aid.[8]

The United States is the world's largest provider of military aid; and that's true even if we do *not* take into account the often staggering financial cost to U.S. taxpayers of U.S. military operations in numerous countries receiving U.S. military assistance.[9] Moreover, the distribution of military aid is even more concentrated than that of economic aid; just ten countries, among more than one hundred total recipients, get over 90 percent of total U.S. military aid.

> The United States is the world's largest provider of military aid, even if we do not take our military operations into account.

It is chilling to note that each of these ten top recipients is a violator of human rights, according to Human Rights Watch.[10]

U.S. Military Foreign Aid: Top Ten Recipients
(US$ Millions)

Afghanistan	9,560
Israel	3,075
Egypt	1,301
Iraq	1,157
Jordan	304
Somalia	145
Russia	102
Colombia	100
Kazakhstan	97
Mexico	92
Top Ten Total:	**$15,932 million (92.5% of total)**
Total Military Aid:	**$17,222 million**

Source: U.S. Overseas Loans and Grants ("Green Book"). Figures are for FY2012.

While most observers as well as the U.S. government classify military aid as part of the foreign aid program, in this book we focus on our economic aid, as we doubt that many people seeking solutions to hunger and poverty would see military aid as part of the solution.

U.S. COUNTRY-TO-COUNTRY ECONOMIC AID

Total U.S. economic aid in the government's fiscal year 2012 was just over $31 billion. Of that, two-thirds, or about $19 billion, was earmarked for specific countries.[11] This is officially called "bilateral aid."

In this chapter:

- We focus on the top 10 recipients of U.S. bilateral economic aid, since among 182 recipients these countries got *half*. Our assumption is that this extreme concentration is likely to reveal a great deal about how U.S. officials define the national interest.
- We briefly outline the major program areas for which the other half of our economic aid—the half going to the remaining 172 countries—is earmarked.
- We examine one aspect of economic aid—food aid. Even though in dollar amounts it is relatively small, we know that for many people food aid comes first to mind when they think of how to help the world's hungry. We hope to clarify the changes in food aid urgently needed—changes we believe most Americans would heartily applaud.
- We probe the rationale and implications for Africa of the New Alliance for Food Security and Nutrition (the "New Alliance"), a recent initiative of U.S. foreign aid in collaboration with other major aid providers and for-profit corporations; we also analyze the Obama administration's initiative "Feed the Future." We and many others believe that the New Alliance and Feed the Future reveal a great deal about the current thrust of U.S. foreign aid.
- We point to contrasting examples in which foreign aid appears to be helping some the world's poor to gain a voice.
- Finally, we close with reflections on the kind of aid supporting the positive power that hungry people are seeking and that simultaneously serves our true national interest.

U.S. Bilateral Economic Aid: Top Ten Recipients
(US$ Millions)

Afghanistan	3,326
Pakistan	1,138
Ethiopia	865
Jordan	832
Iraq	784
Kenya	746
Colombia	544
Haiti	510
West Bank	457
Tanzania	399
Top Ten Total:	**$9,601 million** **(50 percent of total)**
Total Bilateral Economic Aid:	**$19,100 million**

Total Number of Recipient Countries: 182

Source: U.S. Overseas Loans and Grants ("Green Book") website, September 2014. Figures are for FY2012.

U.S. Bilateral Economic Aid Is Concentrated, but Hunger Is Not the Criterion

In the aftermath of 9/11, the "Global War on Terror" has become the centerpiece of U.S. foreign policy, and that is reflected in our country's foreign aid. In the table above of the top ten recipients of U.S. economic aid, the antiterrorism focus is obvious: Afghanistan, Pakistan, and Iraq are among the top five.

The first thing that stands out is that U.S. bilateral—i.e., country-to-country—economic aid is highly concentrated: Again, of the 182 recipients in 2012, just ten received half of the total.

Second, the focus isn't on the countries with the largest numbers of hungry people. Not one of the four countries accounting for a third of the calorie-deficient people in the Global South—India, Bangladesh, Indonesia, and the Philippines—shows up here. In fact, these top ten countries are home to only 16 percent of the world's hungry people, as measured by the FAO's standard of calorie deficiency.[12]

While many Americans would assume that our aid is focused on helping hungry people in Africa, only three African countries are among the top ten. And even there, as we will see, these choices arguably reflect other policy motivations.

What do we know about the countries that receive—or for many years received, as in the case of Egypt—a disproportionate share of our economic aid? Let's see what reports from five of these countries reveal:

Afghanistan. In the aftermath of the U.S. invasion in late 2001, Afghanistan has become far and away the leading recipient of U.S. economic aid. *New Yorker* writer Patrick Radden Keefe noted in 2015 that, according to the Special Inspector General for Afghanistan Reconstruction, since 2002, the United States has spent $104 billion on rebuilding the country—nearly as much in today's dollars as was spent on the entire Marshall Plan to rebuild much of Europe after the devastation of World War II.[13]

And the many billions of U.S. economic aid to Afghanistan are on top of the massive amount of U.S. military aid and the cost of U.S. armed forces there.

But, you may be thinking, surely there were a lot of impoverished and hungry people in Afghanistan when the United States invaded and, hopefully, at least this massive economic assistance—more per person than to any other country and over so many years—has made a difference for them. But just when we too had that thought, the *New York Times* published an article, based on the paper's in-country reporters' visits to the capital as well as to several provinces, headlined "Afghanistan's Worsening, and Baffling, Hunger Crisis."[14] In it, we see heartwrenching photos of emaciated children, including a "5-year-old boy who weighs less than 20 pounds."

The country's severe hunger crisis is captured in these alarming numbers: One in four Afghans suffers from hunger, defined as insufficient calories.[15] As we emphasize in Myth 1, stunting is a stronger measure of nutritional deprivation, and 59 percent of Afghan children are stunted—that's more than twice the world average.[16]

Searching for some explanation, we learned that the U.S. Government Accountability Office (GAO) has issued dozens of reports related to U.S. aid expenditures in Afghanistan. One report in 2010 repeatedly underscored the lack of oversight and of monitoring by the U.S. Agency for International Development (USAID). As a consequence, the GAO notes, "there is a high degree of potential for fraud, waste and misman-agement."[17] Unfortunately, the report was typical of many GAO reports.

As for corruption, Afghanistan is ranked the fourth most corrupt among 175 countries.[18] And a 2012 World Bank evaluation of Afghan economic prospects warned that the large aid inflows—estimated to be about the same size as the country's GDP!—have been "linked to corruption, poor aid effectiveness, and weakened governance."[19] The Center for Strategic and International Studies agrees, describing aid there as "so poorly planned and managed that it had an intensely cor-rupting impact on Afghan officials and the Afghan business sector."[20]

Yet such recurring reports of mismanagement, ineffectiveness, and blatant corruption seem not to have diminished the enormous flow of U.S. aid.

Just as we were completing this book, we read a *New Yorker* article, "Corruption and Revolt: Does Tolerating Graft Undermine National Secu-rity?" based largely on a 2015 book by Sarah Chayes. An American who arrived in Afghanistan in 2001, initially as a correspondent for NPR, Chayes wound up living for years in Kandahar Province, becoming fluent in Pashto, and initially working directly with members of the powerful Karzai coterie. Chayes's observation of the tragic irony in the impact of U.S. monetary aid to Afghanistan, intended to make the country secure from the Taliban, is summed up this way by the *New Yorker* writer:

> . . . systemic corruption became not just a lamentable by-product of the war but an accelerant of conflict. All those bribes and kickbacks radicalized the local population, turning it against the Afghan government and, at least some of the time, toward the Taliban.[21]

A note on corruption: Four of the top ten recipients of U.S. economic aid rank among the bottom quarter—the worst on corruption—out of 175 countries monitored by Transparency International. "Aid can fuel corruption," notes this watchdog organization.[22]

Ethiopia. In 2012, Ethiopia was the third top recipient of U.S. economic aid, with more than half of the funds going for Emergency Response and Population Policies and Reproductive Health.[23] Because the country is internationally known for periodic, severe hunger crises, the choice seems reasonable. Yet Ethiopia is one of fifty-five countries where hunger is deemed "serious" to "extremely alarming," according to the Global Hunger Index.[24]

So what are some of the reasons the United States might choose to emphasize Ethiopia?

Ethiopia is prestigious in much of the continent because of its uniquely successful military resistance to colonization by European powers in the late nineteenth century.[25] Its capital is the site of the headquarters of the African Union. Moreover, the Center for American Progress and the Center for Global Development remind us, "Ethiopia is a key [U.S.] security partner and hosts a U.S. drone base . . . [and] continues to be the backbone of the international force" in Somalia, widely perceived as a failed state.[26]

Kenya. Kenya is a newcomer among the top ten recipients of U.S. economic aid. Its rise parallels its increasing strategic importance to the United States. Ranking sixth among U.S. economic aid recipients, this East African country borders Somalia. It's been struck by numerous terrorist attacks, including the 1998 bombing of the U.S. embassy (attributed to Al Qaeda) as well as the 2013 attack on the capital's upscale Westgate shopping mall. In 2011, Kenyan and Somalian military forces began a coordinated operation in southern Somalia against the Al Shabaab insurgents, considered an Al Qaeda affiliate.

Almost half of Kenya's population lives below the national poverty line, while the political and business elites, less than 1 percent of Kenyans, live lavishly.[27]

And the focus of the aid?

Note first that at least half of Kenyans make their living by farming, yet only 9 percent of U.S. economic aid is for "agriculture." The biggest chunk—40 percent—goes to "population and reproductive health," in part focused on HIV/AIDS-related work. Kenya's HIV/AIDS rates are high. However, the pandemic's spread and the treatment of those infected cannot be effectively addressed as long as the government fails to redress the extreme poverty and inequalities at the pandemic's root. That clearly is not happening in Kenya.[28]

On Transparency International's corruption index, Kenya is among the most corrupt—in 145th place out of 175 countries.[29] Because corruption operates on all strata of a society, we can only wonder how much aid to Kenya ever serves its stated ends. That aid flows on nevertheless makes it is easy to believe that a big, unstated purpose of the support is to satisfy the country's elite in order to keep Kenya in the U.S. political-economic orbit.

Haiti. Aware of Haiti's devastating earthquake in 2010, many Americans might be pleased to see it in eighth place among the top ten recipients of U.S. economic aid. Since the earthquake, the United States has budgeted $3.6 billion in assistance. But even three years after the quake, "USAID and its implementing partners have generally failed to make public the basic data identifying where funds go and how they are spent," reported the Center for Economic and Policy Research in 2013. "The question 'where has the money gone' echoes throughout the country."[30]

On Transparency International's corruption index, covering 175 countries, only 10 scored lower than Haiti.[31] Below we return to Haiti to discuss U.S. food aid.

Egypt. At the time of this writing, Egypt is not among the top ten recipients of U.S. economic aid. We include it here, however, because for three decades it was a leading recipient of U.S. economic aid, in addition to even more substantial military aid. This huge support began—not coincidentally—shortly after Egypt broke ranks with other Arab nations in signing the treaty between Israel and Egypt known as the Camp David Accords in 1978.

Beginning in the late 1990s, the United States began to scale back economic aid to Egypt, while maintaining $1.3 billion annually in military aid to the country's military-dominated regime.[32] For over three decades Egypt and Israel had been the top two recipients of U.S. economic aid, together typically capturing almost a third. Since 2008, however, U.S. economic aid to Israel has been reduced to a trickle. In lieu of direct economic support, Israel receives somewhat more than $3 billion annually in military aid.[33]

Motives for U.S. economic aid to Egypt appear to reflect narrow geopolitical concerns, not the alleviation of poverty and hunger. But have impoverished Egyptians at least benefited?

It would seem not.

The UN World Food Program (WFP) notes that Egypt's "national poverty rate . . . has increased by nearly 50 percent in the last 15 years . . . to 25.2 percent in 2011," with an additional 24 percent who are "near poor" and hovering just above the poverty line.

The Egyptian government operates a costly subsidized food distribution system that appears badly targeted—73 percent of nonpoor households benefit, while 19 percent of poor households do not.[34] And all our economic aid over decades has not prevented Egypt from becoming dependent on imported food. Egypt imports over half of the mainstay of its diet, wheat, making it the world's largest wheat importer.[35] Such acute dependency on a sometimes volatile global market ensures ongoing, detrimental food insecurity.

BEYOND THE TOP TEN

Where did the $9.5 billion in U.S. bilateral economic aid *not* earmarked for the top ten recipients go in fiscal year 2012?

The remaining half of U.S. economic aid went to 172 countries in fiscal year 2012. Most was earmarked for three program categories: health, humanitarian (emergency) assistance, and economic development. At first glance, each sounds potentially quite helpful to poor and hungry people. Yet, even though field-based research on these programs was beyond the scope of this book, we know that too often those examining U.S. foreign aid (including evaluations by the Government Accountability Office) or working within the foreign aid industry (as I, Joseph, have worked) learn to be ready for disappointment. Briefly, a look at the three program areas:

First, "health" made up the largest of these three main categories, at over $5 billion in 2012; and within it, HIV/AIDS was the biggest subcategory.[36] However, diseases such as malaria and tuberculosis, as well as poor public sanitation, affect many more people's health than does HIV/AIDS. So, again, we wonder about the criteria used in targeting assistance.

And, as noted, the HIV pandemic, like hunger, cannot be effectively addressed without fundamental reforms reducing poverty and inequalities.[37] (To begin to grasp this reality: What is the likely response of the impoverished commercial sex worker, perhaps with hungry children to feed, to the customer offering to pay more if he doesn't use a condom?)

Maternal and child health, family planning, and reproductive health are the other prominent subcategories of health-related U.S. aid.

Second, "humanitarian assistance" goes mostly for disaster relief and to aid refugees. Its prominence is not at all surprising given the violent conflicts and increased weather-related catastrophes in today's world. Humanitarian assistance, however, is fraught with challenges, which we spotlight below in discussing U.S. food-aid policies and emergency relief.

Transparency International, the respected "global coalition against corruption," offers this cautionary note on humanitarian assistance:

> Corruption in humanitarian work is among the worst kind. It can mean the difference between life and death. . . . Emergency assistance pumps large amounts of money and goods into damaged economies. . . . Food, water and medical supplies can be stolen and sold on the black market. . . . Aid agencies feel the need for speed. Sometimes this makes them bypass standard anti-corruption measures. The result? Money or goods go missing. . . . Too often, powerful local groups and existing corrupt networks benefit. Those most needing help miss out.[38]

Third is "economic development," said primarily to include "infrastructure" and "agriculture"; but it also covers support for "trade and investment" and the "financial sector." Later in this chapter we discuss disturbing implications of some of our foreign aid in the last two subcategories. As we have seen in previous chapters, and as we'll see again in the following section, "agriculture" could mean either training small-scale poor farmers in sustainable, money-saving practices, or providing a boost to rural elites and agribusiness, to the serious detriment of the poor and hungry.

So, as we've said, care must be taken to avoid assuming too much too readily on the basis of aid program labels.

Moreover, critically important to aid's impact is what is happening on the ground. To evaluate whether external aid has the potential to help the hungry, one has to ask: Are there citizen movements in the recipient country addressing the tight concentration of power over food-and-income-generating resources that previous chapters identify as the root of hunger, along with at least some officials open to change? And could aid support these reform efforts?

If the answers are "no," it's likely that inflows of outside resources will reinforce existing, hunger-creating power structures. If the answers are "yes," outside assistance can sometimes bolster positive change, as we suggest later on under the heading "What, Really, Is Our Nation's Interest?"

BUT ISN'T AT LEAST FOOD AID HELPING THE HUNGRY?

By now you might be thinking that, surely, at least one aspect of U.S. official assistance—food aid—is helping the hungry. It is *food*, after all! When thinking of U.S. food aid, many Americans probably picture ships loaded with our country's food bounty heading off to fill empty stomachs in the aftermath of natural disasters or during violent conflicts.

Once again, the realities are quite different.

Not a Big Piece

Food aid constitutes only a small fraction of the U.S. overall economic aid budget, typically hovering at around 6 percent. Still, the United States is by far the world's largest food-aid supplier, providing over half of all food aid, at a cost of about $2 billion a year.[39]

At Least Half of U.S. Food Aid Is Still "Tied"

Since the inception of the program in 1954, U.S. food aid has predominantly been "tied aid"—meaning that by law the food must be grown, processed, and packaged in the United States and shipped overseas on U.S.-registered vessels.[40] So the biggest winners—and most trenchant lobbyists—for our food aid are a range of private U.S. interests that benefit handsomely from the production, procurement, processing, transport, and distribution of food aid.

The U.S. farm lobby strongly backs tied aid. It's a handy safety valve for producers of rice, soybean oil, nonfat milk powder, and other food commodities when they are abundant relative to people with the money to buy them. At those times, without U.S. government purchases for food aid, prices would sink. Plus, the U.S. government for many years paid an average of 11 percent above normal market prices for the food it procured domestically for aid.[41]

U.S. Food Aid: Top Ten Recipients
(US$ millions)

Ethiopia	307.5
South Sudan	175.5
Sudan	164.9
Mozambique	89.2
Kenya	87.1
Chad	84.4
Afghanistan	83.3
Somalia	79.9
Dem. Congo	68.3
Pakistan	68.1
Top Ten Total:	**$1,208 million**
	(59% of total)
Total U.S. Food Aid:	**$2,039 million**

Source: USAID[42]

Take the American rice industry. Because Americans don't eat much rice relative to many other countries, the U.S. rice industry depends heavily on export markets, where it has to compete with rice-exporting countries whose production costs are lower. No surprise, then, that for the U.S. rice industry, the food-aid program—plus major government subsidies—is a frequent salvation. In 2001, when our supply glutted international rice markets, Thomas Ferrara, chairman of the Rice Millers' Association, made the group's members' needs clear: "Meaningful and immediate increases in food aid now could mean the difference between survival and financial disaster for rice mills in this region," he argued.[43] When the government sought to cut its rice purchases, congressional representatives from rice-producing states convinced the Bush administration to step up its food-aid purchases.[44]

Tied aid is costly. Often shipped great distances, food grown in the donor country typically costs about 50 percent more than if it had been purchased locally, and roughly one-third more than food bought in the region, notes a study by the Organisation for Economic Co-operation and Development.[45]

Despite some reforms in the 2014 Farm Bill, tied food aid continues with only minor improvements. The Obama administration succeeded in opening a small window for responding to some food emergencies more efficiently—with vouchers and cash, which save U.S. taxpayers money and could help farmers with food to sell in or near the affected areas. "But there is still little flexibility," so people on the ground "aren't allowed to use the best tools to meet a crisis," Gawain Kripke, director of policy and research at Oxfam America, explained to us.[46]

Not Driven by Concern for the Hungry

That humanitarian concern is not the primary driver of U.S. food aid becomes even clearer in this apparent paradox: It's often curtailed just as it's most needed. When food commodity prices are rising substantially—precisely when countries dependent on imports face worsening hunger and therefore most need food assistance—shipments of nonemergency food aid have declined.

Why? At such times, strong commercial sales make the food-aid escape route for U.S. producers unnecessary.[47]

Agribusiness giants, including Cargill and Archer Daniels Midland (ADM), are major beneficiaries of the system of U.S. tied food aid. Profiting from grain sales to the government that generally amount to more than half of the food-aid budget, they are—again not surprisingly—lobbyists for continuing the system of tied food aid.[48]

U.S. shipping companies are also major beneficiaries of tied food aid. Advocates for the interests of the hungry have long pushed against tied aid, which for some time meant that 75 percent of food aid had to be purchased domestically *and* transported by U.S.-registered vessels. Over a three-year period in the early 2000s, the U.S. maritime industry made $1.3 billion on transportation contracts with USAID, reports food-aid authority Jennifer Clapp.[49] These contracts are vital to the U.S. shipping industry, since for many years its share of international cargo not used in transporting food aid had been declining steeply, owing

> Almost 60 percent of U.S. food-aid dollars goes to shipping and other overhead costs.

to lower-cost competitors. Overall, according to Oxfam America, almost 60 percent of U.S. food-aid dollars goes to shipping and other overhead costs, while only 40 percent is spent on food.[50]

In 2014, reformers succeeded somewhat, but lobbying by U.S. vested interests kept the tied-aid minimum at 50 percent.[51]

So the lobbying clout of these interests has helped make the United States the world's largest food-aid supplier while keeping our food aid largely "tied." What has been the impact for hungry people overseas? Is this system of tied food aid a "win-win" in which both private U.S. interests and hungry people overseas benefit?

Neither Speedy nor Cheap in Emergencies

The most obvious need for food aid is to meet emergencies, such as those caused by hurricanes, earthquakes, or armed conflicts. By definition, these crises require speedy responses. Yet shipping food over what is often thousands of miles can hardly be speedy. The GAO found that tied food aid from the United States took on average 147 days to reach those in need, whereas food procured locally took on average just 35 days.[52]

> "I've run these operations, and I know that food aid often gets there after everyone's dead."
>
> —Andrew Natsios, former USAID administrator

Andrew Natsios, USAID chief under President George W. Bush, put it quite bluntly: "I've run these operations, and I know that food aid often gets there after everyone's dead."[53]

Another potential consequence of the built-in delay is that, once food does arrive, the consequences can be as harmful as the delay itself.

A Tsunami of American Rice

Some may be surprised to learn that when natural disaster strikes one locality, there's often plenty of food available for purchase elsewhere in the same country or region.

After the tsunami struck coastal regions in Southeast Asia in 2004, the World Food Program estimated that two million people—mainly in Indonesia and Sri Lanka—required emergency food aid. Most needed was the region's main staple, rice. But even though plenty was available within the affected countries, as well as in the region—and bumper rice harvests were anticipated in only a month—the United States shipped U.S. rice across the world.[54]

"If you lived here in the remote southeast of Sri Lanka and hadn't heard of the tsunami," wrote Toronto's *Globe and Mail* reporter Doug Saunders in Ampara, Sri Lanka, at the time, "you might easily convince yourself that people had been displaced and killed by a giant wave of rice. Fifty-kilogram sacks of the white grain have just about overwhelmed this community."

One result is that U.S. food aid ends up undermining the livelihoods of local farmers and farmworkers, many themselves living on the edge of hunger. "When millions of tons of free food are shipped into your neighborhood," Saunders reported from Sri Lanka not long after the tsunami, "what happens to the food you were going to sell? Its price plummets, of course."[55]

"Oversupply and declining prices on local markets when the aid does arrive . . . can be harmful to local farmer incentives to plant in the following season," observes Clapp. "This happened in Guinea in 1988, in Ethiopia and Somalia in 2000, and in Mozambique in 1992–93."[56]

Deepening the Disaster

Haiti's devastating 2010 earthquake, killing two hundred thousand, spurred an unprecedented outpouring of international support. In addition to massive funds for reconstruction, the United States spent $68 million delivering food aid to Haiti less than a month after disaster struck.[57] But only 28 percent of it was procured locally or regionally.[58] As a result, Haiti's entire local rice-supply chain—from farmers to local wholesalers to small resellers and street vendors—was hard hit (or simply put out of business) by the massive injection of free rice.[59]

Not surprisingly, then, we find many countries in which the need for "emergency" food aid becomes chronic.[60] In these cases, the "benefits of food aid in addressing short-term food insecurity may be offset by the cost of reducing long-term food security," concludes a World Bank review of the experience of four of the largest food-aid recipient countries.[61]

Food Aid Stifling Local Agriculture

The GAO has repeatedly evaluated the U.S. food-aid program and has found that the two agencies in charge, USAID and USDA, often neglect to sufficiently assess local markets beforehand in order to help avoid food aid's potentially adverse impact on local farmers. Moreover, the GAO finds that these agencies have exceeded even the volume limits set by their own market assessments.[62]

Two examples tell the story.

Afghanistan. Former USAID administrator Andrew Natsios has written about his "painful experiences" with U.S. food aid in Afghanistan. To improve wheat crop yields and farmers' incomes and to combat worsening famine conditions, USAID had begun distributing better-quality, drought-resistant seeds to farmers in the early 2000s. When in 2002 the country then experienced the best rains in many years, Afghan farmers produced the largest wheat crop in the country's history. And soon prices collapsed to 20 percent of their normal level. But instead of purchasing this local bounty for Afghan refugees and war-displaced people, thereby buoying prices, USAID instead imported several hundred thousand tons of U.S. wheat. Local wheat prices remained so low that many farmers didn't even harvest their wheat crops; and, sadly, the following year they returned to cultivating poppies for heroin production.[63]

Thus, the "rigidities of our food-aid program beholden" to vested U.S. domestic interests "indirectly encouraged heroin production in war-torn Afghanistan."[64]

Ethiopia. For many years Ethiopia has been by far the top U.S. food-aid recipient—at the time of this writing, receiving almost one-fifth of the total—and it is a country many people associate with real need. But a scholarly review of the data from 1996 to 2006 finds that "[f]ood aid is widely regarded as a 'necessary evil': necessary to avert hunger in places where household food security has been compromised, but evil because it is suspected of undermining incentives for local production, thereby creating structural dependency on food aid."[65]

"We really appreciate it," said thirty-five-year-old farmer Jerman Amente, but he added that he and other Ethiopian farmers are "sad and discouraged that the U.S. government buys surplus grain from American farmers and sends it halfway around the world . . . instead of first buying what Ethiopians produce," reported *The Wall Street Journal*. "Some of his [Amente's] own grain has sat for eight months," along with what the farmer described as at least a hundred thousand metric tons of other locally produced foods languishing in warehouses.[66]

And what does it say that in 2012 the U.S. government spent nearly $300 million buying and shipping U.S. food to Ethiopia, but less than a third of that amount to assist Ethiopia's own agricultural development?[67]

Ethiopia and other major recipients of U.S. food aid, including Pakistan, Bangladesh, and Afghanistan, have become chronic recipients, receiving food aid year after year, with no end in sight. A U.K. foreign aid agency's 2006 policy brief drew this conclusion: "Long-term food aid . . . will encourage people and government officials to externalize responsibility/accountability and, consequently, delay the seeking of solutions, while more and more people suffer."[68]

Food Aid: Three Lessons

From decades of experience, many knowledgeable observers argue that three key changes are needed to enable food aid to work for hungry people.

Use food aid primarily for short-term emergencies. To benefit the hungry, while not undermining small-scale farmers overseas, many of whom are poor, food aid must not become chronic.

"Untie" food aid. The United States should quickly catch up with the European Union, which as far back as the mid-1970s began shifting away from tied food aid and had fully "untied" it by 1996. Australia partially untied its food aid in 2004 and fully ended the destructive practice in 2006.[69]

Food aid is best provided not bilaterally, where business interests in the donor country come into play, but rather via food or, better yet, cash grants to the multilateral United Nations World Food Program (WFP). With cash grants from donors, WFP can respond to emergencies by procuring food locally or, if that's not feasible, regionally. In 2013, WFP bought more than $1 billion worth of food, of which 50 percent was purchased in the regions where the aid was needed.[70]

And, since 2008, WFP has sought to ensure that its emergency food *purchases*, and not just its distribution, have positive impacts. Under a five-year pilot project called Purchase for Progress (P4P), WFP experimented with new ways of buying food staples from small farmers, as locally as possible. Its efforts in twenty countries have resulted in the purchase of more than four hundred thousand tons of food from hundreds of thousands of small-scale farmers. Additionally, by providing training and equipment to farmer organizations, WFP has enabled their members to sell to a larger number of buyers, even beyond WFP. Other successes include higher crop yields: P4P farmers in El Salvador and Ethiopia improved corn yields over the five-year pilot, unlike the control group of farmers not in the program. The program has also contributed to the advance of women farmers. Their participation in P4P farmer organizations has increased from 19 to 29 percent over the five-year pilot, with their incomes boosted by sales to the WFP and others.[71]

Stop allowing charities to sell food aid to fund their operations. Some U.S. charitable organizations have been allowed to resell U.S. food aid in countries in which they operate and to then use the proceeds to finance their development projects. It's called "monetization." About a third of U.S. food aid is used in nonemergencies, and of that, roughly 60 to 70 percent is monetized.[72] The practice is wasteful and works against the interests of hungry people, in part by undermining local farmers, who, as we have repeatedly noted, are themselves often among the hungry. The 2014 Farm Bill claimed to make monetization more efficient, but Oxfam America's Gawain Kripke points out that "it is monetization itself that's inefficient." On principle, CARE is phasing out all monetization, but several other major charitable organizations have lobbied to continue the practice, reports Kripke.[73]

ADVANCING HUMAN INTERESTS OR CORPORATE INTERESTS?

Aid Opens the Door for a Corporate Scramble for Africa

In 2012, the United States leveraged its presidency of the Group of Eight (G8)—governments of eight of the world's largest economies—to launch the New Alliance for Food Security and Nutrition (the "New Alliance"). In it, according to our government, the G8 and global-reaching corporations will partner with sub-Saharan African governments to lift fifty million African people out of poverty and hunger by 2022.[74]

G8 governments have pledged $22 billion in aid, and more than seventy corporations have committed to invest $3.75 billion in the region's agriculture over a period of a decade.[75] Global corporations, including agrochemical and seed giants DuPont, Monsanto, and Syngenta, as well as giant grain trader Cargill, are prominent among the participants.

The corporations' eagerness is understandable, since the World Bank has estimated that sub-Saharan Africa's agribusiness market will be worth $1 trillion by 2030.[76]

In 2014, General Electric's CEO Jeff Immelt sounded downright giddy about jumping on board. "We kind of gave Africa to the Europeans first and to the Chinese later, but today it's wide open for us," he said during U.S.-Africa Leaders Summit.[77] Was Immelt not aware of what his words could imply: that the "us" to which he refers—i.e., giant global companies—is the new colonial power?

The United States' pledge to the New Alliance is in the ballpark of what we might give to a major recipient, but its impact could be vastly greater, because the New Alliance seeks to alter the rules that ultimately determine who has access to land and what gets produced.

By 2014, twelve sub-Saharan countries had signed on to the New Alliance.[78]

And whose interests and voices have shaped this multibillion-dollar consortium of foreign donors, private corporations, and African governments? Certainly not the announced beneficiaries, those fifty million poor Africans—for smallholder farmers, their families, and their organizations have been excluded from the planning.

The initiative's design and negotiations with African governments have happened behind closed doors.[79]

Conditional Aid, Corporate Style

New Alliance aid and investments are set up with striking similarity to earlier "structural adjustment programs" imposed by global agencies such as the World Bank. In those programs, help came with strings attached, as recipient governments had to satisfy very specific conditions. In Myth 6 we reported the negative impact on hunger and poverty of this earlier go-round.

In today's reprise, aid and corporate investment are tied to "tough policy reforms by African governments that . . . build investor confidence," including:

- *Protecting intellectual property rights*, particularly relevant for genetically engineered seeds

- *Strengthening private property contracts*, which could privilege corporations over small farmers
- *"Simplifying" tax codes*, which might be helpful or could further reduce public resources
- *"Reforming" seed laws*, which could benefit corporations selling patented seed varieties at the expense of farmers
- *Selling off enormous swaths of public land*, which could further displace small farmers
- *Jettisoning agricultural subsidies and tariffs*, thus weakening governments' capacity to encourage local production[80]

By 2013, six African countries had published national agricultural investment plans.[81] In all, ten African countries have now made more than two hundred policy commitments in order to align with the New Alliance "cooperation frameworks," including changes to laws and regulations affecting agricultural investments.[82]

Many are precisely the type of pro-corporate "reforms" discussed in Myth 6 that often have worked against the interests of the hungry—including those shifting farming toward reliance on chemical inputs and hybrid seeds that are controlled by a handful of foreign corporations. Examples, by country, include:

Ethiopia. The Ethiopian government, USAID, and the U.S.-based chemical and biotechnology giant DuPont have launched a program to distribute hybrid corn seeds to small farmers that will reportedly boost yields significantly.[83] For this program, Ethiopia has passed a new policy that "incentivizes international seed companies to operate in Ethiopian seed markets" and "incentivize[s] the private sector to commercially multiply and distribute seed."[84]

But, as Myth 4 recounted, many Ethiopian farmers are already achieving comparable success using ecological practices that enhance their independence and long-term food security. Also, Ethiopia seemingly did not contemplate Malawi's attempt at a similar strategy, summarized in Myth 3, which resulted in debt-strapped farmers but not the increased output promised.

Burkina Faso. The government hopes to lure the Norwegian fertilizer giant Yara to build a fertilizer plant in the country, at a cost of $1.5 billion to $2 billion. If it is to succeed, Burkina Faso will first have to "reform" its laws in accordance with the New Alliance's framework so as to "facilitate private [business] participation in fertilizer supply contracts."[85]

Mozambique. To comply with the stipulations of the New Alliance, reforms have halted the distribution of free, unpatented seeds.[86] This step was taken after Syngenta, the world's second-largest biotechnology firm, with headquarters in Switzerland, announced that it will invest $500 million in New Alliance recipient countries—but warned that it will *not* invest where Syngenta would have to compete with government-subsidized seed programs.[87]

Olivier De Schutter, while serving as the UN Special Rapporteur on the Right to Food, stressed the tremendous influence that global corporations wield over African governments: "There's a struggle for land, for investment, for seed systems, and first and foremost there's a struggle for political influence."[88]

That influence, bolstered by the promise of new foreign aid funds, means that under the New Alliance, corporations are buying up public lands. Critics say it's nothing less than a giant "fire sale" of Africa's agricultural assets. Its scope is debated, because information is not public; but key details are coming to light:[89]

Ivory Coast. This West African country—where more than 40 percent of citizens are impoverished—has pledged to reform its land laws to facilitate the inflow of private agricultural investment, especially in rice cultivation, under its New Alliance cooperation framework. In exchange, the Ivory Coast receives hundreds of millions of dollars in donor aid and promises from at least eight foreign companies to invest nearly $800 million in large-scale rice plantations.[90]

In early 2013, the French grain-trading giant Louis Dreyfus gained access to roughly five hundred thousand acres for rice cultivation, and two other investors are seeking up to more than a million additional acres. These three deals alone are expected to displace tens of thousands of small-scale rice farmers and to destroy the livelihoods of thousands of local food traders. But while hunger stalks the Ivory Coast, since 2010 its rice *exports* have climbed dramatically.[91]

Tanzania. In this East African country, under the New Alliance framework, Agro EcoEnergy—a local subsidiary of a Swedish company— is investing $500 million to create a sugarcane plantation and an electric power and ethanol factory near the historic coastal town of Bagamoyo.[92] The government offered the Tanzanian company almost twenty thousand acres to grow sugarcane in exchange for shares in the company. In 2013, thousands of villagers, living in the area for decades, officially protested the government's plan to evict them forcefully and hand over their land to the investor.[93] Despite these protests, EcoEnergy's Bagamoyo office opened with government support in May 2014. Forced resettlement of families in the villages began soon after.[94]

Ethiopia. To speed land transfer under the New Alliance framework, Ethiopia has allocated almost 7.5 million acres to corporate investors and will establish a one-stop service for investors to cut through the red tape. To live up to its New Alliance commitments, the country must improve its score in the World Bank's "Doing Business" index. How? One, by increasing new private investment in agriculture; and two, by generating a greater percentage of private investment in the commercial production and marketing of seeds.[95]

So, despite rhetoric about helping small-scale farmers, the New Alliance is promoting precisely the failing model of farming we describe in Myth 3. In this African context, Oxfam America describes what's happening as "large-scale land acquisitions by U.S. investors and investment funds [that] appear to promote monoculture commodity agribusiness at the expense of greater biodiversity and more sustainable agricultural practices."[96] The report warns: "Smallholders risk being sidelined while private investors garner public support and access."[97] In 2014, just six months after publishing this report, Oxfam withdrew from the Leadership Council of the New Alliance. The organization charged that the New Alliance had "inadequately integrated" the voices of smallholders and, in doing so, "created a form of global governance which is exclusive and potentially self-serving."[98]

Here we see clearly that, once officials define our national interest as changing rules in favor of corporate interests, our nation's diplomatic power—along with our honor and our tax dollars—lines up against poor and hungry people and in opposition, we believe, to the values and common sense of most Americans. Throughout our book we present evidence that a very different kind of change—involving a profound democratization and thus fairer access to resources—is necessary to end hunger. It is not possible for the United States and its G8 allies to be supporting the very opposite changes—as in the power-concentrating policies outlined here—and be for the hungry at the same time.

Feed the Future

In step with the G8's New Alliance, in 2009 the Obama administration launched Feed the Future, pledging $3.5 billion over five years with the goal of "giving millions of people a pathway out of hunger and extreme poverty." As of 2014, Feed the Future had projects in nineteen countries.

Feed the Future claims that in 2013 alone it assisted nearly seven million farmers in the use of new technologies and management practices on more than ten million acres for "sustainable intensification" of farming on small plots.[99]

But what's the on-the-ground evidence?

A search of USAID's "Evaluation Showcase" and Feed the Future websites reveals very few, if any, serious evaluations. Moreover, multiple whistle-blowers at USAID charge the agency's inspector general with removing critical details from internal audits before turning them into public reports.[100]

Oxfam America, fortunately, has carried out independent research on Feed the Future in Tanzania and Haiti. As we noted, both countries are among the largest recipients of U.S. economic aid, and they are two of the largest recipients of Feed the Future funds.

And what do we learn from these independent evaluations?

- In Tanzania, the stated objective is to help bring 834,000 Tanzanians, mostly smallholder farmers and their families, out of hunger and poverty. However, the 2013 Oxfam field investigation of how $54 million was used over five years concluded that "[p]articipation is limited to a relatively few farmers well-endowed with resources—access to water and investment capital." Failure to provide enough time to transfer technical knowledge was another problem Oxfam noted. Thus a women's farming co-op inaugurated by Secretary of

State Hillary Clinton in mid-2011 was in near-collapse two years later. The women told Oxfam investigators they felt they'd received the "support in order to put on a good show for Secretary Clinton."[101]

- In 2011 in Haiti, Feed the Future brought under its umbrella an existing USAID project, "Watershed Initiative for National Natural Environment Resources," with the stated goal of raising the incomes of poor, smallholder farmers. To manage the project, USAID then contracted with Chemonics, a for-profit company based in Washington, D.C., and the largest USAID contractor worldwide. Visiting field sites of the project, Oxfam concluded that the "project does not target farmers who are the most poor." Also, "many participating [farmers] . . . did not see the technologies as relevant to their problems." Oxfam's field report also noted concern about the project's sustainability, questioning whether donors would "continue to pour millions of dollars into Haiti with little to show for it once the projects end."[102]

WHAT, REALLY, IS OUR NATION'S INTEREST?

Once we grasp that our own well-being is intimately linked to that of the world's poor— a reality we explore more fully, as we have already noted, in the next chapter—we gain clarity on some key questions we must ask to measure any attempt to help, including official foreign aid:

Does it help people gain power over their lives? Does it help create new, more equitable power relationships?

Here we use "power" in its root meaning: the ability to act. For throughout our book we've sought to make clear that hunger cannot be solved by charity, as its roots lie deep in a society's power relationships that determine who has a voice and who does not, who can lay claim to food and who cannot.

Then we can ask: Is there such a thing as truly empowering aid? Aid that contributes to people's gaining confidence, courage, and skills that make democratic "system change" more possible? And by "system change" we mean altering the rules and norms so that they are fairer and more inclusive, and thus life-enhancing.

> Hunger cannot be solved by charity, as its roots lie deep in a society's power relationships that determine who has a voice and who does not, who can lay claim to food and who cannot.

Here is a sprinkling of examples that suggest possibility:

Consider what Nepal is achieving via a volunteer network of some fifty thousand village women, each elected by her community, working with the Ministry of Health and Population. Called Female Community Health Volunteers, they build on a Nepali tradition of women in every village helping during pregnancies. Today, these health volunteers bring "maternal and child health information and health services to every community in the country." Quite a feat, given that most of Nepal's twenty-nine million people are rural and live in remote, mountainous areas, far from health facilities.[103] Since 1998, donors including USAID have supported the initiative that's helped train the volunteers.[104]

Since the approach got going in 1988, Nepal's child polio vaccination rate has climbed by about 55 percent to over 90 percent, and the country's child mortality rate has dropped by 70 percent.[105]

> In Nepal, Female Community Health Volunteers bring services to every community in the country, even the most remote. Since 1988, the child mortality rate has dropped 70 percent.
> —drawn from USAID, 2011, and from UNICEF data

Another consequence? The status and voice of village women are enhanced, and families are gaining greater power in the form of control over their health, their children's survival, and the size of their family. Such profound, personal shifts enable new aspirations.

In 2000, USAID provided some of the early support for research on what are now called "savings groups," a simple tool by which poor people use group solidarity to save for tough times and new opportunities. As we describe in Myth 10, members of savings groups (mainly women) receive back all they save plus a share of the interest, often during the "lean season" when money is scarcest in the village, reports one of the movement's founders, Jeffrey Ashe.[106] Donors have seen their early, modest investment in people mushroom in a decade into ten million savings group members worldwide, he reports.

Of course, the approach cannot itself end hunger, but it empowers some of the world's poor to better protect themselves during times of weather-damaged harvests or ill health, to increase income through small enterprises, and to create solidarity for community-building work.

Part of our government's foreign aid supports the work of United Nations agencies.

An example is the quarter of the budget of the Food and Agriculture Organization of the United Nations (FAO) that the United States provides.[107] Over the years, the FAO has been criticized for, among other things, supporting the kind of destructive model of farming we describe in Myth 3, but during the last decade it has also been promoting agroecology.[108] One example is the FAO's Farmer Field Schools. Since 1989, the FAO reports, farmers have been trained in integrated pest management in order to cut pesticide use and in other ecofarming practices.[109]

The official aid programs of some other countries also provide examples of aid that can be empowering.

We write in Myths 2 and 10 of regreening being undertaken by small farmers who are spreading the practice of agroforestry in Niger, in Burkina Faso, and elsewhere in West Africa. The Dutch, Swedish, and German governments' development agencies, among others, are helping farmers incorporate this ecological advancement across this drought-prone region.[110] Farmer-to-farmer training in agroforestry is a part of their support.[111] USAID and the World Bank are also investing in the spread of agroforestry.[112]

We hope that these closing, constructive examples suggest another, more accurate, understanding of national interest: one that appreciates that we cannot enjoy security in a world suffering powerlessness, humiliation, and hunger, and that U.S.-based corporate incursions, supported by our government, often contribute precisely to these negative realities.

TRUE AID:
REMOVING THE OBSTACLES TO CHANGE

Understanding that much U.S. foreign aid is not helping the hungry should not lead us to throw up our hands in despair or to fold them in resignation.

Rather, this realization can open our eyes to the many actions we might take that could help to reshape the system-wide roots of hunger. When we come to appreciate how our lives are deeply connected with those suffering from hunger in other countries, we readily see critical actions we can take.

These actions start from the premise that poor people throughout the Global South are hardly passive. In Myth 10, we highlight their courageous and effective efforts, as well as those of people in our own country, to create systemic change to end hunger.

True foreign aid is that which removes the obstacles blocking their paths.

The special responsibility of U.S. citizens is to target those obstacles erected by our government (acting hand in hand with foreign elites, in our name and with our tax dollars) and by corporations (fueled by our consumer spending); the New Alliance is one example. Other obstacles include a deregulated financial industry that unleashes speculative food-price spikes, with disastrous consequences for poor people; trade rules that allow giant companies to undercut food producers in developing countries; our government's long history of undermining efforts in other countries to work toward fairer distribution of land and other sources of economic and political power; foreign aid policies that promote agricultural technologies that trap farmers in debt; and tied food aid that all too often has undercut poor farmers in the Global South.

Taking responsibility for actions being carried out with our tax and consumer dollars, for example, we can stand up as citizens to:

- Oppose economic aid to all countries where the poor and hungry are clearly not benefiting, and are even being harmed by it. As we have seen, this group includes many of the top aid recipients.
- Oppose all military and police aid to governments that are documented violators of human rights.
- Promote a fairer U.S. farm economy in which family farmers can thrive without resorting to income from purchases for the U.S. food-aid program.
- Support efforts to shift purchases of emergency food aid to sources as near as possible to where it is needed; and to shift from donating food to providing cash for buying food locally so that local farmers benefit.
- Demand changes in international law that would end "land grabs" in the Global South, now throwing hundreds of thousands of farmers off their long-held homesteads.[113]
- Speak out against trade and aid agreements serving corporate interests and fostering dependency on imported food throughout the Global South.[114]

- Protest aid that is conditional on a recipient government's changing its laws and regulations to give corporate interests the upper hand.
- Shop critically: whenever possible, purchasing Fair Trade products rather than those of global agribusiness.
- Finally, in order to enable the above, we can support fundamental reform of the U.S. electoral system so that it is no longer driven by powerful moneyed interests.

More broadly, we can overthrow the obstacle of despair itself. In Myth 10, we offer stories of changes under way that are meaningful because they address the roots of powerlessness—the ultimate cause of hunger. Spreading such stories and joining in such efforts, we shed despair through action. And in our concluding essay, "Beyond the Myths," we suggest the possibility of rethinking freedom itself, to replace the dominant, self-destructive framing of freedom with one that releases honest hope.

THE POWER OF EXAMPLE

In tackling such big political and economic shifts within our own country, we amass the greatest power available to help end hunger: *the power of example*. We can fashion our own country into a true democracy in which citizens' values and voices count. Since no one *chooses* hunger, its very existence means that the promise of democracy—a voice for each of us—has been denied. Thus, every act we take, in whatever nation we inhabit, toward truly accountable Living Democracy is evidence that the end of hunger is possible.

It's up to each of us, wherever we are, to provide that evidence.

myth 9

It's Not Our Problem

MYTH: In the United States we do have big problems, for sure. Hunger and poverty persist, and hurt millions of families. But our difficulties are fundamentally different from those of developing countries. Not only are their problems orders of magnitude greater than ours, but the roots of hunger and poverty in developing countries are also quite unlike what we face. Many poor countries lack natural resources we take for granted, like fertile soils or plentiful water. Child labor is common. Plus, while we have some bad apples in high places, hungry countries are rife with corruption at all levels.

And here's an aspect of the divide that may be uncomfortable to acknowledge: Not only are their problems different, but those of us in countries like the United States in some ways actually benefit from the oppressive conditions in which many people live in the Global South. Poor and hungry people have no choice but to accept low wages, but their miserable pay is one reason we can buy jeans for $9 at Wal-Mart—a big boon, especially for low-income citizens.

OUR RESPONSE: While it's certainly true that the extent and severity of hunger differ greatly between the United States and the Global South, there's a deeper reality. In it are powerful parallels and interconnections between our lives and those of hungry people abroad. Exploring them, we've come to see that our own well-being and that of future generations depend on how deeply we grasp this commonality and whether we make choices based on that understanding.

Let's begin with a positive shared reality, one we hope has become clear in previous chapters: that the United States and other "developed" countries don't hold a monopoly on rich natural resources. But to serve our own and others' well-being, we must face deeply troubling parallels as well.

Hunger and poverty. As noted in our opening essay, one in six Americans experiences what our government calls "food insecurity," meaning that a huge proportion of U.S. citizens face ongoing difficulty even ensuring a healthy diet for themselves and their families.[1] It's easy to imagine that such estimates don't show up in elevated mortality rates here, as they do in "hungry countries." But no. Infant death is widely understood to reflect a society's nutritional well-being and poverty, and by this measure the United States ranks fifty-sixth globally—just behind Serbia and Lithuania.[2] Moreover, in both North and South the life span is shortened by poverty and inequality. In the United States men in the bottom half of the population by income live almost six years less than those better off—a gap that's grown fivefold since the early 1970s.[3]

In a dietary parallel, the quality of food is degrading in both North and South. The global study in *Lancet* reported in Myth 1 describes the United States, despite some improvements, as one of the "worst in the world" in terms of consumption of unhealthy foods associated with diseases responsible for most mortality worldwide.[4]

Here in the United States, as in low-income countries, poverty touches not a small share of our people but the majority: By the age of 85, two-thirds of Americans will have spent at least one year of their lives below the poverty line.[5] Officially, more than forty-eight million Americans

> Among thirty-five "economically advanced" countries ranked by their rate of child poverty, the United States would be dead last if not for Romania.

are poor—that's as many people as live in Texas and New York State combined. And, among thirty-five "economically advanced" countries ranked by their rate of child poverty, the United States would come in dead last if not for Romania—barely below us.[6]

All this, yet Americans aren't even counted as "poor" if their annual income is above $11,670.[7] But even in the most affordable U.S. housing markets, paying the median rent for a one-bedroom apartment alone would eat up two-thirds to three-quarters of this income.[8]

Inequality. We imagine inequality to be the marker of countries such as India, Liberia, and Yemen. But today, America's inequality is even more extreme than theirs.[9] In the quarter century before 2004, the average, inflation-adjusted, after-tax income received by the bottom fifth of Americans stayed almost flat—gaining only 6 percent—while the top fifth enjoyed a 69 percent increase, and the top 1 percent saw its income leap almost threefold.[10]

Then it got worse. As noted, between 2009 and 2012, the richest 1 percent of Americans captured an astonishing 95 percent of all income gains.[11] As a result of these trends, four hundred Americans today control vastly more wealth than the bottom half of U.S. households.[12] In 2013 the Pew Research Center found that the wealth gap between middle- and upper-income households had widened to the highest level on record.[13]

A rigid class divide. Beyond inequality itself, many Americans also see poor societies as those with little opportunity to rise out of poverty. We see poor majorities "over there" stuck through the generations in wretched slums next to elites in gleaming office complexes, and think; Glad that's not us. We're the "bootstrap" nation, where anyone who tries hard can make it!

Well, no. Today this always-partial myth is slipping away altogether.

A 2014 Gallup survey found that only about half of Americans view this country as a land of economic opportunity, way down from the 81 percent of 1998.[14] And what they perceive is real. Forty-three percent of Americans who grew up in the bottom fifth by income remain there as adults, while 70 percent remain below the middle fifth.[15] Our neighborhoods, too, increasingly split us between rich and poor.[16] Today, the United States enjoys less class mobility than the U.K.—long known for its rigid class lines.[17]

Exploitation of children. Many in the Global North are appalled by child labor and other forms of child exploitation in Global South countries. They are disturbed to learn, for example, that in sub-Saharan Africa a quarter of all children age five to fourteen work, or to hear of child laborers in Chinese factories that supply our ubiquitous smartphones.[18] Yet many Americans seem unaware that in the United States children as young as twelve, with parental consent, are legally working for hire on farms.[19] A Human Rights Watch investigation found farmworkers in the United States as young as seven. In the Texas Panhandle, the organization documented children being paid only $5 or less an hour, and being cheated by employers who underreport their hours or illegally require them to pay for tools, gloves, and even drinking water. [20]

Corruption and rigged rules. We hear a lot about rampant government corruption in low-income countries where officials demand bribes for just about any public service. We can feel it's a plague that we, thankfully, have escaped. In reality, though, we, too, suffer systemic corruption. Only here much of it might well be called "legal corruption."

Corruption can take the form of distortion of government action by private power. In low-income countries the common form is an explicit bribe paid for a specific favor. In our system, it is unlimited spending by enormously wealthy corporations and individuals in political campaigns. The result is similar: Officials, dependent on moneyed interests to be elected and survive in office, feel they must please their funders first. Sometimes the corruption inherent in that quid pro quo is probably not even conscious.[21] In this way, "economic inequality translates into political inequality and political inequality yields increasing economic inequality," observes economist Joseph Stiglitz.[22]

Without strong rules against the influence of private wealth in politics—which are working to some extent in Scandinavia and Germany, for example—elites with deep pockets co-opt political power to rig the rules of the economic game. This is no secret: Polls in six countries—Brazil, India, South Africa, Spain, the U.K., and the United States—show that most people believe the laws are now skewed in favor of the rich.[23]

Tax dodging by the wealthy is rampant, and escalating. And because the rich make the rules, tax avoidance is largely legal. As a consequence, the public purse is being robbed, and critical resources needed to end hunger and poverty are shrinking. In India, fewer than three in a hundred people pay

The world's richest individuals and companies use tax havens to hide from tax authorities as much as $7.6 trillion. That's more than twice the budget of the United States.

any taxes, according to data from the country's finance ministry.[24] The lost income for public purposes comes to a lot of money. Using a web of tax havens around the world, the richest individuals and companies hide from their countries' tax authorities as much as $7.6 trillion—more than twice the entire U.S. budget.[25]

This loss is just as true here as in the Global South.

In the United States, corporations have gotten so good at exploiting "tax loopholes" that since the late 1980s the effective corporate tax rate—the percent actually paid—has fallen by half, from 30 percent to 15 percent, even though the official rate hasn't changed.[26] So corporate taxes as a percent of GDP are half the level of 1970.[27] On top of that, the tax rate for dividends on capital gains—that is, the return on investment—is 40 percent lower than the tax rate working people pay on their hard-earned income.[28] A legion of well-known and highly profitable corporations—Verizon and FedEx, to name only two—have maneuvered so as to pay low taxes or no taxes at all.[29]

In all, U.S.-based corporations have stashed profits in offshore tax havens in excess of $1.9 trillion, according to a congressional investigation.[30] If instead a tax of 25 percent were paid on those profits, it would cover the entire food stamp program six times over. This vital nutritional support for forty-seven million Americans is a program Congress cut heavily in 2014 in the name of reducing the government deficit.[31]

Clearly, tax injustice afflicts those living in the Global North and South and reflects a common barrier: decision-making systems so rigged to serve private interests that they are unable to solve the world's most pressing problems, including hunger.

THE ILLUSION THAT "WE" GAIN

Now let's continue to explore the views embedded in the myth that "it's not our problem." We've suggested significant commonality in the challenges we face. Now we want to explore how our self-interests also intersect.

Because many in the "better-off world" view our lives as being so different from "theirs," it's easy to understand that many could imagine our self-interests also to be different from those of people living in the Global South, or even that our interests are opposed. Indeed, in the industrial countries we are told—sometimes not too subtly—that we benefit from imported goods made affordable by the very fact of lower wages "over there."

A food blender for under $20? An ultrathin flat-screen TV for only $99? How can prices like that not be great for us? Especially for those struggling to live on low incomes.

Wal-Mart takes in almost one out of every ten retail dollars Americans spend, excluding autos.[32] It offers those cheap goods while at the same time returning so much profit to the founder's six heirs that their wealth now equals the combined wealth of 42 percent of American families.[33]

How is it possible to offer rock-bottom prices while amassing such wealth? And what do these "cheap" goods *really* cost us?

For one, the loss of thousands and thousands of good jobs.

Because Wal-Mart sources much of its production in China, the United States over just a five-year period—2001 to 2006—lost 133,000 decent-paying manufacturing jobs, according to estimates by the Economic Policy Institute.[34] And in their place what kind of jobs does Wal-Mart offer? Wal-Mart "associates" in the United States earn on average not much more than the inadequate federal minimum wage, which itself has lost about 20 percent of its value in the last thirty-five years.[35]

Moreover, beyond the loss of specific better-paying jobs is the overall downward pressure on wages and benefits that's hidden in the perception of "cheap" goods. Globalizing corporations, in effect, require workers here to compete with their counterparts in countries that can keep wages low by suppressing independent unions and failing to uphold safety and environmental standards.[36] Wal-Mart's near-monopoly power means it can pretty much dictate prices to its suppliers everywhere, forcing them to lower their production costs. And that means squeezing workers' wages and benefits.[37]

> "Wal-Mart's low-wage workers cost U.S. taxpayers an estimated $6.2 billion in public assistance including food stamps, Medicaid, and subsidized housing . . ."
> —*Forbes*, 2014, based on Citizens for Tax Fairness report

But there's an even less visible loss to all of us. We the people lose big-time as tax money gets diverted to fill in the gap between what corporations pay and their employees' basic survival needs. "Wal-Mart's low-wage workers cost U.S. taxpayers an estimated $6.2 billion in public assistance including food stamps, Medicaid, and subsidized housing," noted *Forbes* in 2014, drawing on a report by Citizens for Tax

Fairness.[38] Similarly, the U.S. fast-food industry "outsources" its labor costs to taxpayers at a total of $7 billion yearly. That's almost $60 for every household in the country.[39]

With Wal-Mart using our tax-dollar subsidies, little wonder that its prices can be low while its profits remain high.

But let's not assume that companies have to treat workers badly in order to profit. Costco, for example, pays its workers an average of about $21 an hour, roughly twice what Wal-Mart pays.[40] And Costco has flourished, while in recent years Wal-Mart has not.[41]

Today in the United States and elsewhere more and more people are beginning to see—and to act on—the links between poverty in other countries and the increasingly dire straits of workers here. In the following chapter we highlight citizens uniting in common purpose across borders.

"A DANGER TO PROSPERITY EVERYWHERE"

Seventy years ago such awareness of our essential interconnectedness helped launch the International Labor Organization of the United Nations (ILO), whose four founding principles—established in Philadelphia, the birthplace of our Constitution—include: "Poverty anywhere constitutes a danger to prosperity everywhere."[42] Many decades before "globalization" and "outsourcing" had become buzzwords, ILO's constitution was prescient: Its preamble declares that any single nation that fails to adopt humane labor conditions is "an obstacle in the way of other nations."[43]

> "Poverty anywhere constitutes a danger to prosperity everywhere."
> —International Labor Organization, 1944

Today 185 member states make up the ILO.[44] One hundred thirty-eight have signed all eight ILO core conventions protecting workers against discrimination and forced labor and ensuring rights of association (the right to organize trade unions). But the United States is not one of them. In fact, except for some tiny island states, only Brunei has shown as much aversion to supporting ILO's labor-protecting conventions as the United States.

Thus, a legitimate question today is this: Does not the fact that the United States has ratified just two of the ILO's eight core conventions make it—according to the words of this body's constitution—"an obstacle in the way of other nations"?[45]

And at the same time, the United States' resistance to recognizing the rights of labor measurably hurts workers and their families here. On average in the United States, those protected by unions earn $3 to $4 more per hour and enjoy superior benefits compared with nonunion workers.[46]

ILL HEALTH KNOWS NO BORDERS

Beyond these threats to workers' rights and wages are other critical threats that we face in common—and that we can solve only in common—with those everywhere struggling for decent lives.

In Myth 3, we explained the worsening U.S. crisis of infections resistant to antibiotics—in large part because of antibiotic overuse in both humans and livestock. The crisis, we now learn, goes far beyond our shores. A 2014 *New York Times* headline called out to us: "Drug-Resistant 'Superbugs' Kill India's Babies and Pose an Overseas Threat." In just five years an epidemic of infections with antibiotic-resistant germs has spread so widely that 70 percent of all Indian newborns are infected, and has killed tens of thousands annually. People of all ages are endangered.

Authorities blame the poor sanitation and overcrowding of extreme poverty, along with massive antibiotic overuse. India's crisis has created "a tsunami of antibiotic resistance that is reaching just about every country in the world," warned Professor Timothy R. Walsh of Cardiff University.[47] So, as with other health threats of which Americans are increasingly aware, especially owing to the 2014 Ebola outbreak, one lesson emerges. It is the ILO's wisdom on poverty applied to health: "Ill health anywhere is a threat to good health everywhere."

And here is another worrying link between our health and the reality of poverty elsewhere: the contamination of imported food, which makes up 16 percent of what is eaten in the United States.[48] Let's begin with some background about how this common threat arose.

Responding to citizens' concerns, in 1970 Congress passed and President Nixon, a Republican, signed the Clean Air Act and created the Environmental Protection Agency. Two years later came the Clean Water Act. And soon corporations got worried: Insistence by engaged citizens that the U.S. government create and enforce standards to protect them and their country could . . . well . . . cut corporate profits. So corporate scouts began the search, not for greener pastures but for places where citizens are denied power to protect their "green pastures."

And they found them. Soon countries in the Global South began competing with each other to lure foreign investment by lowering their standards. This speeded up the competition that we and many others have called a "race to the bottom," creating a perverse twist to the theory of comparative advantage, described in Myth 7. In it, a country's "advantage" is its willingness to suffer miserable wages along with health hazards and environmental devastation. And the logical conclusion of this race has been "a literal bottoming out of environmental protection," notes political scientist David Konisky.[49]

Many corporations have felt free to relocate operations to "pollution havens" where, unencumbered by laws safeguarding water, clean air, and healthy soils, they can extract natural resources as well as dump toxic waste—all to the detriment of nature, workers, and the public.[50]

> *One out of every six pounds* Americans eat is imported, and imported food was linked to 39 disease outbreaks over a five-year period.
> —drawn from Centers for Disease Control, 2012

This corporate strategy puts us very directly in the hazard loop, in part because millions of tons of food Americans eat is imported—*one out of every six pounds.*[51] And much of it comes from countries where governments are too weak and compromised to enforce safe standards, so soils and water are likely to be contaminated.

Some imported fish in North Carolina's supermarkets contained the known carcinogen formaldehyde, a 2013 North Carolina State University study found.[52] In Alabama, almost half of sampled catfish from Asia tested positive for a powerful antibiotic used to treat tuberculosis and pneumonia that has been banned since 1997 in the United States.[53]

So we should not be surprised that, during one five-year period in the 2000s, contaminated food in the United States imported from fifteen countries was implicated in thirty-nine disease outbreaks and well over two thousand illnesses, as reported by the Centers for Disease Control.[54]

Then, think of the health implications of this fact: 80 percent of the fish Americans eat is imported, yet the FDA tests just over 1 percent for any contaminants at all. And for fruits and vegetables? An even smaller percentage is tested.[55]

Our point here is simple: It is easy to slip into the belief that our nation's wealth sets us apart from the one-quarter of all people worldwide who experience nutritional deprivation. What we have learned in responding to this myth should remind us of what we do share: common needs and common threats from tightly concentrated power.

FREEDOM FROM WANT, FREEDOM FROM FEAR

Throughout this chapter we've suggested many parallels and interconnections regarding the well-being of those living in the Global North and the Global South. But we've not directly mentioned one commonality—the experience of violence, including violence against civilians, and the fear it engenders.

So let us close by briefly exploring the connections among poverty and hunger, violence and fear.

During the Second World War, President Roosevelt awarded a medal to every soldier. Inscribed on its back were the words "Freedom from Fear and Want. Freedom of Speech and Religion." By "want" Roosevelt meant hunger and poverty. The soldiers' medals state what many at the time believed—that we were fighting not only to defeat fascism but to establish peace, and that enduring peace depended upon securing the Four Freedoms.[56]

A "basic essential to peace is a decent standard of living for all individual men and women and children in all nations," Roosevelt said in his 1944 State of the Union address.[57]

What Roosevelt saw in his day as the connection between the denial of these freedoms and violence seems equally clear today. As long as steep and worsening inequalities leave at least two billion of us struggling to survive on less than $2 a day, it's easy to understand that violence, including violent terrorism, is likely to continue to erupt.[58]

Our elected leaders often seem to suggest we can end such violence by force alone. As we were composing this chapter, for example, a *New York Times* front-page headline read: "Destroying ISIS May Take 3 Years, White House Says."[59] Yet three presidents have tried to wipe

> "Freedom from fear is eternally linked with freedom from want."
> —Franklin Delano Roosevelt, 1944

out terrorism since the first World Trade Center bombing in 1993. And all have failed.

So how can we create a world free from want of food and other necessities as well as free from fear of violence?

Our answer touches all the themes of this book. We begin by digging deeply—all the way, in fact, to what we believe are humanity's essential needs, even beyond the physical. Our grounding is this observation: To thrive, most human beings need self-respect and the respect of others—perhaps best captured in the concept of "dignity." And we posit that in order to experience dignity people need three things: *connection* with others, *meaning* in our lives, and *power*. We use "power" here to suggest a sense of agency in creating our own destinies.

Yet, in today's world of extreme inequalities, many people are denied opportunities for meeting these three human essentials. And *what happens if people are blocked from constructive avenues to achieve dignity*?

Some will resort to destructive means to meet this need, and many innocent people will die. In most cases, there are obvious triggers, such as territorial disputes or actions perceived as racial or religious bigotry. But the means often become violent, we hold, because of underlying inequities in power that deny people a voice in righting the perceived wrongs.

Researchers in the United States and abroad studying terrorists, for example, offer observations that reinforce this perspective. By and large, terrorists are not pathological, according to an analysis published by the American Psychological Association.[60] Nor do even the suicide bombers appear to be religious fanatics, concludes Riaz Hassan, an authority on Islam and society. Instead, Hassan explains, their complex motivations include humiliation and revenge in response to a perceived injustice.[61] They commonly "identify with perceived victims of the social injustice," adds the analysis of the American Psychological Association.[62]

Terrorism typically emerges from an "environment of hopelessness and feeling of lack of inclusion," notes a research fellow at the Institute for National Strategic Studies, writing for a think tank based in the Middle East. Many of its perpetrators see no avenue to power other than violence to effect real change.[63]

Thus, to free ourselves and others from the common scourge of violence, we must address the roots of "want" and its dignities, as President Roosevelt called us to do more than seventy years ago. Stating it most simply, Pope Paul VI advised us in 1972: "If you want peace, work for justice."

This chapter's myth—that the problem is not ours but rather that of somebody else "over there"—blinds us to the ways that so many people, there *and* here, are blocked from constructive means to achieve freedom from want and to experience dignity; and it blinds us to the forces supported by economic and political structures here—many outlined in this book—that help to deny equity, basic opportunity, and thus dignity to many Americans as well as to people the world over.

With this in mind, we have a much-needed compass. We can satisfy the need for the deeply connected freedom from want and freedom from fear as we, citizens of the North and South, move together in a direction of greater fairness in both economic and political spheres, thereby lessening humiliation and deprivation. If this deeper framework feels daunting, note that human beings have proved we can find within ourselves tremendous energy and courage *if* we feel the direction is worthy—no matter how long the journey seems. In our next chapter, we hope to persuade our readers that movement in this direction is under way in every part of our world. And that, by taking part, each of us can meet our own needs for connection, meaning, and power.

FROM WHERE HAVE WE COME—AND WHERE ARE WE HEADED?

As we explore these common challenges and shared human needs, let us make explicit an assumption that guides us: Centrally important is not just what a society has achieved but the direction in which it is headed. It is not enough for a society to congratulate itself on achieving better lives for most of its citizens compared with societies still considered "underdeveloped." Every society must ask: Are we moving toward ensuring that everyone can enjoy the basic essentials of human dignity, *or* are we moving toward the life-stunting conditions associated with "hungry countries"?

Clearly, these are not one-time-only questions. Change can happen quickly.

In late-nineteenth-century America, the Progressive Era succeeded, for example, in passing the 1890 Sherman Antitrust Act, countering monopolistic power. Then, beginning in the 1930s and 1940s, the United States moved rapidly in the direction of greater fairness and opportunity. Measures to ensure economic equity—Social Security; the GI

Bill, providing education and training, job-location assistance, and other benefits for veterans—were bolstered by a strong and growing organized labor movement.[64] (One result? Today, Social Security alone keeps twenty-two million Americans out of poverty.)[65]

Decline in the poverty rate began, and then quickened with the War on Poverty—including the Food Stamp Act of 1964. In little more than a decade, from the early 1960s to the early '70s, the United States cut its poverty rate in half, to about 11 percent.[66]

For social advancement, this speed is impressive.

Baby boomers born in the aftermath of World War II grew to adulthood during a time when every income level was advancing significantly. But the poorest fifth of Americans enjoyed the biggest gains—their real household income more than doubled.[67]

This period of freedom-expanding social advances lasted for roughly forty years.

Then came the Great Reversal. By the 1980s, those with ideological and financial interests in taking a different direction had led many citizens to believe that government was the problem, enabling corporate power to grow unchecked.[68] Quickly and dramatically, the trend toward greater opportunity and equity reversed. And the proportion of Americans in poverty climbed by a quarter after 1974.[69]

Our point here, which we now take up in our response to the final myth, is that none of the negatives with which this chapter begins is a given. Each results from human choices. Each generation much choose the direction in which its society is heading.

Myth 10

Power Is Too Concentrated for Real Change—It's Too Late

MYTH: Wealth is so concentrated in most countries that, of course, many people go hungry. Many lack land to farm or can't find work, or they're paid so unfairly they don't have enough money to buy food. It feels naive, and perhaps even dangerous, to suggest that we can end hunger. Given the tight grip of corporate power and so many oppressive and corrupt governments, maybe the best we can do is to relieve as much hunger as we can by giving more generously to programs helping the poor. Suggesting anything more could lead to false hope and ultimately to more despair.

OUR RESPONSE: It's certainly no myth that wealth is concentrated in a few hands. In the previous chapter we noted that today wealth is moving increasingly toward the top 1 percent; and concentrated economic power translates into political power, betraying the democratic accountability that's essential to ending hunger.[1] Nonetheless, and although it is easy to miss, transformational change is under way. In this chapter, striking proof of the power of citizens working together to create truly democratic solutions may surprise you.

Of course, we do understand why—given today's extreme concentration—some feel the prospects are dim that humanity will

231

seize this momentous time to uproot hunger's root causes. But consider this question: If a doctor were to tell you that the likelihood was small that you could make a difference in helping a loved one to survive an illness, what would you do?

Would you fold your hands and say good-bye? We doubt it. And the evidence in this book convinces us that, in effect, the future of all our loved ones is at stake—a realization that can energize, not stultify.

And here is a point to ponder: In the eyes of some, we've returned to feudalism—but in corporate form. So let's ask, *How did feudalism end*?

Once, listening to a learned historian speak, I (Frances) perked up when she asked this very question and then paused. Quickly, I grabbed my notebook, eager to get down all the interesting details. When I looked up, the historian answered her own question: "People stopped believing in it," she said.

Could we be in such a moment, we wonder—one in which people the world over stop believing in the economic and political structures that make so many of us feel powerless? Of course, we do not mean that this is all that's needed. But we *are* suggesting that such a shift—no longer accepting the inevitability and legitimacy of the current order—is a prerequisite to real solutions. And it is happening.

OUR POWER

In making this shift, rethinking the meaning of power itself is helpful.

What scientists in a range of disciplines—from physics to neuroscience to ecology—are seeing ever more clearly is that the nature of existence is continuous change, with all elements influencing the shape of all other elements.[2] The implications for the meaning of power—and even of hope—are huge. In the world we grew up in, power was understood as a "thing" you either had or did not have. It was fixed, with only so much to go around: If you have it, the other guy must not.

But from the emergent "systems view of life," what *is* power? It comes from a Latin root meaning "to be able." It is our capacity to act. And our capacity to act is not fixed. It grows and it shrinks in response to innumerable forces, including our own creativity, insight, fortitude, knowledge, capacity to empathize, desire, and connection with others. Power is so much more than money and guns.

Over and over again through-out history, human beings have proved what was before believed to be unthinkable: whether it was the four-minute mile or the first democratic republic. James Madison said of this newborn nation: America has been "useful in proving things before held impossible."[3]

> Awareness of the reality of continuous change—responding to many forces, seen and unseen—means to us that it is simply not possible to know what's possible.

Awareness of the reality of continuous change—responding to many forces, seen and unseen—means to us that it is simply not possible to know what's possible. When discouraged, we count the numerous occasions throughout our lives in which what most people assumed to be impossible actually happened. Think of the fall of the Berlin Wall. Think of Germany, in our lifetimes, moving from global pariah to a leading world democracy. In fact, we ourselves admit that, when we began decades ago, had someone described to us what you are about to read, we might have said, "Oh, no, that's not possible."

So maybe a bit of humility is in order! Perhaps it's the certainty that entrenched power is unmovable that is the real hubris.

LIVING DEMOCRACY

Another ingredient is needed, however, in order to believe that we can be part of historic change. We must have an idea of where we are headed. Not a road map, but a vision grounded in real-life examples.

Since the word "democracy" is now applied to authoritarian and corporate-dominated societies, a new term is needed to capture the vision. Throughout this book we've used Living Democracy to suggest not a fixed structure but an evolving culture characterized by a wide dispersion of power, transparency, and mutual accountability: neither one-way blame nor one-way power. Living Democracy means not a set system but a *set of system values*, including fairness and inclusion, the essentials of human dignity.

In Living Democracy, democratic decision making extends far beyond the ballot box and infuses economic and cultural life as well. This shift in our understanding of democracy reshapes perceptions of what "ordinary" people can do. And, most important, it meets the deep human need for connection, meaning, and power.

So we fill this chapter with surprising stories of Living Democracy arising, some vital enough to dissolve the roots of hunger. We begin with tales of people-to-people movements that refused to be trapped by the thought that "power is too concentrated." We also look at the new rules they are creating, as well as more democratic forms of economic life and of governance. And we note rules still desperately needed.

In a very real sense, this chapter thus responds to each of this book's ten myths, which help to perpetuate the stultifying notion that what we have now—despite its increasingly obvious failings—is the best we can do. Here we hope to demonstrate the opposite.

THE HYACINTH PRINCIPLE

How does the lovely hyacinth fill up a pond so fast? It reaches out. Each plant sends out lateral "runners" that create daughter plants, so hyacinths are able to double their numbers in weeks. For sure, some experience this humble plant as nothing but an invasive nuisance; but, hey, how about seizing the hyacinth's approach for positive, liberating, exponential growth? It might then be a great metaphor for the power of citizens to make big change by sharing with others what they're learning.

So, let's now bust open the assumptions behind this myth with new sightings of relational power emerging in some unlikely places and on a scale that, we imagine, will leave a few readers wondering why they've never before heard these stories.

India's Ecological Farming Solutions—from the Ground Up

We begin in India because, as noted in Myth 3, in the 1960s India was the primary large-scale testing ground for industrial agriculture in the Global South—what has been called the Green Revolution.

A Village Says "No" to Pesticides and Ripples Spread

The rural areas in the southern state of Andhra Pradesh—India's fourth largest before its 2014 division into the two states of Andhra Pradesh and Telangana—have seen a lot of misery in recent decades. Farmers' production costs—especially for purchasing chemical pesticides—climbed while inflation-adjusted prices for their crops stagnated or fell.[4] By the 1990s, some had dubbed the state the "pesticide capital of the world." A registered medical aide working in a small group of villages reported fifty to sixty pesticide poisonings each season, killing and disabling people.[5] In addition, suicide by debt-laden farmers was taking too many lives.

Then came something quite unexpected.

A spark ignited in the small village of Punukula in about 2000. Burdened by debt and suffering ill health from exposure to toxins in agricultural chemicals, first one Punukula farmer, then another, said "no" to pesticides. And the local civil-society organization SECURE, together with the Hyderabad-based Centre for Sustainable Agriculture (CSA), helped farmers learn a different approach they now call Non-Pesticidal Management.

Farmers turned to homemade pest-control potions, especially those made from the neem tree, along with garlic and chilies. Plus, many began actively helping others achieve what they were enjoying—the increased income from not having to buy pesticides and better health from no longer being exposed to chemicals.

From heartbreak came courage, and by 2004 villagers of Punukula—who before had spent collectively $80,000 to almost $100,000 on pesticides each season—formally declared their village pesticide-free. They told pesticide pushers to just stay away.[6]

> In 2004, the villagers of Punukula formally declared their village pesticide-free and told the pesticide pushers to stay away.

From this small, courageous village grew a movement away from pesticides and toward agroecological practices.

Unlike the U.S. government, which puts its weight largely behind corporate-driven agrochemicals and patented seeds, the Andhra Pradesh Ministry of Rural Development stepped up with a different agenda: It's helping farmers to shift toward healthier, more profitable practices.[7] An affiliate of the state government, the Society for Elimination of Rural Poverty—collaborating with CSA—initiated what it calls Community Managed Sustainable Agriculture, which includes the nonpesticide movement.

Its power is growing through a knowledge-dissemination team in each locality, including a "Village Activist" and a "Cluster Activist," both of whom are paid small stipends, along with technical consultants and coordinators. They're trained to oversee village Farmer Field Schools that encourage farmers to try new methods, first on just a portion of their land—and from there, the farmers report, the "results can speak for themselves."

Now 12,500 Village Activists spread knowledge of healthy farming practices, and each Cluster Activist teaches Farmer Field School courses in about half a dozen villages. Farmers learn about compost, biogas, cover-cropping, intercropping, vermiculture (worms), rainwater catchment, and seed saving.[8]

To get training, a farmer has to make commitments, too: Each agrees to pay a fee of about thirty-three cents, collect neem seed, attend all trainings, keep careful records, and forgo pesticides.[9] And farmers are encouraged to ensure that their teachers uphold their responsibilities. If the Activists don't show up to teach, for example, the villagers who are counting on gaining key knowledge in return for making good on their own commitments can use their mobile phones to report an Activist's absence.[10] This practice of mutual accountability, we believe, is vital to democracy at any level.

Through this knowledge-expanding network, hundreds of Farmer Field Schools have trained tens of thousands of farmers. So the first breakaway village, Punukula, has now been joined by roughly twelve thousand villages on their way to becoming pesticide-free.[11] By 2015 nonpesticidal practices had spread to two million small farms cultivating 15 percent of the arable land in the two states, with women's self-help groups in the lead.[12]

Using Non-Pesticidal Management, including composting, a farmer renting land can realize a net financial gain five to ten times greater than what would be obtained with chemical inputs.[13] As families' health care expenditures go down, they have more disposable income. So they need to rely less on their children and can keep them in school full-time. Then, as children get more education, they are qualified for better-paying jobs. Farmers also can invest in improvements like chicken coops. As farmers work more acres, more farmworkers are hired.[14] New jobs are also created for those collecting and processing the neem seed's natural pesticide.

The Society for Elimination of Rural Poverty also offers farmers small loans to create businesses preparing and selling the botanical pest-control extracts. Some earn $500 to $800 yearly, in the ballpark of this state's average per capita income.[15]

Overall, these efforts have cut chemical pesticide use in the state by half in the ten years from 2004 to 2014. The ultimate goal of this movement? Eliminating all synthetic chemical inputs.[16]

"What We've Most Gained Is Courage"

In 2012, to get a firsthand look, I (Frances) visited the same state, Andhra Pradesh, to meet farmers within the Deccan Development Society (DDS), a parallel, self-organized network of five thousand women in seventy villages who have for thirty years been embracing and teaching organic practices.

After a few hours' drive from Hyderabad, I sat on a straw mat surrounded by a dozen DDS women in brilliant saris. I asked about their lives before the DDS. "We were so poor that in the rainy season our hut floors would turn to mud and we had to pile up branches to sleep on," the women told me. "We were always hungry. We depended on government ration cards. Sometimes the big landowner would pay us for a job with some grains and that would be the only food for our children. It was a dark time."

Gazing at beautifully arranged mounds of diverse seeds from the women's own fields, I asked the obvious: "And what changed?"

"We started meeting and talking. Every week, at nine in the evening our *sanghams* [self-help groups of women] come together and make decisions together. Through the *sanghams*, we've reclaimed the land. We don't use any chemicals. We grow as many as twenty crops on an acre or two. Every family in our village has food security now."

The changes go well beyond food, I learned. "We tell each other our problems. If someone was abused, all of us go together to confront the abuser," they said. "And now if there is any kind of conflict in our village, they call on us."

"We make pledges to each other" in the *sanghams*, the women told me. In a public ritual, all members stand in a circle, arms outstretched, committing to three actions: to forgo agrochemicals, to reject genetically modified seeds, and to share their seeds and what they learn.

Walking the next day in a field of diverse crops—from lentils to oilseeds to greens—I noticed the dry earth. As with two-thirds of Indian farmland, there's no irrigation here, so rain matters a lot. I asked, "Aren't you worried about climate change bringing more drought?"

"No. We know what to do," a farmer told me. "If rainfall is cut by half, we know which seeds will work." Once a year, the Society's colorful seed caravans—accompanied by music and dancing—travel to fifty or more villages creating festivals that teach the art of seed saving.[17]

> "From the *sanghams* [women's groups] what we've gained most is courage."
> —Deccan Development Society member, 2012

A few years ago, DDS calculated that the women's leadership has meant the production of almost three million extra meals each year, as well as almost 350,000 additional days of employment in their villages.

But what most struck me were the women's parting words: "From the *sanghams*, what we've gained most is courage." And with it, the women of DDS are working to create a federation of village *sanghams* with, as you will see, a strong voice in state and national policy.

The Seed of Freedom

The Indian scientist Vandana Shiva spent much of her youth among trees. Her father's job for the Indian government was protecting forests in the northern state of Uttarakhand. While a young woman, in 1977 Shiva joined village women in a protest against deforestation, an event that ultimately shaped her life.

That fateful day, officials arrived near Shiva's home to begin tree cutting and declared, "You foolish women, how can you prevent tree felling by those who know the value of the forest? Do you know what forests bear? They produce profit and resin and timber." Shiva recalls that in response the local women sang back to them, "What do the forests bear? Soil, water, and pure air. Soil, water, and pure air sustain the Earth and all she bears."[18]

Shiva began volunteering in the movement; and from there, she writes, protecting "biodiversity and biodiversity-based living economies" became her life's mission.[19] In 1987 she founded Navdanya, meaning "nine crops that represent India's collective source of food security."

Soon Shiva's family homestead had become Navdanya's full-blown rural study center—Bija Vidyapeeth (Earth University). And, by 2014, Navdanya had trained four hundred thousand men and women—from farmers to government officials—in organic farming practices and how to protect biodiversity.

Since 1991, Navdanya's Seed Freedom campaign, known as the Bija Satyagraha Movement, has organized farmers to actively resist the corporate control of seeds and promote the tradition of saving and sharing seeds. In 2000, the Navdanya farmers' network sent thousands of postcards to the European patent office, aiding the historic defeat of a powerful U.S. corporation's attempt to patent the ancient Indian neem tree. (Neem serves many purposes; it is used in everything from toothpaste to the natural pesticide noted earlier.)

Navdanya's efforts have conserved more than three thousand rice varieties from all parts of India, including rice adapted over centuries to meet a range of ecological conditions, plus seventy-five varieties of wheat and hundreds of millets, legumes (mainly lentils), oilseeds, vegetables, and multipurpose plant species, including medicinal plants. Navdanya has also created a network of 111 community-run seed banks in varied eco-zones across the country.

So in 1999, when a supercyclone hit the state of Orissa (now called Odisha), Navdanya was ready. Victims were able to use the salt-resistant seeds Navdanya had conserved to get on their feet again.

More than half a million Navdanya farmer-members are taking the Seed Freedom movement to neighboring villages, in part via Navdanya's seed fairs, where farmers learn about saving and sharing seeds.[20]

Today, Navdanya's home state, Uttarakhand, calls itself an "organic state" because it is publicly committing to moving to organic methods.[21]

Farmers Leading the Way to Agroecology, Some Surprises

Related, striking transitions are happening in other states of India, each with its own special twist.

In the poor northern state of Bihar, several hundred thousand poor farmers have shifted to the ecological approach not dependent on herbicides, with the backing of the state government.[22] It's the System of Crop Intensification, first used with rice and described in Myth 4. In 2013, using this method, Sumant Kumar, described by the press as a "shy young farmer," was shocked to haul in a world-record rice yield of 22.4 tons from one hectare (about two and a half acres). The deposed record-holder, a Chinese agricultural scientist known as the "father of rice," cried foul.[23] But the Indian government had verified Kumar's yield. His fellow villagers had also increased their yields significantly.

"My whole life has changed," said Kumar. "I can send my children to school."[24]

And then, a big surprise. The state of Kerala, home to thirty million people, in 2010 officially declared the goal of becoming 100 percent organic within ten years.[25] By 2013, fifteen thousand farmers were in the process of securing organic certification.[26]

So in India, the country in the Global South most associated with industrial agriculture, more and more farmers are leading the transition to agroecology. They are benefiting from the power of their growing knowledge and that unleashed by sharing it widely, along with the power of making public commitments to each other and holding each other accountable for fulfilling them.

Campesino a Campesino:
Farmers Learning from Farmers in the Americas

Now jump across the world to Central America, where more than thirty years ago, a time of many crises—from civil wars between wealthy elites and the poor to economic distress—gave birth to the farmer network Campesino a Campesino (Farmer to Farmer).

Arising first among the Kaqchikel Mayans in the highlands of Guatemala, the network today includes hundreds of thousands of farmers in more than a dozen countries in Latin America.

Campesino a Campesino is a social movement based on the belief that farmers "are capable of developing their own agriculture," emphasizes Eric Holt-Giménez, executive director of Food First/Institute for Food and Development Policy, which the two of us founded in 1975.[27]

And key to its success? Like many others, the movement started small and then shared its farmer-generated techniques and farmer-to-farmer knowledge as widely as possible.

In Myth 2 we told the story of farmers in Central America who in 1998 weathered that century's most destructive hurricane by employing ecological practices. Many were part of Campesino a Campesino, and these farmers suffered less soil erosion and fewer crop losses than neighbors who didn't use the practices.

From such experiences, farmers learned that protecting the environment is critical, notes Holt-Giménez. They therefore work together to manage their watersheds and protect biodiversity, making on-farm success possible.[28] They work incrementally to eliminate dependence on purchased, synthetic inputs and to protect and enhance the health of their local ecosystems. Innovations that the network has spread include planting velvet bean and other "green manures"—crops that "fix" nitrogen from the air and thus serve as natural fertilizers—and various soil and water conservation practices.[29]

Instead of one-to-one farmer education, Campesino a Campesino uses "farmer-promoter teams" in which farmers mentor their peers. Local nongovernmental organizations also offer the assistance of agricultural technicians, who hold field-demonstration days, study sessions, and workshops for farmers, and sometimes provide logistical and economic support as well.

But there's another dimension that helps to explain the movement's vitality, one that could be lost on those who assume a farmer's motivation lies simply in greater crop yields and income. Although these farmers are achieving both those gains, Holt-Giménez told us, they are motivated by "deeply held beliefs in the divine, in family, in nature and community."[30] He emphasizes that

> [p]art of farmers' enthusiasm for developing agriculture comes from the sense that they are actually contributing to and shaping society. This subjective, but very powerful, motivational force has been nurtured through cross visits, *"encuentros"* (farmer gatherings . . . sometimes similar to scientists' symposia) and the inclusion of farmer-promoters in workshops held by national and international agencies for agricultural development.[31]

As seems to be the case with every example in this chapter, Campesino a Campesino is succeeding not just in addressing hunger for food and for economic security. It is also meeting the deep human need to know that one's voice counts within a community of meaning.

Farmers Regreen the Earth: Africa

In Myth 2 we celebrated peasant farmers in Niger who are turning back the desert and establishing "agroforestry" fields by interspersing trees and crops—yet another sighting of the "hyacinth principle" at work.

In neighboring Burkina Faso is another striking story addressing hunger and climate change at the same time. In the 1970s, when drought hit the Sahel—the Sahara Desert's southern flank, covering six countries—many families fled. One who did not leave was Yacouba Sawadogo. Knowing his ancestors had survived droughts in this arid land, he trusted that they had devised practices to conserve water and nourish the soil. He was ridiculed, and worse.

But Sawadogo was determined. He adapted his ancestors' technique of digging shallow, circular pits, called *zai*, in his fields. They keep rainfall from running off the hardpan soil's surface. In the *zai* he put dung for nutrients and termites that create tiny tunnels, allowing the water to penetrate. Between these tricks and his use of small earthen dams to trap rainfall, it worked. Others saw his success and began experimenting. Soon he and other trailblazers began classes for neighbors. They became the informal "extension agents" for a new (old) way of farming.

These ingenious methods have regreened about half a million acres in Burkina Faso. If yield increases reach their potential, in just a few decades farmers sharing their knowledge will have created enough additional food to feed half a million people.[32]

In Niger, Burkina Faso, and Mali, the successes of farmers teaching farmers agroforestry have sparked a vast vision—"a network of villages with regenerated agroforestry systems that will span some 15 countries and involve millions of people in creating a more prosperous evergreen agriculture," reports the World Agroforestry Centre, headquartered in Nairobi, Kenya.[33]

Imagine agroforestry's contribution as well to absorbing excess atmospheric carbon, thereby helping address the world's climate crisis.

Strength Across Borders

Breakthroughs we capture here, from India to the Americas to Africa, demonstrate that courage is indeed contagious. People's movements gain and grow power as they spread laterally, reaching out and sparking others' growth.

That's the method behind the power of La Via Campesina, the international peasants' movement founded in 1993. It has grown over twenty years to include 164 organizations in 79 countries, and its slogan is

"Globalize the Struggle, Globalize Hope." By 2009 the organization was a recognized global player invited to join the civil society "mechanism" of the UN's Committee on World Food Security and to deliberate on policies determining the future of farming and food as a human right.

In 2013, La Via Campesina came to a formal "agreement of cooperation" with the director general of the FAO to acknowledge the role of small farmers in ending world hunger, emphasizing the importance of young people and women in food production.[34]

Some living in industrial countries might be surprised that this farmers' movement chooses the term "peasants" to describe its members. But for many small-scale farmers, a "peasant" is not just someone who works the land, but a person embodying a rich culture of values, knowledge, and power.

Strengthened by connectedness through one global network, La Via Campesina member organizations have taken leadership nationally for agroecological farming. Two stories capture this growing power of people whose voices have too often not been heard.

In the autumn of 2014, La Via members in Guatemala won a surprising victory. Earlier that year, the Guatemalan congress passed—without public discussion or participation by the most affected farmers—a law that would have opened up the market for genetically modified seeds. According to La Via Campesina, this law threatened indigenous seeds and seed diversity and disadvantaged local producers in a country where about 70 percent of the population engages in small-scale agricultural work.

Demonstrations against the law began in the Mayan communities of Sololá, a mountainous region not far from the capital. Protesters blocked several main roads. The Mayan people and social-benefit organizations argued that the new law violated both the constitution and the Mayan people's right to traditional cultivation in their ancestral territories. Indigenous people, social movements, and trade unions, along with farmers' and women's organizations, joined in ten days of all-out street protests.

And the congress listened. It repealed the law that "would have given exclusivity on patented seeds to a handful of transnational companies," reports La Via Campesina.[35]

Just to the north is Mexico—corn country, and home to thousands of indigenous varieties also threatened by contamination by genetically modified seeds. There, a year earlier in 2013, a collective hunger strike by La Via Campesina members helped to galvanize farmer support for a lawsuit filed by fifty-three citizen plaintiffs, including farmers, to halt large-scale planting of GMO corn.

It worked. Members kept up the pressure, and the following year their efforts helped to move a judge to uphold the injunction against further testing or commercial planting of GMO corn, citing "the risk of imminent harm to the environment."[36]

Taking to heart the understanding that knowledge is power, La Via Campesina has not just worked to block threats to farmers and the land but created regional agroecology training schools in a half dozen countries, with more in the works. Plus, in its political leadership academies, peasants learn how to pressure governments effectively for pro-peasant policies.

ECONOMICS OF LIVING DEMOCRACY

Land Reform: "For a Brazil Without *Latifundios*"

In Latin America, peasants are also proving that concentrated power, even centuries old, is not as solid as it appears. Only a few decades ago, we imagined Brazil—an extreme example of land concentration—to be among those countries least likely to become a leader in transformational change. We were *really* wrong.

With the end of Brazil's dictatorship in 1985 came the emergence of arguably the largest social movement in the Western Hemisphere: the Landless Workers' Movement, known by its Portuguese acronym MST, first mentioned in our opening essay. The movement was a response to what its founders saw as indefensible: less than 2 percent of Brazil's landowners controlled about half the land, often gained illegally, and left much of it unplanted, while roughly ten million rural workers struggled with hunger because they had too little land or none at all.[37]

One MST member told us about her life before the movement: "You see," she said, "before, we weren't just landless, we were everything-less." From its birth, MST's goal has been nationwide land reform— "Brazil without *latifundios*" (large landholdings).[38]

Faced with such extreme injustice, the MST chose a strong but risky tactic: acts of civil disobedience in which landless workers occupy idle land and demand government help in gaining legal rights to it.

Fortunately, standing with these courageous farmworkers have been a range of social movements, as well as the Catholic Church's Pastoral Land Commission and other religious groups.

Then, as noted, in 1988 Brazil's new constitution provided the movement some legal grounds by affirming government's obligation and power to "expropriate . . . for purposes of agrarian reform, rural property . . . not performing its social function."

To compel the government to uphold this constitutional commitment, the MST has carried out roughly 2,500 occupations on the unused land of large estates—arguing that, because it fails to "perform its social function," the land should be expropriated and transferred to the landless.[39] In this long struggle, almost 1,500 MST members have lost their lives, mostly at the hands of angry landowners and as a result of illegal police action.[40]

Through extraordinary commitment and sacrifice, more than 370,000 families are now building new lives in their own communities on about twenty million acres, spanning almost every state of Brazil. The MST influence, moreover, extends well beyond its own organization. Inspired by its work, wider social mobilization has since 1994 led to a total of more than a million landless families resettled on land they can farm.[41]

In MST communities, children attend one of the two thousand schools the MST has created, serving 150,000 kids and 3,000 adult students.[42]

MST families gain the legal right to live on, and to farm, a parcel of land—and to pass it on to their children. But they do not own this land. That "would just privatize the plots," the MST leadership wrote to us, "allowing people to sell it like any other private property."[43] The right to work the land as one sees fit but not to sell it prevents the reconsolidation of farmland into *latifundios*. (This principle of land held in community trust can also be seen in the 250 "community land trusts" in the United States in which families enjoy security of tenure via very long-term leases, plus a portion of the home's increased value when the lease ends. The land itself, though, is held in community trust in perpetuity so that homes remain affordable.)[44]

Note, however, that because Brazil exists within national and global economies driving in the opposite direction, over these three decades land concentration in Brazil as a whole has not diminished.[45]

Following land occupations, families often live for years in encampments—typically protected from the elements only by plastic-covered wood frames—waiting for legalization of their claims. Currently, there are 180,000 people waiting in these harsh conditions. Some MST families use their time in part to train in ecological farming.[46]

The MST works to resist genetically modified seeds and has created Brazil's first organic seed line, Bionatur, which offers more than ninety organic plant varieties.[47] In 2004, it built the Chico Mendes Agroecology Center in Ponta Grossa, Paraná—ironically, on land that biotech giant Monsanto formerly used to grow genetically engineered crops. The center produces organic seeds native to Brazil for MST farmers and trains farmers in agroecology.[48]

MST farmers are among the many family farmers in Brazil adopting agroecological practices. In southern Brazil, some of these farmers increased average yields of black beans by 300 percent and corn by 100 percent. In the north, the practices increased resilience to irregular weather.[49] Also motivating farmers is the financial gain they enjoy: Those MST farmers making the effort to shift to agroecology have seen their costs drop from $200–$285 per acre to only $11, reports the U.K.'s War on Want.[50]

The MST has also spawned almost one hundred food-processing cooperatives, indirectly benefiting about seven hundred rural towns, along with about five hundred other co-ops providing technical assistance, credit, marketing, and more.[51]

We wanted to know more about *why*—why the MST has led the way in promoting agroecological farming throughout its settlements. And Jacir Pagnussatti, a young MST member of a community near Curitiba, explained, "It's not just that farming without using pesticides means less hazard and lower costs for us. Why would we go to all this trouble and risk to grow food that's just going to hurt people? We are concerned about the people in the cities, too."[52]

Why Has the MST Achieved What Others Have Not?

For many generations the fight for fair access to land in Brazil has been met with violent repression. So what explains MST's power?

In writing our book *Hope's Edge*, my (Frances's) daughter Anna and I asked this question of a movement founder, João Pedro Stédile. His answer stressed the MST's commitment to group action through participatory power. It begins in the movement's settlements, where members organize in groups of ten families making decisions together, and it extends all the way to a national congress, held every five years. The most recent drew fifteen thousand members from every corner of Brazil. Women made up half the delegates, and one in ten members of Brazil's parliament considered this gathering to be important enough that he or she also participated.[53]

For the MST, gender equality is a core value. In MST communities the coordinators must be one man and one woman. The goal is for women to make up half of the participants in all MST education and training courses, as well as in leadership in the organization's national bodies. Joint land-use titles and the right to credit in the names of both members of a couple protect the rights of women.

MST is building relationships of mutual accountability, not one-way control. And at the same time, it is working to remake relationships with the Earth by farming in ways aligned with nature. All of this requires huge changes within. As cofounder Stédile explained to us:

"The first step is losing naïve consciousness—no longer accepting what you see as something that cannot be changed," he told us. "The second is reaching the awareness that you won't get anywhere unless you work together. This shift in consciousness, once you get it, is like riding a bike, no one can take it from you," he declared. When "you forget how to say 'yes, sir' and learn to say 'I think that . . .' This is when the citizen is born."[54]

> When "you forget how to say 'yes, sir' and learn to say 'I think that . . .' This is when the citizen is born."
> —João Pedro Stédile, an MST founder

"The Most Astonishing Thing"—Workers Unite

Even with this brief history, we can appreciate advances made by courageous landless workers in Brazil—but what about the three million to five million farmworkers and their families in the United States who are *our* landless workers? [55]

As a class, they've long been excluded from even the most basic protections of U.S. labor laws. Yet their work is hazardous. Very hazardous. Every year in the United States, sixty-one thousand farmworkers are injured so badly they lose work time.[56] Surprisingly to many, the rate of nonfatal work injuries is higher among agricultural workers than in mining occupations.[57]

But injury is just one risk. Compared with the general U.S. population, farmworkers suffer from higher rates of a range of serious diseases, including certain cancers: lymphomas; leukemia; and brain, cervix, prostate, and stomach cancers.[58] Women farmworkers also suffer high rates of sexual abuse.[59]

And here's another risk that will shock most Americans: Farmworkers are vulnerable to being caught in "involuntary servitude," a euphemism for slavery. Since 1997, in nine cases, more than a dozen labor contractors in the southeastern United States have been prosecuted for holding over one thousand farmworkers against their will.[60]

In the 1970s, the United Farm Workers, led by Cesar Chavez, awakened Americans to the human rights violations suffered by farmworkers. But after his death in 1993—in part, some say, due to the health toll taken by three monthlong hunger strikes years earlier—the farmworkers' struggle faded from the news.

Thus, few Americans were aware that in the same year Chavez died a new flame ignited: In a church in south Florida, a handful of tomato pickers began talking about how to change the intolerable. From there, the Coalition of Immokalee Workers (CIW) arose to protect the well-being and dignity of workers in Florida's tomato fields, which supply almost all of the domestically produced tomatoes Americans eat during the winter months.[61]

And just two decades later, in 2014, former president Bill Clinton publicly proclaimed the CIW's work to be the "most astonishing thing politically in the world we're living in today."[62]

So what grabbed Bill Clinton?

Backed by student, faith, and community organizations—its key allies making up the Alliance for Fair Food—the CIW has achieved what many believed impossible: It has convinced some of the world's most powerful corporations to make legally binding commitments respecting workers' rights.

During the 1990s, CIW members risked their livelihoods and more by carrying out strikes, including a thirty-day hunger strike in 1998 and a 230-mile march through Florida in 2000, with the simple demand that the tomato growers sit down at the table with farmworkers to address decades-old farm labor abuses and subpoverty wages. The workers won raises, but success brought the piece rate for tomato picking back only to what it had been before 1980—still well below the poverty rate.

So in 2001 the CIW shifted its primary target. Launching the Campaign for Fair Food, the CIW called out major corporate players in the food industry, demanding that they step up to improve wages and working conditions in the Florida tomato fields.

Starting with Taco Bell in 2001, the CIW eventually secured "Fair Food Agreements" with twelve major corporate buyers—including Trader Joe's, McDonald's, and Wal-Mart. They included a penny-per-pound increase paid to farmworkers and adherence to an enforceable, worker-designed Code of Conduct.

By 2010, the CIW's original target, the Florida Tomato Growers Exchange, signed on—expanding the agreement to over 90 percent of the Florida tomato industry.

Then the CIW set its sights even higher.

Following the landmark agreement with the growers exchange, the CIW created the Fair Food Program, a unique partnership among farmworkers, Florida tomato growers, and participating retail buyers. Buyers that sign on to the FFP agree to "zero tolerance" for forced labor and sexual assault. The CIW—on company time—is allowed to educate farmworkers about their rights and responsibilities. And, new on-farm health and safety committees give workers a "structured voice" in their working conditions. Finally, the companies allow monitoring and enforcement by an independent, third-party organization, the Fair Food Standards Council. The program also still includes the penny-more-per-pound premium paid by corporate buyers.

The package adds up to big change.

During the first three and a half years of the Fair Food Program, Florida tomato pickers received roughly $15 million in Fair Food Premiums, paid out as bonuses on workers' checks. Fear of enslavement has eased as well. Prior to the Fair Food Program, the CIW helped in prosecuting seven cases of modern-day slavery, freeing

> During the first three and a half years of the Fair Food Program, Florida tomato pickers received roughly $15 million in Fair Food Premiums, paid out as bonuses on workers' checks.

over a thousand workers effectively held captive. No new cases have arisen on participating farms since the program started.[63]

The men and women of CIW embody the courage and integrity at the heart of Living Democracy. So we close our salute to them by pointing to the obvious: The protections the CIW is gaining will be ensured for *all* workers only as we as citizens step up and strengthen federal labor laws and enforcement.

To bring that point home, consider the biggest corporation to sign on to CIW's Fair Food Program, Wal-Mart. In early 2014, the National Labor Relations Board—set up in the 1930s to protect workers—officially issued a complaint and notice of hearing against Wal-Mart for disciplining and illegally firing more than sixty employees organizing against the company's abusive labor policies.[64]

Just two days later, Wal-Mart—on its own with no CIW campaign targeting the company—signed on to the Fair Food Program, stating that "Wal-Mart [is] committed to strong ethical sourcing standards and . . . fair treatment for workers."[65] We were pleased, but we also wonder how many observers noticed that Wal-Mart's action did nothing at all for its *own* 1.3 million U.S. employees.

Our point? That we citizens should be both encouraged by CIW's remarkable story and challenged to step up for *society-wide* rules essential to fairness in the workplace.

As we see here, and throughout our book, the power of common action can indeed dissolve the mind-set—and the reality—that concentrated power is immovable. Sustaining the widening and fluid dispersion of power—essential to ending hunger—requires our learning to create fair economic rules, as well as to hold each other accountable. It's true from the village council to governments negotiating international agreements.

Democratizing Food Power, U.S.-Style

Despite the grip of monopoly power within the U.S. food industry, dramatic change is moving in the opposite direction: It is democratizing power.

Consider Community Supported Agriculture (CSA). It's a model of farmer-consumer collaboration in which consumers pay farmers up front to cover their costs and get great produce all season long.

The idea took off twenty years ago, and already there are an estimated six thousand CSA farms nationwide.[66] Farmers markets have grown fast—multiplying nearly fivefold in twenty years.[67] At the turn of the twentieth century there were seventy-five thousand school gardens, but they disappeared. Now they are returning. In the 1990s, California set the goal of a garden in every school, and already there are two thousand five hundred. Across the country more and more children, with hands in the soil, are learning about our Earth.[68]

Also democratizing access to land and food across North America are eighteen thousand community gardens that now offer eaters the opportunity to become producers, too—often using reclaimed lots or public urban space. Community gardens—where city dwellers have access to small plots—increase access to fresh food and green spaces, and have even been shown to help reduce neighborhood crime.[69]

And do these developments address hunger? In some places, yes— quite directly.

A great example is Growing Power, launched in Milwaukee. Across nearly twenty locations in Wisconsin and Illinois, the initiative now provides over a million pounds of fresh fruits and vegetables to low-income neighborhoods every year. Community members enjoy the bounty through schools, restaurants, co-ops, farm stores, and farmers markets, and even via low-cost CSA food baskets delivered throughout the neighborhood.

On its initial two-acre site in Milwaukee, Growing Power packs in fourteen greenhouses, where herbs and vegetables grow almost up to the ceiling. Out back are pens full of chickens, goats, ducks, turkeys, and beehives. Plus, an ingenious symbiotic system annually raises ten thousand tilapia and perch. Water is pumped upward for sprouts growing above a long tank. The sprouts' dirt and gravel filter the water, which trickles down, pure and aerated, to the fish.

Demonstrations and trainings engage neighbors in urban agriculture. In Milwaukee, on any given day you can learn the ins and outs of composting, participate in a "From the Ground Up" workshop to learn how to develop and run food initiatives in your own community, or, if you're a young person, take part in trainings that range from agricultural basics to leadership and entrepreneurial skills. And Growing Power is helping to set up similar projects across the United States, including projects in Arkansas and Mississippi.[70]

Now, let's refocus this chapter's lens of connectedness and change to examine the power of economic rules—rules shaping the systems we live in, and thus shaping us.

NEW RULES FOR DEMOCRATIC ECONOMIES

A Right to Eat

One international agreement is gaining ground: the human right to food.

As we lay out in Myth 6, for us a "right" is that which citizens agree together to ensure for one another: what no one can be denied. And, today 164 nations have accepted international law committing them to the right to food, and virtually all endorse the 1948 Universal Declaration of Human Rights that identifies food as a right.[71]

In addition, in their own constitutions, more than two dozen nations now include an explicit right to food.[72] One is India. In 2013, it passed a National Food Security Act clarifying that its public distribution of grain at a subsidized price—

> More than two dozen nations now include an explicit right to food in their constitutions.

aiming to reach two-thirds of the people—is not a welfare approach to fighting hunger but rather "right based."[73]

Another sign that food as a human right is being taken seriously? In 2000, the United Nations announced its first Special Rapporteur on the "right to food," signaling a growing global acceptance of the view that hunger is due not to lack of food but rather to lack of access to food because it's not protected as a human right.[74]

But these facts, though encouraging, only begin the conversation. A right means nothing without rules to protect it. What therefore matters is the system of rules we together both create *and* enforce—laws and public policies that make real the right to food.

And after all, food as a human right is hardly a new idea!

We closed the previous chapter recalling that roughly seventy-five years ago President Roosevelt passionately advocated for freedom from want as one of the "four freedoms," to lay the groundwork for lasting peace. Roosevelt's goal was a "Second Bill of Rights" defining economic and social rights.[75] Roosevelt did not live to realize his goal of a Second Bill of Rights, but during the three decades after his death in 1945 Americans proved that we could move in the direction of "freedom from want." As noted, the poorest fifth of our people doubled its real family income.[76]

How did this happen?

Americans together created new rules, for example, to better align taxes with the ability to pay, to provide public support for veterans' education, to protect laborers' freedom to organize unions, and to ensure that the minimum wage was a living wage.[77]

If today we've lost confidence, let us remind ourselves both of what America has achieved and of what so many in our world are creating today as they gain confidence and join together.

On the right to food, consider what has happened in Brazil.

A Right to Eat, the Experience—Brazil and Beyond

In 2010, Brazil added the right to food to its constitution. But unlike people in most countries, Brazilians were already on the path to making the "right to adequate food" real. As you read of the accomplishments below, keep in mind that as of this writing some are under attack—all the more reason, we believe, to understand, defend, and further proven approaches to make real the right to food.

From the mid-1980s onward, freed from the terror of a military dictatorship, religious leaders, workers, students, academics, farmers, householders—all pressed hard. By 2002 they were key in electing a former metalworker as president, Luis Inácio "Lula" da Silva. Everyone calls him simply "Lula."

His stated mission? That by the end of his term every Brazilian would eat three meals a day. To make the point, one of his first presidential acts was suspending purchase of six fighter planes and redirecting $760 million to fighting hunger.[78]

Lula did not achieve his three-meals-a-day mission. But under his leadership, the *Fome Zero*—Zero Hunger—campaign was born. Its emphasis from the beginning has been not just on food security, but on what is consistently referred to as "food and nutritional security." Seemingly, those driving Zero Hunger are aware of what we previously noted—that increasingly "food" does not mean "nutrition."

Today Zero Hunger's thirty-one programs attack economic barriers to access to healthy food while also supporting family farming.

Brazil has taken the lead, for example, with what's become known internationally as the "conditional cash transfer," called *Bolsa Familia* (family payment). Launched in 2003 to bring the poorest families above the poverty line, the monthly electronic payments from the federal government average about $35 per family, but vary depending on the family's size and financial situation. More than 90 percent of the payments go directly to women, on the condition that their kids are in school and vaccinated.[79]

While many assume that such "handouts" discourage people from working, women who receive the benefit have a 16 percent higher rate of employment than those who don't.[80]

Bolsa benefits a quarter of Brazilians directly, and indirectly all Brazilians gain as more women acquire a degree of financial independence and more children become educated and stay healthier.

Its cost? Just 2.5 percent of total government expenditures.[81] Plus, in Brazil every one dollar in *Bolsa* benefits generates almost twice that amount in economic activity.[82] Similar cash transfer programs have spread to over forty countries.[83]

Systemically addressing hunger's roots, Brazil's Family Farming Procurement Program directly supports small family farmers by guaranteeing a market for what they grow. Government purchases replenish the country's food stocks or go to institutions that help to reduce hunger, such as schools offering free meals or the "people's restaurants" that we described in Myth 6.

Other economic policies have enabled jobs with benefits to grow three times faster than jobs without during the 2000s, and the buying power of the country's legal minimum wage nearly doubled.[84] Skeptics attribute such improvement to economic influences unrelated to Zero Hunger. In any case, inequality in Brazil—among the world's most extreme—declined somewhat during the first decade of this century, in contrast to the trend in much of the rest of the world.[85]

Dozens of other programs are helping to make real the "right to food" in Brazil's inner cities, as we highlighted in Myth 6. In the first six years of Zero Hunger, the number of Brazilians living in poverty—using the international standard of $1.25 a day—fell by 42 percent. And, from 2000 to 2012, Brazil cut its child death rate—a prime indicator of hunger—in half, a rate of improvement that might well be unprecedented.[86]

In agriculture, Brazil is directly supporting agroecology by offering farmers who supply school food a 30 percent price premium if they use ecological practices.[87]

Brazilians insist that none of this could have happened without one ingredient: wide participation. Forming community-based action groups from the 1970s through the '90s, civil society worked long and hard for the right to food. Then, in 1998, about a hundred social-benefit organizations, social movements, academic institutions, and religious and other groups came together in São Paulo to create a national forum on food security, preparing the way for Lula in 2003 to establish the National Food and Nutrition Security Council. Its initial seventeen ministers of state and forty-two civil society representatives directly advised Lula on implementing and monitoring *Fome Zero*.[88]

"Make like Lula—participate. This is a story we're creating together." These are words on a Lula campaign T-shirt that we spotted in Pôrto Alegre, Brazil, in 2003. To us, it seems that they turned out to be much more than a great political slogan. Brazil, with all its remaining challenges, is a story from which we all can learn.

Lest readers see Brazil as a solitary leader in the right to food, consider Bolivia.

After the 2005 election of labor leader Evo Morales—the country's first democratically elected president from the long-oppressed indigenous majority—a 2009 national referendum overwhelmingly approved a new constitution, with over 90 percent of the registered electorate voting. It enshrined the right to food.[89]

Since then, Bolivia's multiple food security policies have come together in a national development plan involving all relevant ministries working with civil society groups. It's called the Patriotic Agenda to fight hunger. Already, almost 90 percent of schoolchildren benefit from free school meals, and Bolivia's conditional cash transfer goes to millions of elderly people, poor families with children, and pregnant and nursing women. Like Brazil, Bolivia has cut by half the calorie-deficient percentage of its population since 1989.

Critical to these advances, among many more, have been "participatory processes," observes the FAO. The next challenge, it notes, is making sure the approach becomes anchored in local, collaborative governance.[90]

New Rules for the World's Biggest Food Safety Net—with Big Impacts

Many, many other public policies—or, more simply, "new rules"—are addressing hunger around the world as a result of power-generating movements.

In India, the women of the Deccan Development Society (DDS) you've just met rejected the empty-calorie, polished rice sold at subsidized prices in India's half a million government Fair Price Shops—the world's largest food-subsidy program.[91] It left them weak and vulnerable to disease, they said. Then, as they began growing a mix of nutritious crops, especially traditional millets, their health improved—for some millets contain roughly ten times more iron and calcium than white rice. (Vitally important, given that 50 to 90 percent of adolescent girls and half of all women in India suffer from anemia.)[92]

The women of DDS were therefore understandably troubled as they watched India's government continue to distribute polished white rice bought from big, chemical farms in other parts of the country.

Soon, though, the women took a stand, organizing the millet growers in their region into the Millet Network of India, or MINI. The group achieved a major victory in 2011, when for the first time "food security" legislation included "coarse cereals," such as millet, among the foods made available via India's Public Distribution System. This single rule change has vast implications for small-scale farmers' income and everyone's health.

In 2013, MINI members sent 140,000 postcards to Sonia Gandhi, chairperson of an advisory council to the president and responsible for drafting the 2013 Food Security Bill. In part because of such persistent efforts by MINI, the national Mid-Day Meal Program in schools and the Integrated Child Development Scheme now offer nutritious millet to millions of children.

These new farm leaders, long considered among India's most powerless people—including many once considered in the "Untouchable" caste—have also convinced India's Right to Food Campaign to support decentralizing the Public Distribution System, putting it into the hands of local communities.

Ultimately, farmers of MINI have persuaded about four thousand farmers on six thousand acres across four Indian states to reintroduce millet—a much healthier, traditional Indian grain—as part of their biodiverse farming.[93]

Democratizing the Rules of Work and the Economy

One reason that the concentration of economic power can feel so immovable is that many of us absorb the idea that the market mechanism as we've known it—bringing highest return to existing wealth holders—is all there is. So we think: Of course wealth concentrates . . . how could it not?

Fortunately, there's more than one kind of market economy, as we note in Myth 6. A market economy and capitalism—where owners of capital hold virtually all economic power—are not synonymous. A market economy can work beautifully without any capitalists at all. Examples include markets in which worker- or community-owned enterprises are primary, so profits accrue to those who do the work or use the service. In one of the most prosperous areas of Europe, for example, Italy's Emilia Romagna region, over a third of the economic output is generated by cooperatives, mostly small-scale.[94]

The Taste of Dignity—Cooperatively Produced, Fairly Traded

Cooperatives are one key within a wider movement for economic equity, countering an increasingly global market driving power into ever-fewer hands. Called the Fair Trade movement, it arose in its current form in the 1980s because of outrage at how little of the retail price of agricultural products is actually retained by the too-often-poor farmer. For coffee and bananas, it's less than 10 percent of the retail price; for tea, it's only 1 to 3 percent.[95]

So, allies of poor growers in the Global South—including religious organizations and other justice activists—have for decades been creating a new set of trading rules. Fair Trade grew from the belief that a sizable number of shoppers would *choose* to pay more if they knew it would bring fairer return to poor producers. The global movement is coordinated by Fairtrade International, and the U.S. certifier is Fair Trade USA.

To be permitted to carry the Fair Trade label, sellers must offer a fair and reliable price to the producers, as well as meet other labor and environmental-sustainability standards. The movement also encourages and supports farmers' participating in democratic cooperatives.

So far, Fair Trade involves 1.4 million small growers of coffee, cocoa, sugar, tea, bananas, honey, cotton, wine, fresh fruit, chocolate, flowers, and more in roughly sixty countries. In 2012, $90 million in premiums paid by consumers went to small farmer cooperatives—a jump of 52 percent in one year. To take just one example, nearly three thousand farmers in the Indian state of Kerala formed the Fair Trade Alliance Kerala to market their spices, coffee, and cashews, and in part have used the premiums for a school lunch program.[96]

> Fair Trade involves 1.4 million small growers of coffee, cocoa, sugar, tea, bananas, honey, cotton, wine, fresh fruit, chocolate, flowers, and more in roughly sixty countries.

In the Global North, the U.K. has taken the lead on fair trade, with 500 Fairtrade Towns, as well as more than a hundred universities and thousands of churches and schools. All have publicly committed to using Fair Trade goods.[97]

Currently, the Fair Trade movement benefits primarily farmers who produce for export. But that's a sizable number. Potentially, among coffee and cocoa smallholders alone, thirty million families could benefit.[98]

Another Reason to Love Chocolate . . .

Let's focus now on movement in the direction of greater equity and democratic voice for producers right on the farm—what say do they have? In cooperatives owned by the workers, the answer is "a lot."

Cooperatives of all kinds—owned by producers, consumers, service providers—are growing worldwide. In fact, more people today—one billion—are members of cooperatives than own shares in publicly traded companies.[99]

> Worldwide, more people today—one billion—are members of cooperatives than own shares in publicly traded companies.
> —estimated from data from the International Labor Organization, 2012

One particularly moving story of cooperative success caught our attention.

Most Ghana cocoa farmers don't earn enough to lift themselves out of poverty. Yet some are now co-owners of Divine Chocolate, the world's first Fair Trade chocolate company. Beyond fair pay, these farmers—as co-owners—receive a share of Divine's profits, and they have also gained a voice in Ghana's cocoa industry.[100]

This remarkable story began in the early 1990s, when Ghana's cocoa market went from public to private hands and some gutsy visionaries seized the moment for good ends. They organized farmers to set up a company to sell cocoa to the Cocoa Marketing Company, the state-owned, sole exporter of Ghana's cocoa.

Backed in part by the U.K. fair-trade company Twin Trading, Ghanaian farmers joined forces to create the cocoa cooperative Kuapa Kokoo—meaning "good cocoa growers." Its three goals for its members are: dignified livelihoods, increased female participation, and environmentally friendly farming practices.

The co-op sells its members' cocoa to the government export agent and takes care of weighing, bagging, and transporting it to market.

The co-op also handles members' legal paperwork and promises to be "transparent, accountable and democratic." Being accountable is a very big deal. In the past, farmers were frequently cheated by buyers' inaccurate weighing scales.

In 1998, Divine Chocolate Ltd. gave the world the first "fairness-flavored" chocolate bar. The co-op owns almost half of the company, and two co-op farmer representatives serve on Divine's board of directors. A share of profits from candy sales goes to the farmers. Ghanaian farmers' ownership stake in Divine Chocolate is "a first in the Fair Trade world," boasts the company. (We hope it will not be unique for long.)

Because of all these benefits, Kuapa Kokoo says, membership has climbed to 65,000 growers organized in about 1,400 "village societies." That's almost 10 percent of all cocoa farmers in Ghana.[101]

Beyond Kuapa Kokoo's success, broader cooperative efforts, combined with other economic changes, have helped Ghana cocoa growers garner a much bigger chunk of the cocoa export price—rising from below 20 percent in the 1970s to nearly 80 percent today.[102]

Beyond cocoa, Ghanaian farmers are growing a range of crops cooperatively. In just six years, from 2002 to 2008, agricultural cooperatives in Ghana increased 3.5-fold.[103]

Co-ops Grow Worldwide

Beyond Ghana, another stunning example of the co-op movement is India's Amul Dairy cooperative. Founded in the 1940s, it now includes seventeen thousand village-level co-ops with over three million milk-producer members. Amul has become India's largest milk producer and one of the largest dairy producers in the world. Its democratic processes, including village elections for cooperative officers, are "breaking down social and economic barriers." And the co-ops are also slowly dissolving caste barriers by the simple act of socially equalizing lines at milk collection centers.[104]

Worldwide, cooperatives provide a hundred million salaried jobs—that's 20 percent more jobs than multinational corporations provide.[105]

Banking on Trust

"Occupy Wall Street" in 2011 was only the most visible sign of citizen anger at the big banks that triggered what became known as the Great Recession. While many in the industrial world continue to feel powerless to reform the banking industry, millions of poor women in the Global South are discovering a simple process by which they can become their own bankers.

The microfinance movement hit the international marquee when Muhammad Yunus won the Nobel Peace Prize in 2006 for creating the Grameen Bank in Bangladesh. In this model, groups of villagers—shut out by banks—can get small loans without material collateral because participants share responsibility for each loan. Today two hundred million poor borrowers participate in microfinance worldwide.

The Grameen Bank was officially established in 1983. Then, in the late 1990s, came another breakthrough: The Women's Empowerment Program in Nepal, and a few others elsewhere, launched "savings-led" microfinance. Going into debt can add stress to already stressful lives; so "Savings Groups" start with . . . well . . . savings. The approach is as simple as its name. It is unlike the Grameen model, in which a loan is granted by an outsider; in Savings Groups the women themselves create the fund that becomes the source of their loans.

Members meet weekly—often about twenty sitting in a circle under a favorite tree—and put what they can, which might be twenty cents, into a communal pot. Then, at set intervals, the group decides together who will get a loan at an interest rate the group has agreed on. It might be a leg up for a member to stock her small business, or it might enable a family to get by when a loved one is sick with malaria.

Then, once a year, timed for when need is greatest—typically the "lean season" between harvests—the pot is divvied up according to how much each member saved, plus her share of the collected interest. Note: In Savings Groups, unlike most microfinance formats, every cent plus interest is returned to members. "Why pay them [the bankers] when we can pay ourselves?" one Nepali member told Jeffrey Ashe, coauthor of *In Their Own Hands: How Savings Groups Are Revolutionizing Development*.

Savings Groups will not lift people out of poverty, Ashe emphasizes. But they change poor people's sense of possibility. "I have seen that the groups grow in size . . . and . . . launch their own initiatives—training groups for their children, buying grain when the price is low to better survive the lean season, and launching collective enterprises," reports Ashe, founder of Oxfam America's Saving for Change Initiative.

After six years, during which a coalition of five development organizations gave support, 2014 saw membership in Savings Groups jump from one million to ten million. In social innovation, this movement may have set a speed record. For us, the takeoff confirms this book's core themes: that innovations meeting deep human needs for power, meaning, and connection spread fast; and that solutions lie in the experience of poor people themselves.

Ashe's oft-repeated lesson is simple: "They know how."

GOVERNANCE FOR LIVING DEMOCRACIES

Now we turn from economic life to political life—but not simply the electoral kind. We look at new forums for citizen voices in public choices and at examples of remaking the rules of electoral systems so as to keep private interests out.

Looking through the lens described in Myth 4 as the "systems view of life," we ask, What characterizes governance that enables all elements of a food system, and wider society, not only to end hunger but to nourish everyone well?

> **Living Democracy**
>
> *Three system-characteristics are proving to work in creating solutions:*
>
> • Wide sharing of power
>
> • Transparency in decision making
>
> • Mutual accountability in which participants share in holding themselves as well as others accountable for solutions

The complex lessons of this book yield what we feel are some fairly simple answers to that question. Three characteristics stand out: the wide sharing of power; transparency in decision making; and mutuality, meaning that all participants share in holding themselves and others accountable for solutions. From a systems view of life, we are all connected and thus we are all implicated, whether the crisis is hunger, poverty, or climate change.

A systems view of life also means spotting leverage points that, if shifted, can trigger a cascade. As this chapter opened, we noted that the biggest block to ending hunger may be despair itself. And what releases us from such despair? One release is a vision of systemic solutions—what we've called Living Democracy—so we can see that our actions aren't just "random acts of sanity" but can potentially alter root forces.

Below, in a brief tour of democratic governance—a voice for each of us—we start with new forums of face-to-face community decision making.

The Citizens' Jury—Farmers Gain a Voice

As a part of what some call "democratizing the governance of food systems," Citizens' Juries are arising in South Asia, West Asia, and the Andean region of Latin America.[106] They involve several dozen randomly selected citizens, with diverse experiences and views, coming together over several days to hear evidence on a critical issue, to dialogue, and then to arrive at an agreement on how to proceed.

In 2002, in India, a Citizens' Jury of small and marginal farmers—mostly women—upset a lot of powerful people. It directly challenged the Andhra Pradesh government's own master plan, backed by the U.K.'s aid agency and the World Bank, to modernize farming in the state via large-scale, industrial agriculture that would push small farmers and pastoralists to cities. The Citizens' Jury chose a different future: an agroecological pathway based on smallholder, biodiverse farming.

In the end, a female jury member traveled to the U.K. parliament to deliver the Jury's verdict to the minister of British aid and the MPs. It created "quite a storm," reports development specialist Michel Pimbert. The Citizens' Jury process turned out to be a "defining historical moment for the claiming of food sovereignty and participatory democracy in Andhra Pradesh," he told us.[107] Some of the positive outcomes of this defining moment we shared earlier in this chapter.

In West Africa, representatives from fifteen organizations—including farmers' groups as well as government officials, academics, and more—also designed a Citizens' Jury process. A farmer-led assessment of public agricultural research arrived at recommendations that fed two Citizens' Juries in early 2010. In each, roughly forty jurors from four West African countries weighed diverse evidence and views and came to clear recommendations.

One was: Include us! Farmers want a "central role" in choosing the focus of public research. Another: Avoid genetically modified and hybrid seeds (which can't be effectively saved for replanting). Focus instead on improving local varieties, spreading ecological practices, and promoting both the use and the exchange of local seeds.

The Juries' findings—favoring exactly the approach to agriculture that's best for addressing climate change—got a lot of media attention across West Africa.[108]

Citizens Participate in Power: Kerala, India

In 1996 the Indian state of Kerala—featured in Myth 1 for its success in achieving levels of literacy and health comparable to those in industrial countries—launched a participatory planning effort. In what was called the People's Campaign for Decentralized Planning, citizens gained unprecedented authority, as 40 percent of the state's budget was transferred from traditionally powerful state-level departments to around nine hundred village planning councils.[109] Citizens were at the table in the planning process through village assemblies; and citizen committees helped to design projects with a say in how development funds were spent.

"The decentralization of a wide range of developmental responsibilities" to more than 1,200 government bodies down to the village level could "transform dramatically the everyday practice of democracy for Kerala's thirty-one million inhabitants," write professors Archon Fung and Erik Olin Wright in *Deepening Democracy*.[110]

One in four households in Kerala participated in village assemblies in the first two years of the campaign. Hundreds of thousands of citizens were trained in planning and budgeting. And, thanks to special efforts to engage them, women made up 40 percent of those attending village assemblies—a rate unique in all of India.

And the results? Hundreds of outcomes that included advancements in housing for the poor, small-scale irrigation, local roads and other infrastructure, health and education services, and projects especially beneficial to women and those who under the Indian caste system were once known as "Untouchables."[111]

People's Voice in the Public Purse

Launched in Pôrto Alegre, Brazil, in the 1990s, Participatory Budgeting gathers citizens in a series of open assemblies each year to determine the use of a significant portion of a city's annual budget. After three decades, the approach has spread to 1,200 cities, from South Africa to the suburbs of Paris.

In many cases, the impact has been dramatic, shifting public spending toward poorer communities that under the prior, more centralized and secretive system had less clout. In Belo Horizonte,

Brazil, for example, homeless people gained a new housing complex. Participatory Budgeting also led to a "health center, schools, paved roads, public squares and parks inside and in the vicinity of the housing complex."[112]

In Myth 6, you also read of Belo Horizonte, known as a leader in realizing the right to food. There, over its first fifteen years, Participatory Budgeting led to a thousand public works projects.[113]

And as citizens see more positive impacts of their tax dollars at work, confidence in government grows, making surprises possible: Participatory Budgeting in some cases is associated with improvement in tax collection, and voters have even approved property tax rate increases.[114]

New Food Forums in the Global North

Starting in the 1980s, citizens in the United States and Canada began creating new food forums—public bodies bringing diverse voices together to examine local food needs and resources and then develop agendas for systemic change. They're called Food Policy Councils, and from the beginning one primary goal has been to help more people access healthy food.[115] They do not have governing power—but their influence is felt in government.

Now, roughly two hundred cities, states, and towns in North America boast Food Policy Councils, bringing together stakeholders from chefs to farmers to citizens to policymakers.[116] In 2012, more than 3,500 people in forty-five communities in North America participated in the councils.[117]

Illustrative results?

The Council of Muscogee (Creek) Nation in Oklahoma achieved new procurement policies so that tribal groups can access locally grown fruits and vegetables. In Hartford, Connecticut, the council helped secure nutrition assistance benefits for four thousand poor women and children.[118]

With a view to guiding policies, some U.S. states are also developing "Food Charters." Created with public input, they present a region's unique vision of a just and sustainable food system and guide food-related policy decisions. Minnesota's Food Charter—the fifth nationwide—emerged from the input of over 2,500 people and 144 events across the state.[119]

A Voice at the Global Table

All these developments notwithstanding, here's a big question: Since citizens are struggling to be heard even in the halls of their own national assemblies, how could they possibly gain influence in global governing forums, where corporate interests speak so loudly?

The answer is fascinating. Food-concerned citizen movements have indeed begun to make their voices heard on the global stage.

And their first focus? On the Rome-based FAO, for an obvious reason: Its mandate is "achieving food security for all." Referring to small-scale and often poor farmers in 2014, Director-General José Graziano da Silva—formerly the head of Brazil's Zero Hunger—affirmed that "meaningful engagement . . . of those whose voices have been all too often marginalized cannot be stressed enough."[120] Unfortunately, this view was not at all evident two decades earlier when rural people began clamoring to be heard.

In her 2015 book, *Food Security Governance: Empowering Communities, Regulating Corporations*, Nora McKeon offers an

> Farmers from communities with direct experience of hunger were finally speaking for themselves in a global forum whose explicit purpose is ending hunger.

insider's tour of the journey. The shift to include voices of the poor people themselves, she notes, began during FAO'S World Food Summits of 1996 and 2002.

There, for the first time, speaking at the closing session on behalf of civil society was an Indian peasant, rather than a spokesperson from the charities and advocacy groups that tended to "dominate the scene" in similar forums.[121] In other words, farmers from communities with direct experience of hunger were finally speaking for themselves in a global forum whose explicit purpose is ending hunger.

For the two of us, this reality feels like a different world from what we experienced in 1979 on the occasion of an FAO conference addressing agrarian reform. We had helped organize an alternative gathering in a nearby converted monastery where peasants, ignored by FAO officialdom, were invited to share their stories. Our events were totally separate from the UN-sanctioned deliberations, though virtually across the street.

By 2003, however, those previously voiceless had built their organizational strength from the local level up and formed the International

Planning Committee for Food Sovereignty (IPC), which describes itself as "a self-organized global platform of small-scale food producers, rural workers' associations, grassroots/community-based organizations and social movements."

Its purpose? To "facilitate dialogue and debate among actors from civil society, governments and others actors working on the food security and nutrition agenda at the global and regional level."[122] Today, the IPC involves more than eight hundred organizations whose members include some three hundred million small-scale food producers.

The FAO's commitment to engage with civil society was evident in the official letter it cosigned in 2003 with this new channel for the voice of rural people. In it, the FAO agreed to work together on the four priorities IPC members had identified: the right to food, agroecological approaches to food production, local access to and control of natural resources, and agricultural trade and food sovereignty.

And so they did. The guidelines for applying the right to food at a national level—adopted by all the member governments of the FAO in 2004—are just one striking example of the impact that civil society has had.

The story picks up following the 2008 shock to the global food system that we described in Myth 6. It painfully revealed humanity's vulnerability, as hunger, fear, and anger led to political uprisings in many countries. Among social movements, it sparked insistent calls for a real voice in a forum that had been created in 1974. It's called—get ready for another acronym!—the Committee on World Food Security (CFS), and it's part of the UN's food policy system noted above.

Created on the heels of an earlier famine crisis, the CFS started life as a body whose membership was strictly limited to governments; and over time it had become something of an "ineffectual talk shop," notes McKeon. But in response to the 2008 crisis, civil society groups allied with some governments eager for change began an "audacious effort to transform" the CFS into "an authoritative global policy forum."

The reformed CFS's leadership took the "unusual step of opening up to all concerned stakeholders," and for the first time small-scale farmers sat around a table with government delegates to discuss the shape of reform. Today the CFS enjoys broad participation, involving: representatives of UN member states and its agencies and bodies; international agricultural research centers; the World Bank; the World Trade Organization; associations of corporations and philanthropies;

and, finally, civil society, rural people, and nongovernmental organizations and their networks.

All these voices at one "table"?

Yes. And the catchphrase for this mélange: a "multistakeholder model."

If the CFS has a fighting chance of fulfilling its tough mission, it is due in good part to the fact that those groups whose right to food is most consistently violated can speak on the same footing as the governments that ought to be defending them but often don't, says McKeon. It's an exception to the norm for representatives of "people's" organizations—those, for example, of small-scale farmers—to have such recognition, she observes.[123]

This shift helps to explain how it was possible in 2012 for the reformed CFS to arrive at the first guidelines ever negotiated globally that seek to ensure local people's access to the land and other resources they need to produce food today and provide a future for their children tomorrow. These guidelines, she notes, are now being used by communities around the world to support their struggles to stay on their land.[124]

Yet strong concern continues about the reformed CFS.

Many civil society participants argue that it is not appropriate for associations pursuing private, corporate interests to have standing alongside government representatives mandated to serve the common interest. Some claim the "Principles for Responsible Investment in Agriculture and Food Systems," hammered out within the CFS over two years, place corporate profit ahead of the human right to food.[125]

And does the CFS have power?

In one sense, the answer, relative to any governing body, is fairly easy to discern. One just asks: Do its decisions yield results? And how do we know?

So it is worth noting that the CFS is now emphasizing "stocktaking"—officially and transparently "taking stock" of the consequences of its actions. Its chairs prepared a ten-year stocktaking on CFS's Voluntary Guidelines "to support the progressive realization of the right to adequate food in the context of national food security." The meetings have been webcast and are thus transparent.

"The reform of the CFS was a victory for political decision-making," McKeon says, "informed by effective evidence-giving by those most affected and allied to accountability." But lest anyone become complacent, she adds that "[m]aking this happen in practice is a constant battle and a work-in-progress."

McKeon appreciates that all this could be seen as a "timid beginning" for what is a "paradigm-changing" approach to furthering accountable global food governance. But, she stresses, it *is* nonetheless a real beginning.[126]

Elected Governments Answering to People, Not to Money

From this brief tour of new forums that citizens are creating—from villages in West Africa and India to deliberative bodies within the United Nations—we see the emergence of democracy understood as a living practice far beyond elections. In closing, however, we turn to the leverage point that many see as the "mother of all issues": the power of concentrated wealth over public decision making within formally elected governments.

In the United States, our political rules have allowed elections to reach the point that in 2014 just 0.04 percent of Americans supplied two-thirds of all federal candidate, party, and PAC money.[127] It was hard to imagine a more skewed system . . . until the following year. In 2015, just three people—the Koch brothers, whose riches derive from fossil fuel—announced their intent to pour almost $1 billion into the next presidential election, a greater sum than was spent on President Obama's entire reelection campaign.[128]

In what has been termed "legal corruption," campaigns increasingly rely on a handful of big spenders to whom those elected feel beholden.[129] So we end up with what might aptly be called "privately held government."

> Almost 90 percent of Americans believe corporations have too much power in Washington.

In response, "Money out of politics" is becoming a rallying cry across the political spectrum. Almost 90 percent of Americans believe corporations have too much power in Washington.[130] And that power was mightily reinforced by the 2011 Supreme Court decision commonly referred to as "Citizens United," the name of the conservative lobbying organization that brought suit. The decision threw wide open the floodgates of corporate and private political contributions. Eighty percent of Americans believe the decision was wrong.[131]

Since then, considerable citizen energy has been pushing forward a constitutional amendment to clarify that corporations are not persons, money does not equal speech, and limits on campaign spending do not, therefore, infringe on the right to free speech.[132] Unfortunately, a constitutional amendment is highly unlikely in the foreseeable future. All it takes is a majority in thirteen state legislatures to kill it.

But citizens' hands are not tied! The challenge is to build a compelling, transpartisan Living Democracy movement, one energized by what we can do *right now*.

A broad-based, money-out-of-politics movement is beginning. This "anticorruption" campaign is "fiercely transpartisan," according to its lead organizer, Josh Silver. Its strategy is a series of victories in municipalities, then in state legislatures, then in Congress. (Note that "corruption" here focuses largely on the legal variety: the power of private wealth over elected officials and those running for office.)

Its first success was in Tallahassee, Florida, in 2014. With two-thirds of the vote, citizens supported a measure backed by both Common Cause and the Tea Party and amended the city charter. As a result, contributions in races for the city commission are now capped at $250, with refunds going to citizens who donate. A new Ethics Board is empowered to ensure that public officials are not "inappropriately influenced" by those seeking benefits from the city.[133]

Moving forward, the legislation sought by the anticorruption movement would end all secret political donations, shut the "revolving door" now allowing an elected official to leave office and immediately begin lobbying (a powerful incentive for the legislator to be kind to a potential employer), and a provision making possible small-donor, citizen-funded campaigns. Given the fast-changing landscape, we suggest the website *Represent.us*, a great gateway to join in this movement and to put despair behind us.

To bolster our sense of the possible, let's note what citizens in Europe have accomplished.

Since the mid-1960s, many Western European countries have been shaping new rules, shifting political funding from private to public financing because big private campaign money is "perceived as having a pernicious influence" on democracy. So reports the Institute for Democracy and Electoral Assistance in Stockholm.

Better rules, formal and informal, are also increasing transparency so that voters know who's backing which candidates. The Dutch Social Democratic Party, for example, posts all its accounts online. Many European countries ban anonymous donations altogether. To increase voters' access to all views, the vast majority of countries offer free or subsidized media access to political parties as well. In some countries, including Iceland, small campaign contributors get a tax break to "incentivize grassroots donations," notes the Stockholm institute.[134]

And do such rules make any difference when it comes to addressing hunger?

Apparently so.

Researchers asked more than 1,400 experts on electoral systems to rank 107 countries with elected governments by the strength of their "electoral integrity," using more than a dozen measures, such as fairness and transparency in campaign financing and access to media. In 2014, among the top five were four industrial countries—Norway, Sweden, Germany, and the Netherlands. The United States, however, was way down below Mexico in thirty-fifth place.[135] (The fifth country in the top group is Costa Rica, which we chose not to use in our comparison below only because its situation doesn't seem comparable to that of much wealthier industrial countries.)

To answer the question about how electoral integrity relates to hunger, we then picked a recognized indicator of a nation's nutritional well-being—the infant death rate—and compared the average rate of the four industrial countries ranked highest on electoral integrity with the United States'. Our finding? In the United States infants die at a rate more than twice that of these democratic exemplars.[136] Fully half of American infants, we must also note, depend on federally subsidized food supplements.[137]

We return to this core theme of the triumph of citizenship over cynicism in the closing reflections to follow.

THE PATH AHEAD

In this quick scan of movements democratizing economic and political life, we hope that we've shaken the myth of immovable power. We hope we've encouraged openness to the possibility of transformational change—because, as we noted in opening this chapter, within a systems view of life characterized by continuous change, it is not possible to know what's possible.

And the advantage of such humility?

We remain open. And we can also appreciate something long observed about our species, which has survived plagues and world wars: It is not a big problem that defeats the human spirit—it's feeling futile. Humans thrive when we feel connected to others, when we have meaning and purpose in our lives, and when we feel that we can *do* something—that we can make a difference.

If this is true about us, what a perfect time it is to be alive. Because those deepest of needs are met by joining together to transform what's proved so clearly *not* to work into what we now know does work.

Humanity is for the first time facing a challenge to life not only for our species but also for many others with whom we share this planet. Unprecedented extremes of both social inequality and climate change put us in a new place. And what *else* is new? For the first time, communications link even the remotest villages so that many more of us can actually see what is happening globally.

As we noted earlier, such a "moment of dissonance"—when old ways of seeing no longer serve us—can be terrifying. But it can also be liberating. A fast-spreading, shared perception that a faulty foundation is cracking could—just maybe—break the spell of our disempowerment.

And this thought brings us back to the theme of power with which we began. We humans model ourselves on one another. How many times in our lives do we think: Hmm, if he or she could do that, maybe, just maybe, I could, too—whether it's taking our first crack at Rollerblading or speaking out in a group that intimidates us. We see others doing it, so why not try?

Thus, key to our power is to grasp—truly internalize—the breadth and depth of breakthroughs that people like us are making all over the world. Among them are the people you've met in this chapter, and throughout our book, who'd never have imagined themselves as leaders blazing new trails—until they did.

Ultimately, ending hunger and addressing the climate crisis—which are inseparable—can happen only as citizens everywhere let go of the defeatist notion that power is a static grip we can't break.

So what's next for each of us who wants to end hunger once and for all time?

To answer, we now invite you to our concluding reflections on the lessons we two have learned during decades of questioning, as well as on the outer and inner work—the good work—ahead for each of us. We also encourage you to take advantage of the connection-for-action resources that follow these thoughts.

Beyond the Myths of Hunger:
The Takeaways

Our motivation for writing this book has grown from the realization that approaches to world hunger often elicit guilt (that we have so much), or fear (that they will take it from us), or despair (that there's nothing we can do to make a real difference). Other approaches imply impossible trade-offs: Do we protect the environment *or* grow needed food? Do we seek a just *or* an efficient food system? Do we focus attention on those suffering here at home *or* concern ourselves with those abroad facing even greater hardship?

But we find neither guilt, fear, nor impossible trade-offs very motivating.

Our search for the roots of hunger ultimately led us to a number of positive principles that neither place our deeply held values in conflict nor pit the interests of the well-fed against those who are nutritionally deprived. We offer the following principles as working hypotheses, to be tested through experience:

- Since hunger results from human choices, not inexorable natural forces, the goal of ending hunger is no more utopian than that of abolishing slavery was, not all that long ago.

273

- While slowing population growth in itself cannot end hunger, the very changes necessary to end hunger—the democratization of economic life, especially the empowerment of women—are key to reducing birthrates so that the human population can come into balance with the rest of the natural world.
- Ending hunger does not require damaging our environment. On the contrary, it demands that we protect it by using ecologically sustainable methods that are, fortunately, within the reach of even poor farmers.
- Climate change affects us all and requires big changes, many of which can be positive. If we correct the deep inequities and astounding inefficiencies in our food system, no one need go hungry as we face the climate challenge.
- Greater fairness does not undercut food production. The only path to increased production that can end hunger is to devise food systems in which those who do the work have a greater say and reap a greater reward.
- We need not fear the advance of the poor in the Global South. Their increased well-being is essential to our own.

These and other liberating principles point to possibilities for narrowing the unfortunate rifts we sometimes observe among those concerned about poverty, the environment, climate change, population growth, and world hunger. More than that, they offer practical pathways for those eating well to align with those suffering nutritional deprivation.

In structuring our book around ten beliefs that we see as obstacles to ending hunger, we underscore how strongly we believe that *ideas have power*. If our core assumptions are faulty, no amount of earnest effort can achieve much. Thus, in closing we now turn to our culture's perhaps most formative assumption, that about the meaning of freedom itself. From there, we move to the challenge of taking up the implications of a new understanding of freedom for ourselves as citizens and in our own personal empowerment.

WHAT *IS* FREEDOM?

Certainly, in the United States, "freedom" lies at the core of our national identity. But what does it mean? The two of us believe that only as

Americans grapple together with the meaning of this foundational value can our nation end hunger here—and play a positive role worldwide.

For us, freedom means a whole lot more than the end of hunger and other forms of deprivation denying our humanity. It suggests fulfillment. It involves having choice, in the richest sense—including the capacity to choose how we develop our unique gifts and passions. Political philosopher Harry Boyte captures well this meaning in his definition of freedom as the "liberation of talents."[1]

When we define freedom this way, it becomes clear that societies making real the right to eat, aligned with a thriving Earth, are on the path to freedom. On this path we learn how to create rules together, to hold ourselves and others accountable, and to restore our Earth.

On the other hand, many in American culture hold the view that economic freedom means unlimited material accumulation. By that definition, of course, freedom for some will continue at the cost of hunger for many.

Americans who believe that the right to unlimited accumulation is the guarantor of liberty fail to take in the insight of Yale University economics philosopher Charles Lindblom, who in his classic *Politics and Markets* coolly reminds us that "income-producing property is the bulwark of liberty only for those who have it."[2] In the Global South just a tiny minority have such a bulwark to their liberty. And, as we saw in Myth 9, most Americans don't have it either. About half of Americans have zero or near-zero net wealth, and possess no income-producing assets.[3]

> ". . . a power over a man's subsistence amounts to a power over his will."
> —Alexander Hamilton, *The Federalist Papers*, No. 79, 1788

Fortunately, this understanding of economic freedom was not the vision of many whose philosophy and values shaped the birth of our nation. Some perceived that a link between property and freedom could be positive only when ownership of socially productive property is widely dispersed. In 1785, after a conversation on a country road with a desperately poor single mother trying to support two children with no land of her own, Thomas Jefferson wrote to James Madison, "Legislators cannot invent too many devices for subdividing property." The misery of Europe, he concluded, was caused by the enormous inequality in landholding.[4]

Many of our nation's founders also understood that being able to think and speak for oneself without fear is foundational to the culture of a republic. So "independence" carried the meaning of being free from dependency on others for survival—the kind of dependency characteristic of the aristocracies our founders had rejected. Independence in this sense was yet another key reason for our founders' passion for ensuring a wide dispersion of wealth.[5]

Americans today might easily think such views vanished with the powdered wig. But no. In 2014, two professors surprised a lot of people with research revealing that Americans' ideal level of economic inequality would be a society in which an average CEO's pay is no more than seven times that of an unskilled worker. They estimated the actual U.S. gap to be 30 to 1, but they were *way* off. In fact, in America's real divide, CEOs receive on average more than 350 times what the unskilled worker receives. Among sixteen industrial countries, this huge U.S. pay gap is more than twice that of the two countries with the next highest inequality, Switzerland and Germany.[6]

In America's desire for more equal reward is more evidence that a sense of, and strong desire for, fairness lies deep in the human psyche, to be tapped and nurtured as we move toward a world free of hunger.

> In a 2014 survey, Americans on average said their ideal level of wage inequality would be a seven-fold difference between the CEO's pay and the unskilled worker's, and estimated the actual gap to be thirty-fold. In reality, U.S. CEOs make 354 times more than average employees.
> —drawn from Kiatpongsan and Norton, *Perspectives on Psychological Science*, 2014

Security: Not a Threat to Freedom but the Ground of Freedom

Not only did many of our forebears view a wide dispersion of wealth as the basis of freedom; they also grasped that security is foundational to freedom, not a threat to it.

Many political theorists of Jefferson's day believed that if people were economically dependent—that is, had no economic security, not even ensured access to food—they could not think and act independently, and this would make true democracy impossible.

Economic insecurity constrains our freedom of action both as citizens and in our private lives. Franklin Roosevelt summed up this insight when he declared, "Necessitous men are not free men."[7] More recently, University of Maryland philosopher Henry Shue has offered a helpful exploration of precisely why the right to that which is essential to life itself—particularly the right to an adequate diet and health care—is basic to freedom:

> No one can fully, if at all, enjoy any right that is supposedly protected by society if he or she lacks the essentials for a reasonably healthy and active life. Deficiencies in the means of subsistence can be just as fatal, incapacitating, or painful as violations of physical security. The resulting damage or death can at least as decisively prevent the enjoyment of any right as can the effects of security violations.[8]

Shue argues convincingly that the right to the essentials of survival may be even more basic than the right to be protected from assault: It is much easier to fight back against an assailant than to challenge the very structure of the social order that denies one access to food. This understanding of freedom—rooted in security to allow one's full functioning as a human being—builds, we believe, upon humanity's diverse religious and cultural heritage.

Freedom so understood is not finite. My artistic development need not detract from yours. Your intellectual advances need not reduce my ability to develop my own intellectual powers. And, most pertinently for this chapter, assurances of my protection from physical assault, including my right to nutritious food, need not prevent you from enjoying equal protection. This is true because, as we have shown throughout our book, sufficient resources exist to guarantee the fulfillment of food rights for everyone.

Not only is freedom so defined not a zero-sum situation and not only does your freedom to develop your unique gifts not have to limit my expression, but my development in part *depends on* your freedom. The failure of our society to protect the right to basic security means that all of its members are deprived of the intellectual breakthroughs, artistic gifts, and athletic achievements of those whose development has been blocked by poverty and hunger. When we are denied the potential inspiration, knowledge, example, and leadership of those who are directly deprived, all of us experience a diminution of our freedom to realize our own fullest potential.

We hope this book will contribute to a renewed appreciation of the positive link between economic security and freedom, so that people who love liberty will want to expand it by safeguarding the right of every citizen to the resources necessary to live in dignity.

Freedom as Participation in Power

Another vein in many Americans' understanding of freedom is that it means being free from interference, especially from government. The gist is the absence of something negative.

> A concept of freedom strong enough to end hunger must be active. It must include the freedom to have a say—to participate in power itself, creative power, the power to determine our common destiny.

But a concept of freedom strong enough to end hunger must also include a positive, active dimension, people featured in this book demonstrate. It must include the freedom to have a say—to participate in power itself, in creative power, the power to determine our common destiny. And this freedom, as we have noted, depends on the boldness of citizens stepping up to remove the grip of money over public decision making so that government becomes accountable to them.

Freedom Links Ownership and Responsibility

Ending hunger also requires a fundamental rethinking of the meaning of ownership, certainly when applied to the productive resources on which humanity's diet depends. In this rethinking, we believe Americans would be well served by once again going back to our roots, to the concept of property-*cum*-responsibility held by the original claimants to these soils—the American Indian nations—and by many of our nation's founders.

"Man did not make the earth," declared Thomas Paine, whose writings in the months before our Declaration of Independence are credited with stirring the new colonies to claim their freedom. Then, he added: "[and Man] had no right to locate *as his property* in perpetuity any part of it."[9] The meaning we take from Paine is that humans hold land in trust. Because the community endures beyond the lifetime of any one individual, the Native American concept of community tenure carried within it an obligation to future generations as well.[10]

Ownership of productive resources, rather than being an absolute to be placed above other values, can serve us if it becomes a cluster of rights and responsibilities supporting our deepest values. Such a notion of freedom is neither the rigid capitalist concept of unlimited private ownership nor the rigid statist concept of state ownership.

We then ask ourselves, What would be required to achieve a market responding to human needs rather than narrowly to the demands of wealth? For us, part of the answer lies in rethinking property rights—just as in Myth 6 we recast the market mechanism as a device to serve higher values, not as an end in itself.

RECLAIM DEMOCRACY AS A LIVING PRACTICE

As we've underscored throughout our book, all of the above requires that citizens step up to end the power of concentrated private wealth—corporate and individual—over our political systems.

Our nation's founders perceived full well the seriousness of the threat to liberty. For "early generations" of Americans, writes legal scholar Zephyr Teachout, corruption meant "excessive private interests influencing the exercise of public power." Fears of "conspiracy against liberty . . . nourished by corruption" were at the "heart of the Revolutionary movement," notes Harvard historian Bernard Bailyn.[11]

But here we are today. And clearly the "best democracy money can buy" isn't good enough, nor does it have to be. In much of Western Europe, as we've noted, citizens have built significant—though, of course, imperfect—barriers to corporate domination of the electoral process. Restrictions on paid political ads in Germany, for example, enable voters to learn about issues mainly from media interviews and discussions with politicians, notes an overview in the Law Library of Congress.[12]

At this writing, as we note in Myth 10, creative steps to tackle money in politics are taking shape in the United States that do not require the lengthy and perilous constitutional amendment process. We can join in this essential campaign for the heart of democracy.

In it, we can speak up to deflate the myth that limiting campaign spending would violate the First Amendment because spending equals speech. Spending is not speech. Unlimited spending doesn't level the playing field—essential to democracy—but rather, tilts it wildly.

Equal vote and voice are the essence of democracy. By "voice" we mean the right to be heard and to hear diverse voices in order to hone one's own views. It is of little meaning to have a right to speak but no right to be heard. By way of analogy: In a crowded hall of bellowing voices—including some with electronic megaphones—those speaking normally can never be heard, even when they are the majority. Unfortunately, that's the reality of political discourse in America today. The megaphones of great wealth are drowning out the vast majorities.

Living Democracy, however, means ensuring that one's voice is not overwhelmed by the resources of special interests to whom those elected are then beholden.

ENDING HUNGER: THE MOST PERSONAL QUESTIONS

To believe that we can end hunger is to believe in the possibility of Living Democracy, and thus transformational change in which people now most disadvantaged are themselves participant leaders. But how is it possible to believe that hungry people, who have so much working against them, can challenge their exclusion and construct better lives?

Maybe there's only one way: Maybe it's possible to believe in the capacity of others to change only as we experience *ourselves changing*. With this realization, the crisis of world hunger becomes the personal question, How can I change myself so that I can contribute to ending hunger?

In part, the answer lies in dozens of often mundane choices we make *every day.* Only as we make our choices conscious do we become less and less victims of the world handed to us, and more and more its creators. As we align our life choices with the vision of the world we are working toward, we become more powerful and more convincing, to ourselves and others.

HOW DO WE BEGIN?

A first step is learning from independent sources. As we hope to have demonstrated, as long as we get news of our world only from corporation-sponsored television and Internet sources, our vision will remain clouded by myths. That's why Connection & Action Opportunities at the end of our book includes useful sources that continually challenge prevailing dogma.

Then we put that new learning to work. Whether we realize it or not, each and every one of us is an educator—we teach friends, coworkers, and family. What do we teach? With greater confidence born of greater knowledge, we can speak up effectively when others repeat myths that hurt us all.

Organized public actions count too. You can be sure someone will be watching. (Dr. Benjamin Spock once remarked that it was seeing Women Strike for Peace protesters standing in the rain with placards in front of the Kennedy White House that led him to a leading role in halting nuclear testing.) Letters to the editor, comments on blogs, letters and calls to our representatives, letters and petitions to corporate decision makers—all count, too. You can be sure someone will read your words, even if the impact can never be fully known.

And we can take care to share "solutions stories" that counter despair, a primary obstacle to deep change today. In today's world, sharing a solutions story is a revolutionary act. But given the preponderance of negative news, we deliberately have to seek out such stories. We hope our book is a start, along with the resources that follow.

Another critically important step, determining whether we will be part of the solution to world hunger, is our choice of work; and, if we have little choice, how we find other ways to confront, rather than accept, a status quo in which hunger and poverty are inevitable.

To have greater choice of career path and to contemplate involvement in democratizing societal change, it helps to get clear on what level of material goods we require for happiness. Studies show that in the industrial countries, above a fairly modest income more wealth typically brings no added happiness.[13] Millions of Americans are discovering the emptiness of our society's pervasive myth that material possessions are the key to satisfying lives. They are learning that the less they feel they need, the more freedom of choice they have about where to work, where to live, and what learning experiences are possible.

Moreover, in every community in America, people go hungry and lack shelter. Through our faith-based organizations, community groups, trade unions, and local governments, we can both help address immediate needs to uproot this needless suffering and participate in generating a new understanding of democracy not simply as a vote one casts every few years but as active participation in initiatives for more and better jobs, affordable housing, safe neighborhoods, and environmental protection.

Where and how we spend our money—or don't spend it—is also a vote for the kind of world we want. We can boycott companies that are unfair to their workers and companies that destroy precious rain forests, for example. And as part of the Fair Trade movement we can "buycott" (visit buycott.com)—and selectively choose, for example, markets and food stores that offer less processed and less wastefully packaged foods, and stores where workers have organized unions or that are managed by the workers themselves instead of conglomerate-controlled supermarkets. Savings are another "power opportunity" to shape the market, if we put ours in community banks and invest in life-supporting companies—for example, in green energy or healthy food companies.

In all, though, little is possible by oneself. The people whose work we share in our book embody for us this core lesson: We need others to inspire us, to push us, to console us, and to hold us accountable to commitments we make. Social movements brought women the vote and breakthrough civil rights legislation. They are essential to ending hunger.

But because continuing hunger touches all aspects of economic and political life, what's needed now is a "movement of movements": an energizing cross-cultural, pro-democracy campaign providing a canopy of hope over the many single issues. Because it's ever more obvious to more and more people that we can't address any of our planet's biggest crises without democratic governance, we believe such an unprecedented campaign is now possible. From the stories of people in action in this book, we see the global Zero Hunger movement as a vital element in what can become a strong and inclusive Living Democracy movement.

In our Connection & Action Opportunities at the end of the book are some of the key organizations that offer living proof of the transformation of power necessary to end hunger. So we each can reach out, connect, and believe in possibility.

THE ESSENTIAL INGREDIENT

As we close, let us share a final lesson of our exploration to understand what it takes to end hunger. Whether we seize these possibilities depends in large measure on a single ingredient.

You might expect us to suggest that the needed ingredient is compassion for the millions who go hungry today, for compassion is indeed a profoundly motivating emotion. That comes, however, relatively easily. Our ability to put ourselves in the shoes of others makes us human. Some even say that it's in our genes and that failing to express our innate compassion puts us in emotional and physical peril.

But there is another ingredient that's harder to come by. It is courage.

At a time when the dominant ideas are clearly failing, many cling even more tenaciously to them. So it takes courage to cry out, "The emperor wears no clothes! The world is awash in food, and all of this suffering is the result of human decisions!"

To be part of the answer to world hunger means being willing to take risks, risks many of us find more frightening than physical danger. We have to risk being embarrassed or dismissed by friends or teachers as we speak out against deeply ingrained but false understandings of the world. It takes courage to ask people to think critically about ideas so taken for granted as to be like the air they breathe.

And there is another risk—the risk of being wrong. For part of letting go of old frameworks means grappling with new ideas and new approaches. Rather than fearing mistakes, courage requires that we continually test new concepts as we seek to learn—ever willing to admit error, correct our course, and move forward.

From where does such courage come?

Surely from the same root as our compassion, from learning to trust that which our society so often discounts—our innate moral sensibilities, our deepest emotional intuitions about our connectedness to the well-being of others. Only with this new confidence will we stop twisting our values so that economic dogma might remain intact while millions of our fellow human beings remain needlessly deprived amid astounding abundance.

Only on this firm ground will we have the courage to challenge all dogma, demanding that the value of life be paramount.

Right-to-Food Fundamentals

1. Hunger is an outrage precisely because it is needless.

2. Population growth is not the cause of hunger but a symptom of the inequities at the root of *both* population growth and hunger. Human population can come into balance with the natural world as we remake structures of economic and political power so that everyone has a voice and every voice is heard.

3. Improvements in how we farm and what we eat that best address the climate challenge are fortunately those that also most benefit the world's hungry and the health of us all.

4. Industrial agriculture—concentrating power in the hands of ever fewer corporations, as well as abusing the environment—has proved unable to end hunger; by contrast, agroecological farming is enhancing natural systems and empowering people to end hunger.

5. Greater fairness in farming not only releases untapped productive potential but is essential both to making sustainability possible and to ending hunger.

6. To end hunger, a "free market" must come to mean one in which all people are free to participate fairly to meet their needs.

7. Trade—including the export of agricultural products—is in itself not the enemy of hungry people. But for trade to contribute to ending hunger, its rules must promote the wide dispersion of power, transparency in decision making, and mutual gain.

8. True foreign aid helps to remove the obstacles blocking people's efforts toward self-determination, rather than furthering the concentration of power that fosters corruption and dependency on aid.

9. Freeing ourselves from the illusion that our problems are fundamentally different from those in the Global South, we see that concentrated power harms people *everywhere* and that replacing the power of concentrated wealth with governance accountable to citizens is necessary to end hunger and for all to thrive.

10. Solutions to hunger are emerging as courageous people the world over are demonstrating that Living Democracy—from the village to the global level—can address its very roots.

Connection & Action Opportunities: What Can We Do?

10 THINGS WE CAN DO RIGHT NOW TO CREATE A WORLD FREE OF HUNGER

1. Seize every opportunity to let people know that hunger is absolutely needless! Scarcity of democracy—not of food—creates hunger.

2. Connect with others who give you energy and focus. Throw yourself into an organization that challenges the root causes of hunger in your community, in your country, or around the world. Focus on the piece of the hunger puzzle that most grabbed you in this book—such as shifting public support toward agroecology, reforming trade rules and food aid, or shaping a sane climate policy.

3. Democratize your dollar. Join a cooperative. (Or start one!) Give workers a voice by choosing to support unions as well as businesses that offer fair wages and benefits, good working conditions, and environmentally sound practices. Use tools like buycott.com to make informed shopping decisions.

4. Become an ally and direct supporter of farmer and peasant organizations in your country and in the Global South. Connect through

international organizations such as La Via Campesina or, in the United States, the Coalition of Immokalee Workers. (See Research & Action Organizations below.)

5. Help remove obstacles in the path of hungry people and their allies working for reforms. Oppose military and economic intervention on behalf of repressive governments that keep people poor and hungry.

6. Join the movement to get money out of politics so public policies reflect regular citizens' values and interests. Become active in campaigns supporting pro-democracy measures like publicly funded elections and participatory budgeting.

7. Clean up your kitchen's climate impact. Shift toward a plant-and-planet-friendly diet and reduce food waste. Compost at home and work to make nutrient cycling happen in your community.

8. Eat locally for a healthy climate, the economy, and energy. Supporting local farmers keeps money circulating within your community, and cuts down on the wasteful long-distance transportation of food—all at the same time. Plus, fresh food just tastes better!

9. Travel, if you are so fortunate, and see with new eyes. By getting out of resorts and other tourist traps, you can meet and be inspired by people in other countries working to uproot the forces of hunger.

10. Keep learning and keep sharing your knowledge. Sharing a story of people finding their voices and creating solutions is a revolutionary act. You'll never cease to be surprised by its power!

RESEARCH & ACTION ORGANIZATIONS

In addition to the many organizations highlighted in the book, please check out the following resources. And please send us your suggestions. (Unless otherwise noted, the organization is based in the United States.)

Connection & Action Opportunities: What Can We Do?

Organizations responsible for this book:

- *Food First/Institute for Food and Policy Development*—founded by Frances Moore Lappé and Joseph Collins—researches, publishes, and mobilizes globally to eliminate the injustices that cause hunger.
- *Small Planet Institute*, founded by Frances and Anna Lappé, uses diverse media to move people worldwide beyond powerlessness and despair to "bring democracy to life" to create life-serving societies.

Addressing Hunger & Poverty

- *ActionAid* helps communities in 45 countries fight poverty and promote human rights.
- *FIAN International* exposes violations of people's right to food and stands up against policies preventing people from feeding themselves.
- *Grassroots International* advances the human right to land, water, and food around the world through strategic grant-making and advocacy.
- *International Institute for Environment and Development* (U.K.) uses research and direct links with poor communities to spur sustainable development.
- *Oakland Institute* is a leader in fighting global "land grabs" and promotes debate on social, economic, and environmental challenges.
- *Oxfam America and International* (U.K.) work for practical and innovative ways for people to bring themselves out of poverty and escape injustice.
- *WhyHunger* connects people to nutritious, affordable food and supports grassroots solutions to hunger and poverty.

Addressing Healthy Food, Food Safety & Family Farms

- *Center for EcoLiteracy* helps educators apply ecological thinking and practice in K–12 schools from classroom to garden.
- *Center for Food Safety* exposes harmful food-production practices and promotes healthy alternatives.
- *Food Democracy Now!* boasts a network of 650,000 American farmers and citizens dedicated to reforming policies relating to food, agriculture, and the environment.

- *Food MythBusters* and the *Real Food Media Project* (led by Anna Lappé) provides creative media and teaching tools to take on big myths about food and organizes rapid-response myth-busting actions.
- *Localharvest.org* connects individuals with local farmers markets, community gardens, family farms, and other sources of sustainably grown food.
- *Pesticide Action Network* helps farmers shift from hazardous pesticides to ecologically sound alternatives.
- *Real Food Challenge* unites students for just and sustainable food through campaigns for real food in university campus dining services.
- *Slow Food* links the pleasure of good food with a commitment to community and the environment.
- *Sustainable Table* educates consumers and helps build community to promote local sustainable food and agriculture.

Addressing Agriculture, Environment & Trade

- *Corporate Accountability International* fights abuses by global corporations and holds them publicly accountable.
- *Environmental Working Group* encourages wise consumer choices and promotes civic action based on its research on environmental health, food and agriculture, water, and energy.
- *ETC Group* (Canada) researches social and ecological impacts of technologies on vulnerable people worldwide, promoting empowering, ecological alternatives.
- *Food and Water Watch* works to ensure wholesome food, clean water, and sustainable energy.
- *Friends of the Earth* seeks to focus public understanding and action on the most needed environmental policy changes.
- *Institute for Agriculture and Trade Policy* researches, publishes, and advocates for fair and sustainable food, farm, and trade policies.
- *Land Stewardship Project* helps foster an ethic of stewardship for farmland, promotes sustainable agriculture, and develops sustainable communities.
- *Soil Association* (U.K.) campaigns for healthy, humane, and sustainable food, farming, and land use.
- *USC Canada* promotes vibrant family farms, strong rural communities, and healthy ecosystems around the world.

Connection & Action Opportunities: What Can We Do?

U.S. and International Family Farm & Agroecology Networks

- *National Family Farm Coalition* empowers communities to democratically advance a food and agriculture system that ensures health, justice, and dignity for all.
- *International Federation of Organic Agricultural Movements* (IFOAM) unites and assists the global organic movement.
- *Third World Network* advocates for the needs and rights of peoples in the Global South, fair distribution of resources, and ecologically sustainable development.
- *Via Campesina* links peasant movements globally to promote food sovereignty and sustainable agriculture.

Research Institutes on Food, Hunger & Related Environmental Concerns

- *Berkeley Food Institute* at UC Berkeley works to catalyze transformation in global and local food systems and to promote diversity, justice, resilience, and health.
- *Center for a Livable Future* at Johns Hopkins University shares research about the interrelationships among diet, food production, the environment, and human health.
- *Centre for Agroecology and Food Security* (U.K.) is a joint research initiative of Coventry University and Garden Organic, which focuses on creating resilient food systems worldwide.
- *Food Climate Research Network* supports knowledge-sharing and public education on the intersections of food and climate change.
- *GRAIN* (Spain) researches and publishes reports to support small farmers and social movements working for community-controlled, biodiversity-based food systems.
- *Organic Farming and Research Foundation* focuses on policy, education, grant-making, and community building to promote widespread adoption and improvement of organic farming systems.
- *Rodale Institute* boasts the longest-running side-by-side organic and conventional agricultural methods field study in the United States and promotes organic farming practices through workshops and online tools.
- *South Centre* (Malaysia) is a policy think tank protecting and promoting the interests of Global South nations in the international arena.

- *The Organic Center* conducts and convenes credible, evidence-based science on the environmental and health effects of organic food and farming and communicates findings to the public.

Food-connected Worker Alliances

- *Coalition of Immokalee Workers* unites Florida farmworkers and their supporters to improve wages and working conditions.
- *Food Chain Workers Alliance* is a coalition of worker-based organizations that organize to improve wages and working conditions in the food industry.

Replacing Concentrated Money with Citizen Voices in the U.S. Political System

- *Every Voice* works to run and win campaigns for political candidates who champion the reform of how elections are financed at the federal and state level, so that all Americans have a voice.
- *Of Us by Us* works to pass the Government by the People Act and the Fair Elections Now Act. It is a project of Public Campaign.
- *Represent.Us* pursues a transpartisan strategy to pass specific legislation to create citizen-funded elections, transparency in campaign finance, and the end of the "revolving door" of elected officials and corporate lobbyists.

ONGOING LEARNING OPPORTUNITIES

Some Useful Blogs, News Sites, Magazines

- *Beautiful Solutions Gallery and Lab* is an interactive website for stories, solutions, and ideas that inspire a better future.
- *Civil Eats* provides daily news to provoke critical thought about the American food system and sustainable agriculture.
- *Common Dreams* aggregates news, analysis, and links to independent organizations and information
- *Daily Good* is a portal dedicated to sharing inspiring and positive news from around the world.

- *The EcoTipping Points Project* offers in-depth reports of community-based solutions worldwide.
- *Food Tank* publishes reports and provides other educational resources for safe, healthy, and nourished eaters globally.
- *New Internationalist* (U.K.) exposes crises of global poverty and hunger as well as solutions arising.
- *Resurgence and Ecologist* offers positive perspectives on topics ranging from ecology to social justice.
- *Seedmap.org* (Canada) is an interactive online tool exploring where our food comes from, the challenges facing agriculture today, and strategies to overcome them.
- *Solutions Journal* showcases innovative, positive solutions to ecological, social, and economic problems.
- *Triple Crisis* is a blog promoting open, global dialogue about crises in finance, development, and the environment.
- *Yes!* is a magazine offering a vision and tools to create a healthy planet and vibrant communities.

About the Authors

Frances Moore Lappé is the author or a coauthor of seventeen books, including *Diet for a Small Planet*, which has sold three million copies. In 1975, with Joseph Collins, she cofounded the Oakland-based Institute for Food and Development Policy (Food First). Today, with her daughter Anna Lappé, she leads the Cambridge, Massachusetts–based Small Planet Institute. In 1987, Lappé received the Right Livelihood Award, known as the Alternative Nobel, and in 2008 she received the James Beard Humanitarian of the Year Award. She has been a visiting scholar at the Massachusetts Institute of Technology and the University of California, Berkeley. *Gourmet Magazine* named Lappé, along with Thomas Jefferson and Julia Child, among twenty-five people who have changed the way America eats.

Joseph Collins, Ph.D., brings to this book five decades of researching and writing about issues in international development. In 1975, he cofounded with Frances Moore Lappé the Institute for Food and Development Policy (Food First). He is a consultant in Africa, Latin America, and Asia to UN agencies, USAID contractors, and international nongovernmental organizations. A Guggenheim Fellow, he has been a Distinguished Visiting Lecturer at the University of California, Santa Cruz. He codirected the pioneering program on the socioeconomic context of HIV/AIDS at the United Nations Research Institute in Social Development (UNRISD). Collins makes his home and surfs in Santa Cruz, California.

Notes

Beyond Guilt, Fear & Despair

1. "Hunger Statistics," World Food Programme, last updated 2013, accessed August 1, 2014, www.wfp.org/hunger/stats. Explanation: 3.1 million children die each year from poor nutrition; 66,000 people were killed in Hiroshima; 3.1 million/365 days a year = 8493.15 children die every day; 66,000 Hiroshima deaths/8493.15 child deaths per day = 7.77, which rounds to 8: Every 8 days the child death toll from poor nutrition equals the death toll of Hiroshima.

2. "We are millions," *New Internationalist*, December 12, 2009, accessed May 22, 2012. www.newint.org/features/special/2009/12/01/we-are-millions.

3. FAO, "Women and Rural Employment: Fighting Poverty by Redefining Gender Roles," Economic and Social Perspectives, Policy Brief 5, August 2009, accessed March 12, 2015, www.fao.org/3/a-ak485e.pdf.

4. Ibid.; FAO, "Poverty and Inequality," 112, www.fao.org/docrep/015/i2490e/i2490e02c.pdf, accessed December 30, 2014.

5. GRAIN, "Hungry for Land," May 2014, 2, 4, accessed August 4, 2014, www.grain.org/article/entries/4929-hungry-for-land-small-farmers-feed-the-world-with-less-than-a-quarter-of-all-farmland.

6. Fred Magdoff, "Twenty-First-Century Land Grabs: Accumulation by Agricultural Dispossession," *Monthly Review* 65, no. 6 (2013), accessed September 3, 2013, www.questia.com/magazine/1P3-3120536091/twenty-first-century-land-grabs-accumulation-by-agricultural#.

7. "Data: Female headed households," World Bank, accessed August 4, 2014, data.worldbank.org/indicator/SP.HOU.FEMA.ZS.

8. "Who Are the Hungry?" World Food Program, www.wfp.org/hunger/who-are; "Millennium Development Goals: 1. Eradicate extreme poverty and hunger," UNICEF, www.unicef.org/mdg/poverty.html.

9. FAO, "The State of Food Insecurity in the World 2015: Meeting the 2015 International Hunger Targets: Taking Stock of Uneven Progress" (Rome, 2015), 8, 44, accessed June 10, 2015, www.fao.org/3/a-i4646e.pdf.

10. Ibid., 10; "Joint UN report says rate of world hunger dropping amid wider eradication efforts," UN News Centre, May 27, 2015, accessed June 9, 2015, www.un.org/apps/news/story.asp?NewsID=50978#.VXdbhPlVgSV; Rick Gladstone, "U.N. Reports About 200 Million Fewer Hungry People Than in 1990," *New York Times,* May 27, 2015, accessed June 9, 2015, www.nytimes.com/2015/05/28/world/united-nations-reports-global-hunger-down-since-1990.html?_r=0.

11. FAO, "The State of Food Insecurity,"2015, Table A1, 40.

12. Ibid., 40, 42.

13. FAO, "Poverty and Inequality," 112; Branko Milanovic, "Inequality and Its Discontents: Why So Many Feel Left Behind," *Foreign Affairs,* August 12, 2011, accessed August 1, 2014, www.foreignaffairs.com/articles/68031/branko-milanovic/inequality-and-its-discontents.

14. Deborah Hardoon, "Wealth: Having It All and Wanting More," Oxfam Issue Briefing, January 2015, 2, accessed February 18, 2015, www.oxfam.org/sites/www.oxfam.org/files/file_attachments/ib-wealth-having-all-wanting-more-190115-en.pdf.

15. "2013 The 'Borlaug Dialogue,'" World Food Prize, last modified 2013, 4, accessed December 30, 2014, www.worldfoodprize.org/documents/filelibrary/documents/borlaugdialogue2010_/2013borlaugdialogue/WFP172013PanelHansHerren_FB7895363A7A7.pdf.

16. FAO, "State of Food Insecurity in the World," 2015, 40.

17. U.S. Department of Agriculture, Economic Research Service, Alisha Coleman-Jensen, Christian Gregory, and Anita Singh, "Household Food Insecurity in the United States in 2013," ERR-173, September 2014, 6, accessed October 1, 2014, www.ers.usda.gov/media/1565415/err173.pdf.

18. "PACs, Big Companies, Lobbyists, and Banks and Financial Institutions Seen by Strong Majorities as Having Too Much Power and Influence in DC," May 29, 2012, accessed September 30, 2014, www.harrisinteractive.com/NewsRoom/HarrisPolls/tabid/447/mid/1508/articleId/1069/ctl/ReadCustom%20Default/Default.aspx.

19. Martin Gilens and Benjamin Page, "Testing Theories of American Politics: Elites, Interest Groups, and Average Citizens," *Perspective on Politics* 12 (2014): 564–581, accessed March 2, 2015, doi: dx.doi.org/10.1017/S1537592714001595.

20. Harvey Kaye, *The Fight for the Four Freedoms: What Made FDR and the Greatest Generation Truly Great* (New York: Simon & Schuster, 2014), 75.

MYTH 1: Too Little Food, Too Many People

1. FAO, "The State of Food Insecurity," 2015, 1, 40.

2. M. McEniry, "Infant mortality, season of birth and the health of older Puerto Rican adults," *Social Science and Medicine* 72, no. 6 (2011): 1004-1015, doi:10.1016/j.soscimed.2010.08.026; Andrew M. Prentice and Timothy J. Cole, "Seasonal changes in growth and energy status in the Third World," *Proceedings of the Nutrition Society* 53 (1994): 509–519, journals.cambridge.org/download.php?file=%2FPNS%2FPNS53_03%2FS0029665194000650a.pdf&code=6794c0102e32340ce6bbfbf916a408e1; Sharon I. Kirkpatrick, Lynn McIntyre, and Melissa L. Potestio, "Child Hunger and Long-Term Adverse Consequences for Health," *Archives of Pediatric and Adolescent Medicine* 164, no. 8 (2010): 754, 760, accessed August 4, 2014, doi: 10.1001/archpediatrics.2010.117.

3. FAO, "The State of Food Insecurity in the World: The Multiple Dimensions of Food Security" (Rome, 2013), 16–17, accessed July 29, 2014, www.fao.org/docrep/018/i3434e/i3434e.pdf.

4. FAO, "Voices of the Hungry," 2014, accessed March 24, 2014, www.fao.org/3/aml872e.pdf.

5. "Guidelines for Control of Iron Deficiency Anemia" (New Delhi: Ministry of Health and Family Welfare, 2013), Table 2.2, 6, accessed April 27, 2015, www.pbnrhm.org/docs/iron_plus_guidelines.pdf.

6. Fumiaki Imamura et al., "Dietary quality among men and women in 187 countries in 1990 and 2010: a systemic assessment," *The Lancet* 3 (March 2015): 132–142, www.thelancet.com/pdfs/journals/langlo/PIIS2214-109X%2814%2970381-X.pdf.

7. World Health Organization, "Global burden of noncommunicable diseases," accessed October 23, 2014, www.searo.who.int/entity/noncommunicable_diseases/advocacy/global_burden_ncd_advocacy_docket.pdf.

8. World Health Organization, "Obesity and Overweight," Fact Sheet No. 311, updated January 2015, accessed February 24, 2015, www.who.int/mediacentre/factsheets/fs311/en/.

9. Marie Ng et al., "Global, regional, and national prevalence of overweight and obesity in children and adults during 1980–2013: A systematic analysis for the Global Burden of Disease Study 2013," *Lancet* 384 (May 2014): 766–781, dx.doi.org/10.1016/S0140-6736(14)60460-8.

10. Dr. Kartik Kalyanram, Rishi Valley Rural Health Centre, personal communication with author, October 16, 2013.

11. World Health Organization, Global Health Observatory Data Repository, Joint child malnutrition estimates (UNICEF-WHO-WB), "Global and regional trends by WHO Regions, 1990-2013 Stunting" (Global, Stunting prevalence, 2013; accessed March 24, 2015), apps.who.int/gho/data/node.main.NUTWHOREGIONS?lang=en.

12. FAO, "The State of Food Insecurity in the World 2013," 16–17, 21; UNICEF, "Improving child nutrition: The achievable imperative for global progress," April 2013, 8, accessed February 17, 2015, www.unicef.org/gambia/Improving_Child_Nutrition_-_the_achievable_imperative_for_global_progress.pdf.

13. Kathryn G. Dewey and Khadija Begum, "Long term consequences of stunting in early life," *Maternal & Child Nutrition* 7, no. s3 (2011): 5–18, doi:10.1111/j.1740-8709.2011.00349.x.

14. WHO Global Health Observatory Data Repository, "Childhood Stunting: Context, Causes, Consequences, WHO Conceptual Framework," September 2013, accessed February 17, 2015, www.who.int/nutrition/events/2013_ChildhoodStunting_colloquium_14Oct_ConceptualFramework_colour.pdf.

15. Gardiner Harris, "Poor Sanitation in India May Afflict Well-Fed Children with Malnutrition," *New York Times*, July 13, 2014, accessed January 12, 2015, www.nytimes.com/2014/07/15/world/asia/poor-sanitation-in-india-may-afflict-well-fed-children-with-malnutrition.html; World Health Organization, Annette Prüss-Üstün, et al., "Safer Water, Better Health: Costs, benefits and sustainability of interventions to protect and promote health" (Geneva, Switzerland, 2008), 7, accessed July 25, 2014, whqlibdoc.who.int/publications/2008/9789241596435_eng.pdf.

16. FAO, "State of Food Insecurity in the World," 2015, 16. Calculated from UNICEF, "Improving Child Nutrition," 9.

17. Mercedes de Onis, Edward A. Frongillo, and Monika Blössner, "Is malnutrition declining? An analysis of changes in levels of child malnutrition since 1980," *Bulletin of the*

World Health Organization 78, no. 10 (2000): 1222, www.scielosp.org/scielo.php?pid=S0042
-96862000001000008&script=sci_arttext.

18. FAO, "The State of Food and Agriculture 2013: Food Systems for Better Nutrition,"
(Rome, 2013), ix, accessed July 29, 2014, www.fao.org/docrep/018/i3300e/i3300e.pdf;
"Vitamin A deficiency," Micronutrient Deficiencies, World Health Organization, accessed
March 11, 2015, www.who.int/nutrition/topics/vad/en/; "Micronutrient deficiencies,"
World Health Organization, accessed February 17, 2015, www.who.int/nutrition/topics/
ida/en/; www.who.int/nutrition/topics/vad/en/.

19. FAO, "State of Food Insecurity in the World," 2015, 40.

20. Calculated from FAOSTAT [Production, Production Indices, Country: World + (Total),
Item: Food (PIN) + (Total), Year: 1961–2013, Element: Net Per Capita Production Index
Number (2004–2006 = 100); accessed May 7, 2015], faostat3.fao.org/download/Q/QI/E;
United Nations, Department of Economic and Social Affairs (UNDESA), Population Division,
Population Estimates and Projections Section [World, Population (thousands), medium vari-
ant, 1960–2010], accessed October 23, 2014 esa.un.org/unpd/wpp/unpp/p2k0data.asp.

21. FAOSTAT [Food Balance Sheets, Country: World + (Total), Year: 2011, Food supply (kcal/capita/day), accessed March 1, 2015]. faostat3.fao.org/download/FB/FBS/E.

22. FAO, Henning Steinfeld, et al., "Livestock's Long Shadow: Environmental Issues
and Options" (Rome, 2006), 43, accessed July 25, 2014, www.fao.org/docrep/010/a0701e/
a0701e00.HTM; calculated from FAO, "Food Outlook: Biannual Report on Global Food
Markets" (Rome, June 2013), 1, accessed July 23, 2014, www.fao.org/docrep/018/al999e/
al999e.pdf.

23. Emily S. Cassidy et al., "Redefining Agricultural Yields: From tons to people nourished by hectare," *Environmental Research Letters* 8, no. 3 (2013): 3, accessed July 18, 2014,
doi:10.1088/1748-9326/8/3/034015.

24. FAO, Robert Van Otterdijk, and Alexandre Meybeck, "Global Food Losses and Food
Waste: Extent, Causes and Prevention" (Rome, 2011), 4, accessed July 18, 2014, www.fao
.org/docrep/014/mb060e/mb060e.pdf.

25. World Resources Institute, Brian Lipinski, et al., "Reducing Food Loss and Waste,"
Working Paper, installment 2 of Creating a Sustainable Food Future (Washington, DC,
June 2013), 1, accessed July 18, 2014, www.wri.org/sites/default/files/reducing_food_
loss_and_waste.pdf.

26. FAO, Van Otterdijk, and Meybeck, "Global Food Losses and Food Waste," 5.

27. Barbara Vincenti et al., "The Contribution of Forests to Sustainable Diets," Background paper for the International Conference on Forests for Food Security and Nutrition,
FAO, Rome, May 2013, 2, 5–7, accessed January 14, 2015, www.fao.org/forestry/37132
-051da8e87e54f379de4d7411aa3a3c32a.pdf.

28. *Lost Crops of Africa*, Volume III, *Fruits,* National Academy of Sciences, 2008, accessed
January 2, 2015, sites.nationalacademies.org/PGA/cs/groups/pgasite/documents/
webpage/pga_054647.pdf.

29. Lester R. Brown, "The New Geopolitics of Food," *Foreign Policy*, May/June 2011,
www.foreignpolicy.com/articles/2011/04/25/the-new-geopolitics-of-food/; FAO, Agricultural Development Economics Division, "World Agriculture: Towards 2030/2050, The
2012 Revision," Summary, Figure 1, www.fao.org/fileadmin/user_upload/esag/docs/
AT2050_revision_summary.pdf.

30. "FAO Food Price Index in nominal and real terms," Excel data download from FAO,
"World Food Situation: FAO Food Price Index," last modified February 5, 2015, accessed
March 2, 2015, www.fao.org/worldfoodsituation/foodpricesindex/en/.

31. Calculated from FAOSTAT [Production, Production Indices, Country: World, Element: Net per capita Production Index Number (2004–06 = 100), Item: Agriculture (PIN) + (Total), Year: 1999–2011; accessed March 12, 2015], faostat3.fao.org/download/Q/QI/E.

32. Frederick Kaufman, *Bet the Farm* (Hoboken: John Wiley & Sons, Inc., 2012), Chapter 12.

33. FAO, "The State of Food Insecurity in the World: How does international price volatility affect domestic economies and food security?" (Rome, 2011), 14, accessed May 22, 2014, www.fao.org/docrep/014/i2330e/i23303.pdf.

34. Calculated from FAOSTAT [Production, Production Indices, Country: Low Income Food Deficit Countries + (Total), Element: Net Per Capita Production Index Number (2004–06 =100), Item: Food + (Total), Year: 1990–2012; accessed January 2, 2015], faostat3.fao.org/download/Q/QI/E.

35. FAO, "State of Food Insecurity in the World," 2015, 40, 42.

36. Calculated from FAOSTAT [Production, Production Indices, Country: India, Item: Food (PIN) + (TOTAL), Element: Net per Capita Production Index Numbers (2004–06=100), Year: 1990–2012; accessed May 20, 2014], faostat.fao.org/site/612/DesktopDefault. aspx?PageID=612#ancor; FAO, "State of Food Insecurity in the World," 2014, 42.

37. Calculated from Vikas Bajaj, "As Grain Piles Up, India's Poor Still Go Hungry," *New York Times*, June 7, 2012, accessed July 23, 2014, www.nytimes.com/2012/06/08/business/global/a-failed-food-system-in-india-prompts-an-intense-review.html?pagewanted=all; "Foodgrain Stock in Central Pool (As on 1st Day of Month) for Last Ten Years (2005–2014)," Stocks, Food Corporation of India, accessed March 6, 2015, fciweb.nic.in/upload/Stock/12.pdf. Numbers and assumptions: Rice stocks were 30.07 million metric tons and wheat stocks were 36.96 million metric tons; 1 cup of rice weighs 195 g (rough average); 1 cup of dry rice makes 3 cups of cooked rice; 1 lb of wheat makes 0.7 lb of flour; 1 lb of flour makes 1 loaf of bread; India's population in 2012 was 1,236,686,732 according to data .worldbank.org/indicator/SP.POP.TOTL.

38. Calculated from UNICEF, "Improving Child Nutrition: The achievable imperative for global progress" (New York: UNICEF, 2013), Figure 5, 9, www.unicef.org/publications/files/Nutrition_Report_final_lo_res_8_April.pdf.

39. Harris, "Poor Sanitation in India"; WHO, Prüss-Üstün et al., "Safer Water, Better Health," 7.

40. Calculated from FAOSTAT [Production, Production Indices, Country: Africa, Item: Food (PIN) + (Total), Year: 1990–2013, Element: Net Per Capita Production Index Number (2004–06=100); accessed October 23, 2014], faostat.fao.org/site/612/DesktopDefault.aspx?PageID=612#ancor. Calculated from FAOSTAT [Production, Production Indices, World, Item: Food (PIN) + (Total), Year: 1990–2013, Element: Net Per Capita Production Index Number (2004–06 = 100); accessed October 23, 2014], faostat.fao.org/site/612/DesktopDefault.aspx?PageID=612#ancor.

41. FAO, "State of Food Insecurity in the World," 2015, 40.

42. Calculated from FAOSTAT [Production, Production Indices, Region: Middle, Eastern, Western and Southern Africa, Element: Net Food Per Capita Production Index (2004–06 =100), Items Aggregated: Food (PIN) + (Total), Year: 1990, 2011; accessed January 14, 2015], faostat3.fao.org/download/Q/QI/E; US Department of Agriculture, Economic Research Service, Stacey Rosen and Shahla Shapouri, "Factors Affecting Food Production Growth in Sub-Saharan Africa," ERS Feature: International Markets & Trade, September 20, 2012, accessed January 12, 2015, www.ers.usda.gov/amber-waves/2012-september/factors -affecting-food-production.aspx#.UidzcBukp8p.

43. United Nations Development Programme, "Africa Human Development Report 2012: Towards a Food Secure Future" (New York: UNDP RBA, 2012), 33, 69, accessed March 25, 2015, www.afhdr.org/AfHDR/documents/HDR.pdf.

44. United Nations Development Programme, Nicolas Depetris Chauvin, Francis Mulangu, and Guido Porto, "Food Production and Consumption Trends in Sub-Saharan Africa: Prospects for the transformation of the agricultural sector," Working Paper 2012-011, February 2012, 41, 50, 68, accessed July 24, 2014, www.undp.org/content/dam/rba/docs/Working%20Papers/Food%20Production%20and%20Consumption.pdf; WHO, Global Health Observatory Data Repository, Child malnutrition country estimates (WHO global database), "Children aged <5 years stunted data by country," accessed May 7, 2015, apps .who.int/gho/data/node.main.1097?lang=en.

45. Ritu Verma, "Land Grabs, Power, and Gender in East and Southern Africa: So, What's New?" *Feminist Economics* 20, no. 1 (2014): 52–75, doi: 10.1080/13545701.2014.897739; Magdoff, "Twenty-First-Century Land Grabs."

46. FAO, "Biofuels and Food Security: A Report by the High Level Panel of Experts on Food Security and Nutrition," HLPE Report 5, (Rome: FAO, June 2013), 84, accessed October 23, 2014, www.fao.org/fileadmin/user_upload/hlpe/hlpe_documents/HLPE_Reports/HLPE-Report-5_Biofuels_and_food_security.pdf.

47. Regional Strategic Analysis and Knowledge Support System, Shenggen Fan, et al., "Public Spending for Agriculture in Africa: Trends and Composition," April 2009, 4, accessed January 12, 2015, www.resakss.org/sites/default/files/pdfs/public-expenditure -tracking-in-africa-trends-and-c-42375.pdf.

48. FAO, "Monitoring African Food and Agricultural Policies," MAFAP Brochure (Rome), 2, 5, accessed July 18, 2014, www.fao.org/fileadmin/templates/mafap/documents/MAFAP_Brochure_EN.pdf.

49. Masimba Tafirenyika, "Africa's Food Policy Needs Sharper Teeth," United Nations Department of Public Information, *Africa Renewal*, 2014, 3, www.un.org/africarenewal/magazine/special-edition-agriculture-2014/africa%E2% 80%99s-food-policy-needs -sharper-teeth.

50. Godfrey Mwakikagile, *Post-colonial Africa: A General Review* (Dar es Salaam: New Africa Press, 2014), 365.

51. Institute for Agriculture and Trade Policy, Shiney Varghese, and Karen Hansen-Kuhn, "Scaling Up Agroecology: Toward the Realization of the Right to Food," October 9, 2013, 1, accessed August 6, 2014, www.iatp.org/files/2013_10_09_ScalingUpAgroecology_SV_0.pdf.

52. FAO, "Monitoring African Food and Agricultural Policies," 2, 5; World Bank, "Africa Can Help Feed Africa: Removing barriers to regional trade in food staples," October 2012, 18, 32, accessed July 18, 2014, siteresources.worldbank.org/INTAFRICA/Resources/Africa-Can-Feed-Africa-Report.pdf.

53. Sophia Murphy, Ben Lilliston, and Mary Beth Lake, "WTO Agreement of Agriculture: A Decade of Dumping, United States Dumping on Agricultural Markets," Institute for Agriculture and Trade Policy, Publication no. 1, February 2005, 1, accessed May 20, 2014, www.globalpolicy.org/images/pdfs/02dumping.pdf.

54. Michigan State University, Nicole M. Mason, T. S. Jayne, and Bekele Shiferaw, "Wheat Consumption in Sub-Saharan Africa: Trends, Drivers, and Policy Implications," International Development Working Paper 127 (East Lansing, MI, December 2012), v, fsg. afre.msu.edu/papers/idwp127.pdf.

55. Jean-Yves Carfantan and Charles Condamines, *Vaincre la Faim, C'est Possible* (Paris: L'Harmattan, 1967), 63; Gunilla Andrae and Björn Beckman, *The Wheat Trap: Bread and Underdevelopment in Nigeria* (London: Zed Books, 1985).

56. "Food Security in the U.S.," Economic Research Service, U.S. Department of Agriculture, last modified September 3, 2014, accessed February 18, 2015, www.ers.usda.gov/topics/food-nutrition-assistance/food-security-in-the-us.aspx.

57. USDA ERS, Coleman-Jensen, Gregory, and Singh, "Household Food Insecurity," 6.

58. Marcelo Ostria, "How U.S. Agricultural Subsidies Harm the Environment, Taxpayer, and the Poor," National Center for Policy Analysis, 2013, accessed October 23, 2014, www.ncpa.org/pub/ib126; Michael Pollan, "You Are What You Grow," *New York Times*, April 22, 2007, accessed January 12, 2015, www.nytimes.com/2007/04/22/magazine/22wwlnlede.t.html? pagewanted=all&_r=0.

59. U.S. Department of Agriculture, Economic Research Service, "U.S. Bioenergy Statistics: Overview," Table 5, July 16, 2014, accessed July 28, 2014, www.ers.usda.gov/data-products/us-bioenergy-statistics.aspx#.U9aBV_ldURo.

60. World Bank (Data, by Country: European Union, Sub-Saharan Africa, India, World), Indicator: Population Density (people per sq. km. of land area); accessed May 22, 2014, data.worldbank.org/country.

61. World Bank [Data, Indicator, Population density (people per sq. km of land area, Bangladesh; accessed January 14, 2015], data.worldbank.org/indicator/EN.POP.DNST; FAOSTAT (Browse Data, Food Balance, Bangladesh, 2011, Average; accessed October 24, 2014), faostat3.fao.org/browse/FB/*/E.

62. UN data (Total fertility rate, Region: World, Years: 1950–55; accessed April 27, 2015), data.un.org/Data.aspx?d=PopDiv&f=variableID%3A54; "Country Comparison: Total Fertility Rate," The World Factbook, Central Intelligence Agency, accessed July 24, 2014, www.cia.gov/library/publications/the-world-factbook/rankorder/2127rank.html; "Fact Sheet: The Decline in U.S. Fertility," Population Reference Bureau, last modified July 2012, accessed July 24, 2014, www.prb.org/Publications/Datasheets/2012/world-population-data-sheet/fact-sheet-us-population.aspx.

63. Paul Ehrlich, *The Population Bomb* (New York: Sierra Club-Ballantine, 1968).

64. Garrett Hardin, "Lifeboat Ethics: The Case Against Helping the Poor," *Pyschology Today*, September 1974, accessed May 21, 2014, www.garretthardinsociety.org/articles/art_lifeboat_ethics_case_against_helping_poor.html.

65. United Nations Department of Economic and Social Affairs, Population Division (UNDESA), "World Population Prospects: The 2012 Revision, Highlights and Advance Tables," Working Paper No. ESA/P/WP.228, 2013, Table II.1, 12, accessed December 9, 2014, esa.un.org/unpd/wpp/Documentation/pdf/WPP2012_HIGHLIGHTS.pdf.

66. Population Reference Bureau, "2012 World Population Data Sheet," 2012, 10, accessed December 9, 2014, www.prb.org/pdf12/2012-population-data-sheet_eng.pdf.

67. UNDESA, Population Division, "World Population Prospects," 12.

68. Thomas Merrick et al., "Population Dynamics in Developing Countries," in *Population and Development: Old Debates, New Conclusions,* ed. Robert Cassen (New Brunswick, NJ: Transaction Publishers, 1994), 79–105; Tim Dyson, *Population and Development: The Demographic Transition* (New York: Zed Books, 2010), 42, 54–61.

69. Merrick, "Population Dynamics in Developing Countries," 82; UNdata (Total fertility rate, Country: India and China, Years: 1950–95 and 2010–15; accessed December 8, 2014), data.un.org/Data.aspx?d=PopDiv&f= variableID%3A54.

70. UNDESA, Population Division, "World Population Prospects," 2, 12.

71. Ibid., 11.

72. Ibid., xix.

73. Office of the Registrar General and Census Commissioner of India, "Chapter 3: Estimates of Fertility Indicators," SRS Statistical Report 2012 (New Dehli, 2013), 48, accessed May 21, 2014, www.censusindia.gov.in/vital_statistics/SRS_Report_2012/10_Chap_3_2012.pdf; Calculated from Census Organization of India, "States Census 2011," accessed November 11, 2014, www.census2011.co.in/states.php.

74. Calculated from World Bank [Data, Indicators (Nigeria, Tanzania, Democratic Republic of Congo, Niger, Uganda, Ethiopia, United States, World), Indicator: Population, Total, 2009–13; accessed December 11, 2014], data.worldbank.org/indicator/SP.POP.TOTL.

75. United Nations, Department of Economic and Social Affairs, Population Division, Population Estimates and Projections Section, "World Population Prospects: The 2012 Revision—Definition of Regions," accessed December 11, 2014, esa.un.org/wpp/Excel-Data/definition-of-regions.htm.

76. UNDESA, Population Division, "World Population Prospects," xix.

77. UNDESA, Population Division, "Definition of Regions"; Calculated from United Nations, Department of Economic and Social Affairs, Population Division, Population Estimates and Projections Section, "World Population Prospects: The 2012 Revision—Excel Tables–Fertility Data," Data File, Total Fertility (TFR), accessed December 17, 2014, esa.un.org/wpp/Excel-Data/fertility.htm.

78. World Resources Institute, Tim Searchinger, et al., "Achieving Replacement Level Fertility," Working paper, Installment 3 of Creating a Sustainable Food Future (Washington DC, July 2013), 6, accessed October 21, 2014, www.wri.org/sites/default/files/achieving_replacement_level_fertility_0.pdf.

79. UNDESA, Population Division, "World Population Prospects," 11.

80. Calculated from Guttmacher Institute, Gilda Sedgh, Susheela Singh, and Rubina Hussain, "Intended and Unintended Pregnancies Worldwide in 2012 and Recent Trends," Studies in Family Planning 45, no. 3 (2014): 310, onlinelibrary.wiley.com/doi/10.1111/j.1728-4465.2014.00393.x/pdf; Population Reference Bureau, "2012 World Population Data Sheet," 2.

81. Tomas Frejka, "Long Range Global Population Projections: Lessons Learned," in The Future Population of the World, ed. Wolfgang Lutz (London: Earthscan Publications, 1996), 5, table 1.1.

82. United Nations Educational, Scientific and Cultural Organization, Nicole Bella, and Saïd Belkachla, "VII. Impact of Demographic Trends on the Achievement of the Millennium Development Goal of Universal Primary Education," March 14, 2005, VII–2, accessed May 28, 2014, www.un.org/esa/population/publications/PopAspectsMDG/06_UNESCO.pdf; Mary K. Shenk et al., "A model comparison approach shows stronger support for economic models of fertility decline," Proceedings of the National Academy of Sciences 110, no. 20 (2013): 8045, 8048–8049, accessed May 28, 2014, doi:10.1073/pnas.1217029110; United Nations Population Fund (UNFPA), Information and External Relations Division, "The State of World Population 2011: People and possibilities in a world of 7 billion," 2011, 3–4, 14–15, 44, 62, accessed January 20, 2015, www.unfpa.org/webdav/site/global/shared/documents/publications/2011/EN-SWOP2011-FINAL.pdf.

83. Jenna Nobles, Elizabeth Frankenberg, and Duncan Thomas, "The effects of mortality on fertility: Population dynamics after a natural disaster," NBER Working Paper No. 20448, September 2014, 1, accessed January 10, 2015, www.nber.org/papers/w20448.pdf.

84. FAO, "Growing Greener Cities in Latin America and the Caribbean," 75, accessed June 23, 2014, www.fao.org/3/a-i3696e/i3696e09.pdf.

85. Edward Bbaale and Paul Mpuga, "Female Education, Contraceptive Use, and Fertility: Evidence from Uganda," *Consilience: The Journal of Sustainable Development* 6 (2011): 20–47, accessed January 5, 2015, www.consiliencejournal.org/index.php/consilience/article/viewFile/234/79.

86. Ilene S. Speizer, Lisa Whittle, and Marion Carter, "Gender Relations and Reproductive Decision Making in Honduras," *International Family Planning Perspectives* 31, no. 3 (2005): 131, accessed May 23, 2014, doi:10.1363/3113105; Akinrinola Bankole and Susheela Singh, "Couples' Fertility and Contraceptive Decision-Making in Developing Countries: Hearing the Man's Voice," *International Family Planning Perspectives* 24, no. 1 (1998): 15–24, accessed December 9, 2014, www.guttmacher.org/pubs/journals/2401598.html.

87. Hypothesis is authors' own. Inequality/fertility evidence Klaus Gründler and Philipp Scheuermeyer, "Income Inequality, Economic Growth, and the Effect of Redistribution," University of Würzburg, Department of Economics, October 2014, www.boeckler.de/pdf/v_2014_10_30_gr%C3%BCndler_scheuermeyer.pdf.

88. Tim Dyson, *Population and Development: The Demographic Transition* (New York: Zed Books, 2010), 176; United Nations Development Programme, Kevin Watkins, "Human Development Report 2007/2008, Fighting Climate Change: Human solidarity in a divided world" (New York, 2007), 330, accessed May 23, 2014, hdr.undp.org/sites/default/files/reports/268/hdr_20072008_en_complete.pdf; UNDESA, Population Division, "World Population Prospects," 11.

89. United Nations Population Fund (UNFPA), "Niger-Husbands' Schools Seek to Get Men Actively Involved in Reproductive Health," 2011, 1, accessed December 7, 2014, niger.unfpa.org/docs/SiteRep/Ecole%20des%20maris.pdf.

90. United Nations Population Fund, "'School for Husbands' Encourages Nigerien Men to Improve the Health of Their Families," April 20, 2011, accessed December 7, 2014, www.unfpa.org/news/%E2%80%98school-husbands%E2%80%99-encourages-nigerien-men-improve-health-their-families.

91. Jason Beaubien, "School for Husbands Gets Men to Talk About Family Size," National Public Radio, November 27, 2014, accessed December 7, 2014, www.npr.org/blogs/goatsandsoda/2014/11/27/358113783/school-for-husbands-gets-men-to-talk-about-family-size.

92. Mahendra Dev et al., "Causes of Fertility Decline in India and Bangladesh: An Investigation," Centre for Economic and Social Studies and Bangladesh Institute of Development Studies, 2002, 3–5, accessed January 20, 2015, saneinetwork.net/Files/02_01.pdf.

93. Ibid., 35.

94. "Total Fertility Rate India," Government of India, last modified 2012, accessed May 22, 2014, data.gov.in/catalog/total-fertility-rate-india#web_catalog_tabs_block_10; "Press Note on Poverty Estimates, 2011–2012," Government of India Planning Commission, last modified July 22, 2013, 6, accessed May 22, 2014, planningcommission.nic.in/news/pre_pov2307.pdf.

95. Ruhul Amin, A. U. Ahmed, and J. Chowdhury, "Poor women's participation in income-generating projects and their fertility regulation in rural Bangladesh: Evidence from a recent survey," *World Development* 22, no. 4 (1993): 555–565, accessed May 23, 2014, doi:10.1016/0305-750X(94)90111-2; Syed M. Hashemi, Sidney Ruth Schuler, and

Ann P. Riley, "Rural Credit Programs and Women's Empowerment in Bangladesh," *World Development* 24, no. 4 (1996): 650, accessed May 23, 2014, doi:0305-750X(95)00159-X.

96. Max Fischer, "Why China's one-child policy still leads to forced abortions, and always will," *Washington Post*, November 15, 2013, accessed October 24, 2014, www .washingtonpost.com/blogs/worldviews/wp/2013/11/15/why-chinas-one-child-policy -still-leads-to-forced-abortions-and-always-will/.

97. World Bank, "The Little Data Book on Gender: 13," 2013, 58, accessed May 1, 2015, data.worldbank.org/sites/default/files/the-little-data-books-on-gender-2013.pdf.

98. Calculated from United Nations, Department of Economic and Social Affairs, Population Division, "World Contraceptive Use 2011," last modified December 2010, accessed May 21, 2014, www.un.org/esa/population/publications/contraceptive2011/wallchart_front. pdf; A. Singh et al., "Sterilization Regret Among Married Women in India: Implications for the Indian National Family Planning Program," *International Perspectives on Sexual and Reproductive Health* 38, no. 4 (2012): 187–195, accessed November 21, 2014, doi: 10.1363/3818712.

99. Ellen Barry and Suhasini Raj, "Web of Incentives in Fatal Indian Sterilizations," *New York Times*, November 12, 2014, accessed November 13, 2014, www.nytimes. com/2014/11/13/world/asia/web-of-incentives-in-fatal-indian-sterilizations.html?mo dule=Search&mabReward=relbias%3Ar%2C%7B%222%22%3A%22RI%3A16%22%7D&_r=0; Editorial Board, "India's Lethal Approach to Birth Control," *New York Times*, November 20, 2014, accessed November 21, 2014, www.nytimes.com/2014/11/21/opinion/indias -lethal-birth-control.html?_r=0.

100. Editorial Board, "India's Lethal Approach to Birth Control."

101. Andrew MacAskill, "India's Poorest Women Coerced into Sterilization," *Bloomberg*, June 11, 2013, accessed May 21, 2014, www.bloomberg.com/news/2013-06-11/ india-s-poorest-women-coerced-into-sterilization.html; Andrew MacAskill, "Inside India's Female Sterilization Camps," *Bloomberg*, June 20, 2013 www.businessweek.com/ articles/2013-06-20/inside-indias-female-sterilization-camps; "Women in India Targeted for Sterilization in Population Fix," Bloomberg Visual Data, 2013, accessed January 12, 2015, www.bloomberg.com/infographics/2013-06-11/women-in-india-targeted-for -sterilization-in-population-fix.html.

102. Editorial Board, "India's Lethal Approach to Birth Control."

103. UNDESA, Population Division, "World Contraceptive Use 2011"; United Nations, Department of Economic and Social Affairs (UNDESA), Population Division, "Trends in Contraceptive Methods Used Worldwide," Population Facts No. 2013/9, December 2013, accessed November 21, 2014, www.un.org/en/development/desa/population/publica- tions/pdf/popfacts/popfacts_2013-9.pdf; Male Health Center, "Vasectomy," 2006, November 21, 2014, www.malehealthcenter.com/c_vasectomy.html; G. L. Smith, G. P. Taylor, and K. F. Smith, "Comparative risks and costs of male and female sterilization," *American Journal Public Health* 75, no. 4 (1985): 370–374, PMCID: PMC1646249.

104. Tashya de Silva, "Low Fertility Trends: Causes, Consequence and Policy Options," Institute for Health Policy, Sri Lanka, 2008, accessed November 3, 2014, www.ihp.lk/ publications/docs/lowfertility.pdf; United Nations Department of Economic and Social Affairs (UNDESA), Population Division, Population Estimates and Projections [Sri Lanka, Total fertility (children per woman), medium variant, 1980–2010; accessed October 23, 2014], esa.un.org/wpp/unpp/panel_indicators.htm.

105. Medea Benjamin, Joseph Collins, and Michael Scott, *No Free Lunch: Food and Revolution in Cuba Today* (New York: Grove Press/Food First Books, 1986), 26, 92.

106. UNdata (Total Fertility Rate, Cuba, 1950–55, 2010–15, medium variant; accessed January 20, 2015), data.un.org/Data.aspx?d=PopDiv&f=variableID%3A54.

107. "Himachal Pradesh, Secrets of Success," World Bank, January 28, 2015, www.worldbank.org/en/news/feature/2015/01/28/himachal-pradesh-secrets-of-success. Based on: Maitreyi Bordia Das et al., "Scaling the Heights: Social Inclusion and Sustainable Development in Himachal Pradesh," World Bank, Washington, DC, 2015, openknowledge.worldbank.org/handle/10986/21316 License: CC BY 3.0 IGO.

108. National Health Mission, Ministry of Health and Family Welfare, Government of India, nrhm.gov.in/nrhm-in-state/state-wise-information/himachal-pradesh.html.

109 "Kerala Population Census data 2011," Indian Census, 2011, accessed November 3, 2014, www.census2011.co.in/census/state/kerala.html; Anikita Gandhi et al., "India Human Development Report 2011: Toward Social Inclusion," Institute of Applied Manpower Research, Planning Commission, Government of India, accessed November 3, 2014, www.iamrindia.gov.in/ihdr_book.pdf; World Bank [Data, Indicators, Life expectancy at birth, total (years), United States, 2010-2014, accessed November 3, 2014], databank.worldbank.org/data/views/reports/tableview.aspx; World Bank Data [Mortality rate, infant (per 1,000 live births), United States, 2014, accessed November 3, 2014], databank.worldbank.org/data/views/reports/tableview.aspx#.

110. Richard W. Franke, "Land Reform Versus Inequality in Nadur Village, Kerala," *Journal of Anthropological Research* 48, no. 2 (1992): 81, accessed July 23, 2014, www.jstor.org/stable/3630406.

111. Government of Kerala, "Literacy Rate 2011," www.kerala.gov.in/index.php?option=com_content&id=4007&Itemid=3187; Christophe Z. Guilmoto and Irudaya Rajan, "Fertility at District Level in India, Lessons from the 2011 Census," Centre Population & Développement, June 2013, 16, Table 1, accessed January 20, 2015, www.ceped.org/IMG/pdf/ceped_wp30.pdf; World Bank [Data, Indicators, Fertility rate, total (births per woman, United States, 2010–2014; accessed October 24, 2014], data.worldbank.org/indicator/SP.DYN.TFRT.IN/countries.

112. Rupa Subramanya Dehejia, "Economics Journal: Can Kerala Kick Remittance 'Curse'?" *Wall Street Journal*, April 11, 2011, accessed May 11, 2015, blogs.wsj.com/indiarealtime/2011/04/11/economics-journal-can-kerala-kick-remittance-curse.

113. World Bank [Data, Indicators, Mortality Rate (per 1,000 live births), Costa Rica, 2013; accessed January 2, 2015], data.worldbank.org/indicator/SP.DYN.IMRT.IN; World Bank [Data, Indicators, Life expectancy at birth, female (years), Costa Rica, 2013; accessed January 2, 2015], data.worldbank.org/indicator/SP.DYN.LE00.FE.IN.

114. World Bank [Data, Indicators, Literacy rate, youth female (% of females ages 15–24), Costa Rica, 2011; accessed January 2, 2015], data.worldbank.org/indicator/SE.ADT.1524.LT.FE.ZS/countries; Matthew Kelly et al., "Thailand's Work and Health Transition," *International Labour Review* 149, no. 3 (2010): 376, accessed May 22, 2014, doi: 10.1111/j.1564-913X.2010.00092.x.

MYTH 2: Climate Change Makes Hunger Inevitable

1. Justin Gillis, "Panel's Warning on Climate Risk: The Worst Is Yet to Come," *New York Times*, March 31, 2014, accessed May 22, 2014, www.nytimes.com/2014/04/01/science/earth/climate.html?_r=0.

2. Calculated from Intergovernmental Panel on Climate Change, IPCC Working Group II Contribution to AR5, John R. Porter, et al., "Chapter 7: Food Security and Food Productions

Systems," in *Climate Change 2014: Impacts, Adaption, and Vulnerability: Global and Sectoral Aspects*, 2014, 6, 12, 21, accessed December 16, 2014, www.ipcc.ch/pdf/assessment -report/ar5/wg2/WGIIAR5-Chap7_FINAL.pdf.

3. Ibid., 491, 497, 504.

4. Aiguo Dai, "Increasing drought under global warming in observations and models," *Nature Climate Change* 3 (January 2013): 58, accessed November 20, 2014, doi:10.1038/ nclimate1633; Intergovernmental Panel on Climate Change, IPCC Working Group I Contribution to AR5, Thomas F. Stocker, et al., "Technical Summary," in *Climate Change 2013: The Physical Science Basis*, 2013, 91, accessed November 20, 2014, www.ipcc.ch/pdf/ assessment-report/ar5/wg1/WG1AR5_TS_FINAL.pdf; Intergovernmental Panel on Climate Change, IPCC Working Group I Contribution to AR5, Matthew Collins, et al., "Chapter 12: Long-term Climate Change: Projections, Commitments, and Irreversibility," in *Climate Change 2013: The Physical Science Basis*, 2013, 1079–1082, 1084–1086, accessed January 6, 2015, www.ipcc.ch/pdf/assessment-report/ar5/wg1/WG1AR5_Chapter12_FINAL.pdf.

5. IPCC, WGII, AR5, Porter, et al., "Chapter 7," 504–505.

6. Intergovernmental Panel on Climate Change, IPCC Working Group II Contribution to AR5, Christopher B. Field, et al., "Technical Summary" in *Climate Change 2014: Impacts, Adaptation, and Vulnerability*, 2014, 42, 60, 64, 76, accessed December 30, 2014, ipcc -wg2.gov/AR5/images/uploads/WGIIAR5-TS_FINAL.pdf; IPCC, WGII, AR5, Porter, et al., "Chapter 7," 500.

7. Samuel S. Myers et al., "Increasing CO2 Threatens Human Nutrition," *Nature* 510 (2014): 139, accessed June 3, 2014, doi:10.1038/nature13179.

8. "Fish as Food," Marine Stewardship Council, accessed December 10, 2014, www .msc.org/healthy-oceans/the-oceans-today/fish-as-food; Richard A. Feely, Christopher L. Sabine, and Victoria J. Fabry, "Carbon Dioxide and Our Ocean Legacy," April 2006, accessed December 12, 2014, www.pmel.noaa.gov/pubs/PDF/feel2899/feel2899.pdf.

9. World Food Programme, Martin Parry, et al., "Climate Change and Hunger: Responding to the Challenge" (Rome, 2009), 4, accessed May 28, 2014, documents.wfp.org/stellent/ groups/public/documents/newsroom/wfp212536.pdf.

10. Sonja J. Vermeulen, Bruce M. Campbell, and John S.I. Ingram, "Climate Change and Food Systems," *Annual Review of Environment and Resources* 37 (2012): 195, accessed December 12, 2014, doi: 10.1146/annurev-environ-020411-130608.

11. Bojana Bajželj et al., "Importance of food-demand management for climate mitigation," *Nature Climate Change* 4 (2014): 924–929, accessed December 12, 2014, doi:10.1038/ nclimate235; United Nations Framework Convention on Climate Change, "Report of the Conference of the Parties on Its Sixteenth Session, held in Cancun from November 29 to December 10, 2010," FCCC/CP/2010/7/Add.1 (Cancun: United Nations, 2011), 3, accessed December 12, 2014, unfccc.int/resource/docs/2010/cop16/eng/07a01.pdf.

12. "Who Are the Hungry," World Food Programme, last modified 2014, accessed May 28, 2014, www.wfp.org/hunger/who-are.

13 Joel Mokyr, "Irish Potato Famine," *Encyclopaedia Britannica*, accessed June 10, 2014, www.britannica.com/EBchecked/topic/294137/Irish-Potato-Famine; Eric Vanhaute, Richard Paping, and Cormac Ó Gráda, "The European Subsistence Crisis of 1845–1850: A Comparative Perspective" (paper, XIV International Economic History Congress, Helsinki, Finland, August 21–25, 2006), 10, www.helsinki.fi/iehc2006/papers3/Vanhaute.pdf.

14. John MacHale, *Letter to Lord Russell* (1846), cited in Cecil Woodham-Smith, *The Great Hunger: Ireland 1845–9* (New York: Harper & Row, 1962); Joel Mokyr, *Why Ireland Starved: A Quantitative and Analytical History of the Irish Economy, 1800–1850* (Boston:

Allen & Unwin, 1983); Cormac O'Grada, *Ireland Before and After the Famine: Explorations in Economic History 1800–1925*, 2nd edition (Manchester: Manchester University Press, 1993).

15. UNICEF, "Response to the Horn of Africa Emergency: A Continuing Crisis Threatens Hard-Won Gains," Regional Six-Month Progress Report, April 2012, 4–5, accessed May 28, 2014, www.unicefusa.org/sites/default/files/assets/pdf/Horn-of-Africa-Six-Month -Report-April-2012.pdf.

16. Associated Press, "Somalia: Famine Toll in 2011 Was Larger Than Previously Reported," *New York Times*, April 29, 2013, accessed July 17, 2014, www.nytimes .com/2013/04/30/world/africa/somalia-famine-toll-in-2011-was-larger-than-previously -reported.html?ref=famine.

17 UNICEF, "Response to the Horn of Africa Emergency," 5.

18. FAO, "Drought Emergency—Overcoming the Crisis: Horn of Africa" (Rome, July 2011), 1, accessed June 4, 2014, www.fao.org/crisis/28421-030a401bd1fcec17aa8cd 1cb1f4892185.pdf.

19. Xan Rice, ""Hunger Pains: Famine in the Horn of Africa," *The Guardian*, August 8, 2011, accessed July 17, 2014, www.theguardian.com/global-development/2011/aug/08/ hunger-pains-famine-horn-africa; Christopher Thompson, "Floods and Droughts: How Climate Change Is Impacting Africa," *Time*, November 11, 2009, accessed May 29, 2014, content.time.com/time/specials/packages/article/0,28804,1929071_1929070_1936772,00 .html; Heather McGray et al., "Famine in the Horn of Africa," World Resources Institute, August 24, 2011, accessed May 29, 2014, www.wri.org/blog/2011/08/famine-horn-africa; "Millions facing severe food crisis amid worsening drought in Horn of Africa—UN," UN News Centre, last modified June 28, 2011, accessed May 29, 2014, www.un.org/apps/ news/story.asp?NewsID=38876#.U4dB4CgVe9Q.

20. IRIN, "La Niña blamed for east African drought," *The Guardian*, July 14, 2011, accessed July 9, 2014, www.theguardian.com/global-development/2011/jul/14/east -africa-drought-la-nina.

21. Leo Hickman, "Is it now possible to blame extreme weather on global warming?" *The Guardian*, July 3, 2012, accessed July 10, 2014, www.theguardian.com/environment/ blog/2012/jul/03/weather-extreme-blame-global-warming.

22. Jeffrey Gettleman, "Somalia's Agony Tests Limits of Aid," *New York Times*, November 11, 2011, accessed July 17,2014, www.nytimes.com/2011/11/02/giving/some-aid-trickles-into-somalia-surrounded-by-death-and-disease.html?ref=famine; Save the Children and Oxfam, "A Dangerous Delay: The cost of late response to early warnings in the 2011 drought in the Horn of Africa," January 18, 2012, accessed January 5, 2015, www.oxfam.org/sites/www.oxfam.org/files/bp-dangerous-delay-horn-africa-drought -180112-en.pdf.

23. Calculated from FAOSTAT [Prices, Consumer Price Indices, Regions: Eastern Africa + (Total), World + (Total), Months: January and December, Item: Consumer Prices, Food Indices (2000 = 100), January 2009 and December 2011; accessed December 31, 2014], faostat3.fao.org/faostat-gateway/go/to/download/P/CP/E.

24. Samuel Loewenberg, "Humanitarian response inadequate in Horn of Africa crisis," *The Lancet* 378, no. 9791 (2011): 555–558, accessed July 15, 2014, doi:10.1016/S0140 -6736(11)61276-2.

25. Calculated from FAOSTAT [Food Balance, Food Supply-Crops Primary Equivalent, Countries: Djibouti, Ethiopia, Kenya, Somalia, Element: Food supply (kcal/capita/day), Year: 2011, Item: Grand Total + (Total); accessed April 9, 2015], faostat3.fao.org/download/FB/CC/E.

26. Calculated from FAOSTAT (Food Balance Sheets, Country: Ethiopia, Kenya, Somalia, Djibouti, Year: 2001, 2011; accessed January 20, 2015), faostat3.fao.org/download/FB/FBS/E.

27 Calculated from FAOSTAT (Trade, Crops and Livestock Products, Country: Kenya, Elements: Export quantity, Items: Beans, green, Years: 2009–11; accessed January 20, 2015), faostat3.fao.org/faostat-gateway/go/to/download/T/*/E.

28. Nicole M. Mason et al., "The 2011 Surplus in Smallholder Maize Production in Zambia: Drivers, Beneficiaries, and Implications for Agricultural and Poverty Reduction Policies," Working Paper No. 58, Food Security Research Project, November 2011, 30, accessed December 10, 2014, fsg.afre.msu.edu/zambia/wp58.pdf; "Government-Owned Corn Destroyed in One of Africa's Poorest Countries," Bureau of Investigative Journalism, last modified September 20, 2012, accessed December 10, 2014, www.thebureauinvestigates.com/2012/09/20/government-owned-corn-destroyed-in-one-of-africas-poorest-countries/.

29. Sarah Coll-Black et al., "Targeting Food Security Interventions When 'Everyone Is Poor': The Case of Ethiopia's Productive Safety Net Programme," Working Paper No. 24, International Food Policy Research Institute-Ethiopia Strategy Support Program II, May 2011, 1, accessed March 1, 2015; McGray et al., "Famine in the Horn of Africa."

30. Lester Brown, "The Weakest Link," *Resurgence and Ecologist*, no. 276 (2013): 24–25.

31. Abdolreza Abbassian, Senior Economist Trade and Markets Division (D-804) FAO, e-mail communication with the authors, April 2015.

32. Amartya Sen, *Poverty and Famines* (Oxford: Clarendon Press, 1981), 136–138.

33. Betsy Hartmann and James K. Boyce, *A Quiet Violence: View from a Bangladesh Village* (San Francisco: Food First Books, 1983), 189.

34. Bangladeshi citizen, interview by Michael Scott of Oxfam America, Boston, 1979.

35. "Man Dies: Found In Unheated Home," *Chicago Tribune*, January 22, 1994, accessed December 4, 2014, articles.chicagotribune.com/1994-01-22/news/9401220167_1_hypothermia-unheated-home-fourth-fatality.

36. "Seven Hundred Killed from Hypothermia Annually in United States," National Coalition for the Homeless, January 7, 2010, accessed January 20, 2015, www.nationalhomeless.org/publications/winter_weather/index.html.

37. Christopher B. Barrett, "Measuring Food Insecurity," *Science* 327 no. 5967 (2010): 827.

38. Organization for Economic Co-operation and Development and FAO, "OECD-FAO Agricultural Outlook 2012–2021" (Paris: OECD Publishing and FAO, 2012), 17, accessed January 20, 2015, dx.doi.org/10.1787/agr_outlook-2012-en; IPCC, WGII, AR5, Porter et al., "Chapter 7," 490.

39. FAO, Van Otterdijk, and Meybeck, "Global Food Losses and Food Waste," 10; National Resources Defense Council and Dana Gunders, "Wasted: How America Is Losing Up to 40% of Its Food from Farm to Fork to Landfill," IP:12-06-B, August 2012, 1, www.nrdc.org/food/files/wasted-food-ip.pdf.

40. United Nations General Assembly Sixteenth Session, Human Rights Council, Olivier de Schutter, "Report submitted by the Special Rapporteur on the right to food," A/HRC/16/49, December 2010, 4, accessed December 16, 2014, www2.ohchr.org/english/issues/food/docs/A-HRC-16-49.pdf; United Nations Environment Program, "Food Waste Facts," accessed December 12, 2014, www.unep.org/wed/2013/quickfacts/.

41. Human Rights Council, "Report submitted by the Special Rapporteur," 19.

42. FAO, Van Otterdijk, and Meybeck, "Global Food Losses and Food Waste," 10.

43. Our logic: Today the world produces enough food per person, even after we account for most waste, and provides more than adequate calories. Today's population is roughly 7.2 billion, of which 15 percent is 1.08 billion. Thus, 15 percent more food made available, owing to waste reduction, could theoretically feed roughly 1 billion people.

44. "Villages in Cameroon Find Solution to Yearly Hunger Season," World Food Programme, last modified August 21, 2011, accessed December 10, 2014, www.wfp.org/stories/villages-cameroon-find-solution-yearly-hunger-season.

45. "Reducing Food Waste: Making the Most of Our Abundance," Nourishing the Planet, Worldwatch Institute, last modified June 28, 2011, accessed December 10, 2014, blogs.worldwatch.org/nourishingtheplanet/reducing-food-waste-making-the-most-of-our-abundance/#more-11699.

46. "3 Food Waste Facts About France's Ugly Produce Campaign," Organic Authority, September 3, 2014, accessed November 6, 2014, www.organic authority.com/3-food-waste-facts-about-frances-ugly-produce-campaign/; Rose Prince, "How the French can teach us to love ugly fruit and veg: It doesn't have to be beautiful to taste nice or do us good," Mail Online, August 4, 2014, accessed November 6, 2014, www.dailymail.co.uk/news/article-2716133/How-French-teach-love-ugly-fruit-veg-It-doesn-t-beautiful-taste-nice-good.html#ixzz3H4SckS1K.

47. "Finding Takers for Lonely Leftovers in a Culinary Nook of the Sharing Economy," New York Times, November 26, 2014, accessed December 10, 2014, www.nytimes.com/2014/11/27/world/europe/german-matchmakers-pair-lonely-leftovers-and-rumbling-bellies.html?_r=1.

48. "UK Media and Resources," Love Food Hate Waste, accessed November 6, 2014, england.lovefoodhatewaste.com/content/uk-media-resources; "The Bite," Love Food Hate Waste, December 2014, accessed December 10, 2014, us1.campaign-archive2.com/?u=6534310dd35be920e719fccd&id=2c4c35b0b9.

49. "Food Recovery Challenge," U.S. Environmental Protection Agency, last modified January 28, 2015, accessed January 29, 2015, www.epa.gov/foodrecoverychallenge/; "EPA Recognizes Outstanding Food Recovery Challenge and WasteWise Program Participants," News Releases from Headquarters, Newsroom, U.S. Environmental Protection Agency, January 28, 2015, accessed March 18, 2015, yosemite.epa.gov/opa/admpress.nsf/bd4379a92ceceeac8525735900400c27/9816d3c528ecdb9b85257ddb006163f8!OpenDocument.

50. United Nations Department of Economic and Social Affairs (UNDESA), "World Population Prospects," Table 1.1.

51. Hans Hurni et al., "Key Implications of Land Conversions in Agriculture," in Wake Up Before It Is Too Late, Trade and Development Review 2013, United Nations Conference on Trade and Development (UNCTAD), 2013, 221, accessed January 20, 2015, unctad.org/en/PublicationsLibrary/ditcted2012d3_en.pdf.

52. Mario Herrero and Philip K. Thornton, "Livestock and global change: Emerging issues for sustainable food systems," Proceedings of the National Academy of Sciences 110, no. 52 (2013): 20879, accessed July 25, 2014, doi:10.1073/pnas.1321844111.

53. Paul C. West et al., "Leverage Points for Improving Global Food Security and the Environment," Science 345, no. 6194 (2014): 326, accessed on July 18, 2014, doi: 10.1126/science.1246067l; FAO, "Food Outlook: Biannual Report on Global Food Markets," June 2013, 1, accessed July 23, 2014, www.fao.org/docrep/018/al999e/al999e.pdf; "Soy Facts," Soyatech, accessed December 8, 2014, www.soyatech.com/info.php?id=175.

54. Emily S. Cassidy et al., "Redefining agricultural yields: from tonnes to people nourished per hectare," *Environmental Research Letters* 8, no. 3 (2013): 3, accessed December 12, 2014, doi:10.1088/1748-9326/8/3/034015.

55. Ibid., 3.

56. Ibid., 6.

57. FAO, "Report of the Panel of Eminent Experts on Ethics in Food and Agriculture," 2011, 27, accessed January 20, 2015, www.fao.org/docrep/014/i2043e/i2043e02c.pdf. See following citation.

58. Calculated from U.S. Department of Agriculture, Economic Research Service, Cynthia Nickerson, et al., "Major Uses of Land in the United States, 2007," Economic Information Bulletin No. 89, December 2011, 20, accessed July 2, 2014, www.ers.usda.gov/media/188404/eib89_2 .pdf; U.S. Department of Agriculture, Economic Research Service, "Table 5: Corn supply, disappearance, and share of total corn used for ethanol," from U.S. Bioenergy Statistics, updated June 2014, accessed July 7, 2014, www.ers.usda.gov/data-products/us-bioenergy-statistics.aspx#.U7qnsvldURq; Cassidy et al., "Redefining agricultural yields," 6.

59. Stefan Wirsenius, Christian Azar, and Göran Berndes, "Global Bioenergy Potentials: A New Approach Using a Model-Based Assessment of Biomass Flows and Land Demand in the Food and Agriculture Sector 2030," Second World Biomass Conference: Biomass for Energy, Industry, and Climate Protection, 1 (2004): 471, accessed October 20, 2014, publications.lib.chalmers .se/publication/163454-global-bioenergy-potentials-a-new -approach-using-a-model-based-assessment-of-biomass-flows-and-land.

60. L. B. Guo and R. M. Gifford, "Soil carbon stocks and land use change: a meta analysis," *Global Change Biology* 8, no. 4 (2002): 345, accessed July 23, 2014, doi: 10.1046/j.1354 -1013.2002.00486.x.

61. Nathaniel D. Mueller et al., "Closing yield gaps through nutrient and water management," *Nature* 490, no. 7419 (October 2012): 254–57, accessed June 16, 2014, doi:10.1038/nature11420.

62. Calculated from Jonathan A. Foley et al., "Solutions for a cultivated planet," *Nature* 478 (2011): 339, accessed January 20, 2015, doi:10.1038/nature10452. Based on sixteen major crops.

63. West et al., "Leverage points for improving global food security and the environment," 325.

64. Intergovernmental Panel on Climate Change, IPCC Working Group III Contribution to AR5, Ottmar Edenhofer, et al., "Technical Summary," in *Climate Change 2014: Mitigation of Climate Change*, 2014, 42, Figure TS.1, accessed January 20, 2015, www.ipcc.ch/pdf/assessment-report/ar5/wg3/ipcc_wg3_ar5_technical-summary.pdf.

65. Calculated from FAO, "Greenhouse Gas Emissions from Agriculture, Forestry, and Other Land Use," March 2014, accessed November 6, 2014, www.fao.org/resources/infographics/infographics-details/en/c/218650/; "Emissions from Forestry and Land Use," CGIAR, accessed November 13, 2014, ccafs.cgiar .org/bigfacts2014/#theme=food -emissions&subtheme=indirect-agriculture; "Peatlands," Climate Smart Agriculture, accessed November 13, 2014, www.fao.org/climate-smart-agriculture/82057/en/; Intergovernmental Panel on Climate Change, IPCC Working Group III Contribution to AR5, Ralph Sims, et al., "Chapter 8: Transport," in *Climate Change 2014: Mitigation of Climate Change*, 2014, 603, accessed January 2, 2015, report.mitigation2014.org/drafts/final-draft-postplenary/ipcc_wg3_ar5_final-draft_postplenary_chapter8.pdf.

66. "Agriculture and Food," United Nations Environment Program, accessed March 2, 2015, www.unep.org/resourceefficiency/Home/Business/SectoralActivities/Agriculture Food/tabid/78943/Default.aspx.

67. Intergovernmental Panel on Climate Change, IPCC Working Group I Contribution to AR5, Gunnar Myhre, et al., "Chapter 8: Anthropogenic and Natural Radiative Forcing," in *Climate Change 2013: The Physical Science Basis*, 2013, 714, accessed January 2, 2015, www.ipcc.ch/pdf/assessment-report/ar5/wg1/WG1AR5_Chapter08_FINAL.pdf.

68. Nadia Scialabba, FAO, personal communication with author, December 12, 2014.

69. Vermeulen, Campbell, and Ingram, "Climate Change and Food Systems," 195; "Food Emissions: Supply Chain Emissions," CGIAR and CCAFS Research Program on Climate Change, Agriculture, and Food Security, accessed December 15, 2014, ccafs.cgiar.org/bigfacts/#theme=food-emissions& subtheme=supply-chain.

70. FAO, Natural Resources Management and Environment Department, "Organic agriculture and climate change," accessed January 20, 2015, www.fao.org/DOCREP/005/Y4137E/y4137e02b.htm#96.

71. FAO, "Food Wastage Footprint: Impacts on Natural Resources," Summary Report, 2013, 6, accessed May 28, 2014, www.fao.org/docrep/018/i3347e/i3347e.pdf.

72. Intergovernmental Panel on Climate Change, Robert T. Watson, et al., "Land Use, Land-Use Change and Forestry," IPCC Special Report, 2000, accessed July 14, 2014, www.ipcc.ch/ipccreports/sres/land_use/index.php?idp=98; Earth Policy Institute and Emily E. Adams, "World Forest Area Still on the Decline," Eco-Economy Indicators, August 31, 2012, accessed May 30, 2014, www.earth-policy.org/indicators/C56; Jonathan Foley, "A Five-Step Plan to Feed the World: Step 1," *National Geographic*, May, 2014, accessed July 10, 2014, www.nationalgeographic.com/foodfeatures/feeding-9-billion/.

73. Guo and Gifford, "Soil carbon stocks and land use change," 345–360; FAO, "State of the World's Forests 2012" (Rome: FAO, 2012), accessed July 22, 2014, www.fao.org/docrep/016/i3010e/i3010e.pdf.

74. Calculated from Andreas Gattinger et al., "Soil Carbon Sequestration of Organic Crop and Livestock Systems and Potential for Accreditation by Carbon Markets," in FAO, Organic Agriculture and Climate Change Mitigation: A Report of the Round Table on Organic Agriculture and Climate Change (Rome: FAO, 2011), 16, accessed March 19, 2015, www.fao.org/docrep/015/i2537e/i2537e00.pdf.

75. FAO, P. J. Gerber, et al., "Tackling Climate Change Through Livestock—A Global Assessment of Emissions and Mitigation Opportunities" (Rome: FAO, 2013), xii, accessed January 21, 2015, www.fao.org/docrep/018/i3437e/i3437e.pdf; Gidon Eshel et al., "Land, irrigation water, greenhouse gas, and reactive nitrogen burdens of meat, eggs, and dairy production in the United States," *Proceedings of the National Academy of Sciences* 111, no. 3 (2014): 1, accessed January 21, 2015, www.pnas.org/cgi/doi/10.1073/pnas.1402183111.

76. Dario Caro et al., "Global and regional trends in greenhouse gas emissions from livestock," *Climatic Change* (2014): 210, accessed July 23, 2014, doi: 10.1007/s10584-014-1197-x.

77. William J. Ripple et al., "Ruminants, climate change and climate policy," *Nature Climate Change* 4 (2014): Figure 2, 4, doi:10.1038/nclimate2081.

78. Ibid., 3.

79. "International Decade for Action 'Water for Life' 2005–2015: Water and food security," United Nations Department of Economic and Social Affairs (UNDESA), last modified October 23, 2014, accessed December 15, 2014, www.un.org/waterforlifedecade/food_security.shtml.

80. Julian Fulton, Heather Cooley, and Peter H. Gleick, "California's Water Footprint," Pacific Institute, December 2012, 3, accessed April 17, 2015, pacinst.org/wp-content/uploads/sites/21/2013/02/ca_ftprint_full_report3.pdf.

81. University of California, Division for Agriculture and Natural Resources, Alfalfa and Food Systems Workgroup, Daniel H. Putnam, Charles G. Summers, and Steve B. Orloff, "Alfalfa Production Systems in California," in *Irrigated Alfalfa Management for Mediterranean and Desert Zones*, Publication 8287, December 2007, 7–9, 12 accessed December 15, 2014, alfalfa.ucdavis.edu/IrrigatedAlfalfa/pdfs/UCAlfalfa8287ProdSystems_free.pdf

82. James McWilliams, "Meat Makes the Planet Thirsty," *New York Times,* March 7, 2014, accessed December 31, 2014, www.nytimes.com/2014/03/08/opinion/meat-makes-the-planet-thirsty.html?mwrsm=Email&_r=0.

83. Pacific Institute, Michael Cohen, Juliet Christian-Smith, and John Berggren, "Water to Supply the Land: Irrigated Agriculture in the Colorado River Basin," May 2013, vi, accessed December 15, 2014, www.pacinst.org/wp-content/uploads/2013/05/pacinst-crb-ag.pdf.

84. Mesfin M. Mekonnen and Arjen Y. Hoekstra, "A global assessment of the water footprint of farm animal products," *Ecosystems* 15 (2012): 401–415, accessed June 30, 2014, doi:10.1007/s10021-011-9517-8.

85. Brian Machovina and Kenneth J. Feeley, "Taking a bite out of biodiversity," *Science* 343, no. 6173 (2014): 838, doi: 10.1126/science.343.6173.838-a.

86. FAO, Food and Nutrition Division, "Rice and Human Nutrition" (Rome: FAO, 2004), accessed January 29, 2015, www.fao.org/rice2004/en/f-sheet/factsheet3.pdf.

87. FAO, "Greenhouse Gas Emissions from Agriculture, Forestry and Other Land Use"; University of California at Davis, "Rice agriculture accelerates greenhouse gas emissions," October 22, 2012, accessed December 12, 2014, news.ucdavis.edu/search/news_detail.lasso?id=10382; Kees Jan van Groenigen, Craig W. Osenberg, and Bruce A. Hungate, "Increased soil emissions of potent greenhouse gases under increased atmospheric CO2," *Nature* 475 (2011): 214–216, accessed December 12, 2014, doi:10.1038/nature10176.

88. "SRI Rice," SRI International Network and Resource Center, Cornell University, sri.ciifad.cornell.edu/; Reiner Wassmann, Yasukazu Hosen, and Kay Sumfleth, "Agriculture and Climate Change: An Agenda for Negotiation in Copenhagen—Reducing Methane Emissions from Irrigated Rice," International Food Policy Research Institute, 2009, accessed November 10, 2014, www.ifpri.org/sites/default/files/publications/focus16_03.pdf.

89. Klein Ileleji, Chad Martin, and Don Jones, "Basics of Energy Production Through Anaerobic Digestion of Livestock Manure," Purdue University Extension, August 2008, www.extension.purdue.edu/extmedia/ID/ID-406-W.pdf.

90. World Bank, "Carbon Sequestration in Agricultural Soils," 67395-GLB, 2012, 6, accessed January 21, 2015, documents.worldbank.org/curated/en/2012/05/16274087/carbon-sequestration-agricultural-soils.

91. Worldwatch Institute, "Oceans Absorb Less Carbon Dioxide as Marine Systems Change," accessed May 28, 2015, www.worldwatch.org/node/6323; Feely, Sabine, and Fabry, "Carbon Dioxide and Our Ocean Legacy."

92. Meredith Niles, "Sustainable Soils: Reducing, Mitigating, and Adapting to Climate Change with Organic Agriculture," *Sustainable Development Law & Policy* 9, no. 1 (2008): 21, accessed December 11, 2014, digitalcommons.wcl.american.edu/cgi/viewcontent.cgi?article=1082&context=sdlp; U.S. Environmental Protection Agency, "Solid Waste Management and Greenhouse Gases: A Life-Cycle Assessment of Emissions and Sinks," 3rd Edition, September 2006.

93. World Bank, "Carbon Sequestration in Agricultural Soils," Report No. 67395-GLB, May 2012, accessed February 21, 2015, www-wds.worldbank.org/external/default/WDSContentServer/WDSP/IB/2012/05/18/000333038_20120518003322/Rendered/PDF/673950REVISED000CarbonSeq0Web0final.pdf.

94. Robert J. Zomer et al., "Trees on Farm: Analysis of Global Extent and Geographic Patterns of Agroforestry," ICRAF Working Paper No. 89 (Nairobi, Kenya: World Agroforestry Center, 2009), 12, accessed January 21, 2015, worldagroforestry.org/downloads/publications/PDFs/WP16263.PDF.

95. Jo Smith, Bruce D. Pearce, and Martin S. Wolfe, "Reconciling productivity with protection of the environment: Is temperate agroforestry the answer?" *Renewable Agriculture and Food Systems* 28, no. 1 (2012): 80–92, doi:10.1017/S1742170511000585.

96. Joris Aertsens, Leo De Nocker, and Anne Gobin, "Valuing the carbon sequestration potential for European agriculture," *Land Use Policy* 31 (2013): 584, accessed June 4, 2014, dx.doi.org/10.1016/j.landusepol.2012.09.003.

97. World Bank, "Carbon Sequestration in Agricultural Soils," 47.

98. Ibid., xxi–xxiii.

99. Andrew R. Zimmerman, Bin Gao, and Mi-Youn Ahn, "Positive and negative carbon mineralization priming effects among a variety of biochar-amended soils," *Soil Biology and Biochemistry* 43, no. 6 (2011): 1169–1179, accessed December 12, 2014, doi:10.1016/j.soilbio.2011.02.005; Andrew Crane-Droesch et al., "Heterogeneous global crop yield response to biochar: A meta-regression analysis," *Environmental Research Letters* 8 (2013): 2, accessed December 12, 2014, doi:10.1088/1748-9326/8/4/044049; World Bank, "Carbon Sequestration in Agricultural Soils," xxi–xxiii.

100. S. Krishnakumar et al., "Impact of Biochar on Soil Health," *International Journal of Advanced Research* 2, no. 4 (2014): 933-950, accessed April 9, 2015, journalijar.com/uploads/568_IJAR-2264.pdf; Dominic Woolf et al., "Sustainable biochar to mitigate global climate change," *Nature Communications* 1, no. 56 (2010): accessed December 12, 2014, doi:10.1038/ncomms1053.

101. Rattan Lal, "Managing soils and ecosystems for mitigating anthropogenic carbon emissions and advancing global food security," *BioScience* 60, no. 9 (2010): 718, accessed November 12, 2014, www.jstor.org/stable/10.1525/bio.2010.60.9.8; Rattan Lal, Ohio State University, personal communication with authors, January 5 and 6, 2015.

102. Rattan Lal, "Soil carbon sequestration impacts on global climate change and food security," *Science* 304 (2004): 1623, doi:10.1126/science.1097396; Klaus Lorenz and Rattan Lal, "Soil organic carbon sequestration in agroforestry systems. A review," *Agronomy for Sustainable Development* 34 (2014): 447, doi: 10.1007/s13593-014-0212-y; Aertsens, Nocker, and Gobin, "Valuing the carbon sequestration potential," 585; Lal, "Managing soils and ecosystems," 708.

103. Allan Savory, "How to fight desertification and reverse climate change," TED Talk, 22:19, from TED2013 in February 2013, accessed January 28, 2015, www.ted.com/talks/allan_savory_how_to_green_the_world_s_deserts_and_reverse_climate_change.

104. Based on Rattan Lal, personal communication with author, December 9, 2014.

105. David D. Briske et al., "The Savory Method Can Not Green Deserts or Reverse Climate Change," *Rangelands* 35, no. 5 (2013): 72–74, accessed July 18, 2014, doi: 10.2111/RANGELANDS-D-13-00044.1; John Carter et al., "Holistic Management: Misinformation on the Science of Grazed Ecosystems," *International Journal of Biodiversity* (2014), accessed July 18, 2014, doi:10.1155/2014/163431.

106. Wiebke Volkmann, "Managing Community Based Rangelands in Namibia," in

IFOAM FAO Organic Agriculture: African Experiences in Resiliency and Sustainability, ed. Raymond Auerbach, Gunnar Rundgren, and Nadia El-Hage Scialabba (Rome, May 2013), 38–97, www.fao.org/docrep/018/i3294e/i3294e.pdf.

107. Elinor Ostrom, *Governing the commons: The evolution of institutions for collective action* (New York: Cambridge University Press, 1990).

108. Jonathan L. Batchelor et al., "Restoration of Riparian Areas Following the Removal of Cattle in the Northwestern Great Basin," *Environmental Management* (2015): 930, 935, 938, doi:10.1007/s00267-014-0436-2.

109. Ripple et al., "Ruminants, climate change and climate policy," 2–5; FAO, Gerber, et al., "Tackling Livestock Through Climate Change," estimated from Figure 27A, 54.

110. Tara Garnett, "Where are the best opportunities for reducing greenhouse gas emissions in the food system (including the food chain)?" *Food Policy* 36 (2011): 26, doi:10.1016/j.foodpol.2010.10.010.

111. Jessica Bellarby et al., "Livestock GHG emissions and mitigation potential in Europe," *Global Change Biology* 19 (2013): 3–18, accessed December 15, 2014, doi: 10.1111/j.1365-2486.2012.02786.x.

112. Tony Weis, *The Ecological Hoofprint, The Global Burden of Industrial Livestock* (London: Zed Books, 2013).

113. Calculated from FAOSTAT (Production, Livestock Primary, Item: Meat, Total, Area: World, Year: 1961–2012, Aggregation: average; accessed December 15, 2014), faostat3 .fao.org/browse/Q/QL/E; Calculated from FAOSTAT (Population, Annual Population, World, 1961–2010; accessed December 15, 2014), faostat3.fao.org/browse/O/OA/E; Worldwatch Institute, "Peak Meat Production Strains Land and Water Resources," August 2014, accessed December 15, 2014, www.worldwatch.org/peak-meat-production-strains -land-and-water-resources-1.

114. Ripple et al., "Ruminants, climate change and climate policy," 3–4.

115. David Tilman and Michael Clark, "Global diets link environmental sustainability and human health," *Nature* 515 (2014): 518–522, accessed December 15, 2014, doi:10.1038/ nature13959; An Pan et al., "Red Meat Consumption and Mortality: Results from Two Prospective Cohort Studies," *Archives of Internal Medicine* 172, no. 7 (2012): 555–563, accessed December 15, 2014, doi: 10.1001/archinternmed.2011.2287; Rashmi Sinha et al., "Meat Intake and Mortality: A Prospective Study of Over Half a Million People," *Archives of Internal Medicine* 169, no. 6 (2009): 562–571, accessed December 15, 2014, doi:10.1001/ archinternmed.2009.6.

116. *$11 Trillion Dollar Reward*, Union Concerned Scientists, Executive Summary, 2013, 2, www.ucsusa.org/sites/default/files/legacy/assets/documents/food_and_agriculture/ 11-trillion-reward.pdf.

117. U.S. Department of Agriculture and Department of Health and Human Services, "Scientific Report of the 2015 Dietary Guidelines Advisory Committee," USDA and DSHS, January 28, 2015, accessed February 24, 2015, www.health.gov/dietaryguidelines/2015 -scientific-report/.

118. Calculated from U.S. Department of Agriculture, Agricultural Research Service, "Table 2. Nutrient Intakes from Food: Mean Amounts and Percentages of Calories from Protein, Carbohydrate, Fat, and Alcohol, One Day, 2005–2006," 2008, accessed July 23, 2014, www.ars.usda.gov/SP2UserFiles/Place/12355000/pdf/0506/Table_2_NIF_05.pdf; "Nutrition for Everyone: Protein," Centers for Disease Control and Prevention, last modified October 4, 2012, accessed July 23, 2014, www.cdc.gov/nutrition/everyone/basics/ protein.html#How%20much%20protein; "Adults' Daily Protein Intake Much More Than

Recommended," National Center for Health Statistics, last modified March 3, 2010, accessed June 10, 2014, nchstats.com/2010/03/03/adults'-daily-protein-intake-much-more-than -recommended/.

119. Erin Coleman, "How Many Grams of Protein Are in an Eight-Ounce Top Sirloin?" Healthy Eating, SFGate, accessed February 20, 2015, healthyeating.sfgate.com/many -grams-protein-eightounce-top-sirloin-8184.html.

120. Dietary Proteins, National Institutes of Health, Medline Plus, accessed May 28, 2015, www.nlm.nih.gov/medlineplus/dietaryproteins.html.

121. "Heart Disease and Diet," Medline Plus, last modified July 16, 2013, accessed July 21, 2014, www.nlm.nih.gov/medlineplus/ency/article/002436.htm.

122. Vincent H. Smith, "Bloated Farm Subsidies: Will the 2013 Farm Bill Really Cut the Fat?" Research Summary, 1, January 21, 2015, mercatus.org/sites/default/files/ Smith_FarmBill_RS.pdf; Union of Concerned Scientists and Doug Gurian-Sherman, "CAFOs Uncovered: The Untold Cost of Confined Animal Feeding Operations," April 2008, 29, accessed December 15, 2014, www.ucsusa.org/sites/default/files/legacy/assets/ documents/food_and_agriculture/cafos-uncovered.pdf.

123. U.S. Department of Agriculture, "Census of Agriculture," 2012, Census Volume 1, Chapter 1, "U.S. National Level Data," Table 37, "Specified Crops by Acres Harvested: 2012 and 2007," 31–32, accessed May 28, 2015, www.agcensus.usda.gov/Publications/2012/ Full_Report/Volume_1,_Chapter_1_US/st99_1_037_037.pdf; to access all "Census of Agriculture," Volume 1, Chapter 1 tables, please visit www.agcensus.usda.gov/Publications/2012/ Full_Report/Volume_1,_Chapter_1_US.

124. "Trends in Current Cigarette Smoking Among High School Students and Adults, United States, 1965–2011," Smoking and Tobacco Use, Centers for Disease Control and Prevention, November 14, 2013, accessed September 2, 2014, www.cdc.gov/tobacco/ data_statistics/tables/trends/cig_smoking/; "Apr 1, 1970: Nixon signs legislation banning cigarette ads on TV and radio," This Day in History, accessed September 2, 2014, www.history.com/this-day-in-history/nixon-signs-legislation-banning-cigarette-ads -on-tv-and-radio.

125. U.S. Department of Agriculture, Economic Research Service, "Red meat, poultry, and fish (boneless weight): Per capita availability," data set, downloaded from Food Availability (Per Capita) Data System, last modified February 1, 2014, accessed March 23, 2015, www.ers.usda.gov/data-products/food-availability-(per-capita)-data-system/. aspx#26705.

126. USDA, ERS, "Red meat, poultry, and fish (boneless weight)"; U.S. Department of Agriculture, Foreign Agricultural Service, "Livestock and Poultry: World Markets and Trade," October 2014, 6, accessed April 6, 2015, apps.fas.usda.gov/psdonline/circulars/ livestock_poultry.PDF.

127. "Emissions from Forestry and Land Use," CGIAR; "Agriculture and Food," United Nations Environment Program, accessed January 6, 2015, www.unep.org/resourceefficiency/ Home/Business/SectoralActivities/AgricultureFood/tabid/78943/Default.aspx.

128. United Nations Environment Program, "The Emissions Gap Report 2012" (Nairobi: UNEP, 2012), 41, accessed June 9, 2014, www.unep.org/pdf/2012gapreport.pdf.

129. Doug Boucher, "How Brazil Has Dramatically Reduced Tropical Deforestation," *Solutions Journal* 5, no. 2 (2014): 67; Union of Concerned Scientists, Doug Boucher, et al., "Deforestation Success Stories: Tropical Nations Where Forest Protection and Reforestation Policies Have Worked," June 2014, 13, accessed November 14, 2014, www.ucsusa.org/sites/default/files/legacy/assets/documents/global_warming/

deforestation-success-stories-2014.pdf; "Global meat demand plows up Brazil's 'underground forest,'" *The Daily Climate,* November 10, 2014, accessed December 15, 2014, www.dailyclimate.org/tdc-newsroom/2014/11/brazil-meat-cerrado-deforestation; Vincent Bevins, "Brazil Says Rate of Amazon Deforestation Up for First Time in Years," *L.A. Times,* September 10, 2014, accessed January 12, 2015, www.latimes.com/world/mexico-americas/la-fg-brazil-amazon-deforestation-rises-20140910-story.html.

130. World Bank Institute, Climate Change Unit, "Rehabilitating a Degraded Watershed: A Case Study from China's Loess Plateau," World Bank Group, 2010, 4, 15, 9, accessed July 14, 2014, wbi.worldbank.org/wbi/Data/wbi/wbicms/files/drupal-acquia/wbi/0928313 -03-31-10.pdf.

131. Juergen Voegele, Agriculture Global Practice Director, World Bank, phone interview with author, July 8, 2014; World Bank Institute, Xie, et al., "Rehabilitating a Degraded Watershed," 9.

132. Juergen Voegele, World Bank, phone interview with author, July 8, 2014.

133. World Bank Institute, Xie, et al., "Rehabilitating a Degraded Watershed," 14; World Bank, "Project Performance Assessment Report: People's Republic of China," Report No. 41122, October 4, 2007, 22, accessed July 7, 2014, www-wds.worldbank.org/external/default/WDSContentServer/WDSP/IB/2007/10/31/000020953_20071031102004/Rendered/PDF/41122.pdf.

134. World Resources Institute, Kathleen Buckingham, et al., "Taking Culture into Account in Restoring China's Loess Plateau," December 15, 2014, accessed December 30, 2014, www.wri.org/blog/2014/12/taking-culture-account-restoring-china%E2%80%99s -loess-plateau.

135. World Bank, "Project Performance Assessment Report," 23.

136. Juergen Voegele, World Bank, e-mail communication with author, October 27, 2014; World Bank, "Project Performance Assessment Report," 23; "Greenhouse Gas Equivalencies Calculator," U.S. Environmental Protection Agency, April 2014, www.epa.gov/cleanenergy/energy-resources/calculator.html#results.

137. World Bank, "Restoring China's Loess Plateau," March 15, 2007, accessed July 14, 2014, www.worldbank.org/en/news/feature/2007/03/15/restoring-chinas-loess -plateau; World Bank Institute, Xie et al., "Rehabilitating a Degraded Watershed," 10.

138. Juergen Voegele, World Bank, phone interview with author, July 8, 2014.

139. "Rwanda Report Shows Successes and Challenges of Post-Conflict Sustainable Development," United Nations Environment Program, November 16, 2011, accessed July 22, 2014, www.unep.org/newscentre/default.aspx?DocumentID=2659&ArticleID=8944.

140. Government of India, Ministry of Environment and Forests, "National Mission for a Green India," accessed February 21, 2015, www.naeb.nic.in/documents/GIM_Brochure_26March.pdf; "200 cr trees to be planted along highways: Gadkari," *The Hindu*, June 13, 2014, accessed April 9, 2015, www.thehindu.com/sci-tech/energy -and-environment/200-cr-trees-to-be-planted-along-highways-gadkari/article6111239 .ece?utm_source=RSS_Feed&utm_medium=RSS&utm_campaign=RSS_Syndication.

141. Plant for the Planet, accessed April 2, 2015, www.plant-for-the-planet.org/en/home#intro.

142. International Labour Organization, "Global Unemployment Trends 2014: Risk of a Jobless Recovery?" (Geneva, ILO: January 2014), 11, accessed June 3, 2014, www.ilo.org/wcmsp5/groups/public/---dgreports/---dcomm/---publ/documents/publication/wcms_233953.pdf.

143. World Bank, "World Development Report 2008: Agriculture for Development" (Washington, DC: World Bank, 2007), 90, accessed June 4, 2014, siteresources.worldbank.org/INTWDRS/Resources/477365-1327599046334/WDR_00_book.pdf.

144. U.S. Department of Agriculture, Natural Resources Conservation Service, "Unlock the Secrets in the Soil: 2014 Soil Planner," 2014, accessed April 25, 2014, nrcspad.sc.egov.usda.gov/distributioncenter/pdf.aspx?productID=1019.

145. Andreas Gattinger et al., "Enhanced top soil carbon stocks under organic farming," *Proceedings of the National Academy of Sciences* 109, no. 44 (2012): 18227, www.pnas.org/cgi/doi/10.1073/pnas.1209429109.

146. D. W. Lotter, R. Seidel, and W. Liebhardt, "The performance of organic and conventional cropping systems in an extreme climate year," *American Journal of Alternative Agriculture* 18, no. 2 (2003): 1, accessed May 7, 2014, donlotter.net/lotter_ajaa_article.pdf; Binju Abraham et al., "The System of Crop Intensification: Agroecological Innovations for Improving Agricultural Production, Food Security, and Resilience to Climate Change," SRI International Network and Resources Center (Ithaca, NY: Cornell, 2014), 60–62, accessed May 6, 2014, sri.ciifad.cornell.edu/aboutsri/othercrops/SCImonograph_SRIRice2014.pdf.

147. "EverGreen Agriculture: Re-greening Africa's landscape" (Nairobi: World Agroforestry Center, 2013), 1–6, www.ard-europe.org/fileadmin/SITE_MASTER/content/eiard/Documents/Impact_case_studies_2013/ICRAF_-_EverGreen_agriculture.pdf.

148. Peter M. Rosset and Maria Elena Martínez-Torre, "Rural Social Movements and Agroecology: Context, Theory, and Process," *Ecology and Society* 17, no. 3 (2012): 6, accessed June 4, 2014, www.ecologyandsociety.org/vol17/iss3/art17/#ms_abstract.

149. UN Economics Commission for Latin America and the Caribbean, "Appendix X—Agricultural Losses in Honduras following Hurricane Mitch," in *Handbook for Estimating the Socio-economic and Environmental Effects of Disasters*, 2003, Appendix X, accessed June 4, 2014, www.cepal.org/publicaciones/xml/4/12774/lcmexg5i_VOLUME_IIIb.pdf.

150. Eric Holt-Giménez, "Measuring agroecological resistance to Hurricane Mitch," *LEISA Magazine* 17, no. 1 (2001): 19, accessed November 14, 2014, www.agriculturesnetwork.org/magazines/global/coping-with-disaster/measuring-farmers-agroecological-resistance-to/at_download/article_pdf; Eric Holt-Giménez, "Measuring farmers' agroecological resistance after Hurricane Mitch in Nicaragua: A case study in participatory, sustainable land management impact monitoring," *Agriculture, Ecosystems and Environment* 93 (2002): 93, accessed November 13, 2014, www.bio-nica.info/biblioteca/HoltGimenez2002AgroecolgyNic.pdf.

151. Holt-Giménez, "Measuring farmers' agroecological resistance," 19.

152. Chris Reij, Gray Tappan, and Melinda Smale, "Agroenvironmental Transformation in the Sahel," Discussion Paper 00914 (Washington, DC: IFPRI, 2009), 2, 7, 19, accessed July 9, 2014, www.ifpri.org/sites/default/files/publications/ifpridp00914.pdf.

153. Chris Reij, "Food security and water in Africa's drylands," Africa Re-greening Initiatives, March 8, 2012, africa-regreening.blogspot.com/2012/03/food-security-and-water-in-africas.html; World Bank [Data, Indicator, Poverty headcount ratio at $2 a day (PPP) (% of population), Niger, 2011; accessed January 21, 2015], data.worldbank.org/indicator/SI.POV.2DAY.

154. Reij, Tappan, and Smale, "Agroenvironmental Transformation in the Sahel," 2, 7, 19.

155. "Green Wall for the Sahara: Opportunity Local Regeneration Initiatives?" Both Ends: Connecting People for Change, accessed June 3, 2014, www.bothends.org/en/Themes/Projects/newsitem/87/Green-Wall-for-the-Sahara-opportunity-local-regeneration-initiatives-.

156. Chris Reij, "Learning from African Farmers: How 'Re-greening' Boosts Food Security, Curbs Climate Change," World Resources Institute (blog), June 27, 2013, www.wri.org/blog/2013/06/learning-african-farmers-how-%E2%80%9Cre-greening%E2%80%9D-boosts-food-security-curbs-climate-change.

157. Alex Perry, "Land of Hope," *Time,* December 13, 2010, accessed July 14, 2014, content.time.com/time/magazine/article/0,9171,2034507,00.html.

158. Loewenberg, "Humanitarian response inadequate."

159. Samuel Loewenberg, "The Famine Next Time," *New York Times,* November 26, 2011, accessed April 30, 2015, www.nytimes.com/2011/11/27/opinion/sunday/in-kenya-famines-lessons.html?_r=0.

MYTH 3: Only Industrial Agriculture & GMOs Can Feed a Hungry World

1. Cynthia Hewitt de Alcántara, "The 'Green Revolution' as history: The Mexican experience," *Development and Change* 5, no. 2 (1974): 25, doi: 10.1111/j.1467-7660.1974.tb00655.x.

2. Frank Miller, "Knowledge and Power: Anthropology, Policy Research, and the Green Revolution," *American Ethnologist* 4, no. 1 (1977):191, www.jstor.org/stable/643530; Philippine Institute for Development Studies, Eulito U. Bautista, and Evelyn F. Javier, "The Evolution of Rice Production Practices," Discussion Paper Series No. 2005–14, July 2005, accessed August 5, 2014, www.eaber.org/sites/default/files/documents/PIDS_Bautista_2005.pdf.

3. Peter Rosset, "Lessons from the Green Revolution," March/April 2000, 2, accessed May 26, 2015, www.soc.iastate.edu/sapp/greenrevolution.pdf.

4. Gordon Conway, *The Doubly Green Revolution: Food for All in the 21st Century* (Ithaca, NY: Cornell University Press, 1997), 44.

5. Calculated from FAOSTAT (Production, Crops, Country: India, Element: Area harvested, Item: Wheat, Years: 1967–89; accessed October 24, 2014), faostat3.fao.org/download/Q/QC/E.

6. Oxfam-Solidarity and Stephane Parmentier, "Scaling Up Agroecological Approaches: What, Why, and How?" Discussion Paper, January 2014, 14–17, accessed August 6, 2014, www.ikgroeimee.be/uploads/assets/332/1390912349733-201401%20Scaling-up%20agroecology,%20what, %20why%20and%20how%20-OxfamSol-FINAL.pdf.

7. Calculated from FAOSTAT [Food Balance, Food Supply-Crops, Primary Equivalent, Country: India, Element: Food Supply Quantity (kg/capita/yr), Items Aggregated: Cereals –Excluding Beer + (Total), Year: 1961–1980; accessed March 6, 2014], faostat3.fao.org/download/FB/CC/E.

8. FAO, "Women and the Green Revolution," Women and Population Division, Sustainable Development Department, accessed May 26, 2015, www.fao.org/docrep/x0171e/x0171e04.htm.

9. Calculated from FAOSTAT [Production, Production Indices, Country: India, Item: Food (PIN) + (TOTAL), Element: Net per Capita Production Index Numbers (2004–06 = 100), Year: 1990–2012; accessed May 20, 2014], faostat3.fao.org/download/Q/*/E; FAO, "State of Food Insecurity in the World," 2015, 42.

10. Raju J. Das, "The green revolution and poverty: a theoretical and empirical examination of the relation between technology and society," *Geoforum* 33 (2002): 67, accessed December 1, 2014, is.muni.cz/el/1423/jaro2012/HEN437/um/GreenRevDas.pdf.

11. Donald K. Freebairn, "Did the Green Revolution Concentrate Incomes? A Quantitative Study of Research Reports," *World Development* 23, no. 2 (1995): 265, accessed December 1, 2014, doi: 0305-750X(94)00116-2.

12. Fritjof Capra and Pier Luigi Luisi, *The Systems View of Life: A Unifying Vision* (Cambridge: Cambridge University Press: 2014), xiii.

13. Milton Friedman and Rose Friedman, *Free to Choose: A Personal Statement* (New York: Harcourt, Inc., 1980).

14. USDA ERS, Coleman-Jensen, Gregory, and Singh, "Household Food Insecurity," 6.

15. U.S. Department of Agriculture, Economic Research Service, Alisha Coleman-Jensen, and Mark Nord, "Food Security Status of U.S. Households in 2012," September 2013, www.ers.usda.gov/topics/food-nutrition-assistance/food-security-in-the-us/key-statistics-graphics.aspx#.Uyik4a1dWCI.

16. U.S. Bureau of the Census, "Chapter 2, Farms: Number, Use of Land, Size of Farm," in Census of Agriculture, 1969, Volume II, General Report (Washington, DC, 1973), 14, accessed November 21, 2014, usda.mannlib.cornell.edu/usda/AgCensusImages/1969/02/02/1969-02-02.pdf; U.S. Department of Agriculture, "Census of Agriculture, 2012: United States, Summary and State Data," Volume 1, Geographic Area Series, Part 51 (Washington D.C., 2012), 7, accessed November 21, 2014, www.agcensus.usda.gov/Publications/2012/Full_Report/Volume_1,_Chapter_1_US/usv1.pdf.

17. "Farm Household Income (Historical)," Economic Research Service, U.S. Department of Agriculture, February 11, 2014, accessed February 19, 2014, www.ers.usda.gov/topics/farm-economy/farm-household-well-being/farm-household-income-(historical).aspx#.UwTZmZEupQA; "Glossary—Farm Typology," Economic Research Service, U.S. Department of Agriculture, last modified February 11, 2014, accessed May 12, 2014, www.ers.usda.gov/topics/farm-economy/farm-household-well-being/glossary.aspx#Farmtypology; Timothy A. Wise and Alicia Harvie, "Boom for Whom? Family Farmers Saw Lower On-Farm Income Despite High Prices," Global Development and Environment Institute, Policy Brief No. 09-02 (Tufts University: February, 2009), 2, accessed May 12, 2014, www.ase.tufts.edu/gdae/Pubs/rp/PB09-02BoomForWhomFeb09.pdf.

18. Smith, "Bloated Farm Subsidies," 2.

19. U.S. Department of Agriculture, Economic Research Service, Cynthia Nickerson, et al., "Trends in US Farmland Values and Ownership," Economic Information Bulletin No. 92, February 2012, i, accessed January 9, 2014, www.ers.usda.gov/media/377487/eib92_2_.pdf.

20. Max Kutner, "Death on the Farm," *Newsweek*, April 10, 2014, www.newsweek.com/death-farm-248127.

21. John D. Beard et al. "Pesticide Exposure and Depression Among Male Private Pesticide Applicators in the Agricultural Health Study," *Environmental Health Perspective* 122, no. 9 (2014): 1, accessed January 21, 2015, doi: 10.1289/ehp.1307450; Dan Nosowitz, "Landmark 20-Year Study Finds Pesticides Linked to Depression In Farmers," *Modern Farmer*, November 7, 2014, accessed December 1, 2014, modernfarmer.com/2014/11/landmark-20-year-study-finds-pesticides-cause-depression-farmers/.

22. Ivette Perfecto, John Vandermeer, and Angus Wright, *Nature's Matrix: Linking Agriculture, Conservation and Food Sovereignty* (London: Earthscan, 2009), 48.

23. Adjusted for inflation, calculated from Wen Jun Zhang, Fu Bin Jiang, and Jian Feng Ou, "Global pesticide consumption and pollution: With China as a focus," *Proceedings of the International Academy of Ecology and Environmental Sciences* 1, no. 2 (2011): 126,

accessed March 20, 2014, www.biological-control.org/chemicalpest/Global-pesticide -consumption-pollution.pdf.

24. ETC Group, "Who Will Control the Green Economy?" November 2011, 25, accessed February 22, 2015, www.etcgroup.org/sites/www.etcgroup.org/files/publication/pdf_file/ ETC_wwctge_4web_Dec2011.pdf.

25. U.S. Environmental Protection Agency, Arthur Grube, et al., "Pesticides Industry Sales and Usage: 2006 and 2007 Market Estimates," February 2011, 13, accessed December 2, 2014, www.epa.gov/opp00001/pestsales/07pestsales/market_estimates2007 .pdf.

26. Calculated from IFADATA (Activity: Consumption, Region: World, Year: 1961–2011, Total N+P2O5+K2O; accessed May 5, 2014), www.fertilizer.org/En/Statistics/IFADATA. aspx?WebsiteKey= 411e9724-4bda-422f-abfc-8152ed74f306.

27. T. Vijay Kumar et al., "Ecologically Sound, Economically Viable: Commu- nity Managed Sustainable Agriculture in Andhra Pradesh, India" (Washington, DC: World Bank, 2009), 7, siteresources.worldbank.org/EXTSOCIALDEVELOPMENT/ Resources/244362-1278965574032/CMSA-Final.pdf.

28. Penelope Macrae, "India Faces Huge Job in Giving Bank Accounts to All," *Busi- ness Insider,* August 27, 2014, accessed December 1, 2014, www.businessinsider.com/ afp-india-faces-huge-job-in-giving-bank-accounts-to-all-2014-8#ixzz3C0CL4bo2; World Bank, Agriculture and Rural Development Department, Hans Binswanger, and Shahidur Khandker, "The Impact of Formal Finance on the Rural Economy of India," WPS 949, August 31, 1992, accessed December 3, 2014, econ.worldbank.org/external/default/ main?pagePK=64165259&piPK=64165421&theSitePK=469372&menuPK=64216926&ent ityID=000009265_3961003061022.

29. Calculated from "Farmer depend [*sic*] on private moneylenders," *Times of India*, August 2, 2012, accessed October 7, 2014, timesofindia.indiatimes.com/city/nagpur/ Farmer-depend-on-private-moneylenders/articleshow/15321269.cms; Pushkar Maitra et al., "Financing Smallholder Agriculture: An Experiment with Agent-Intermediated Microloans in India," March 2014, 9, accessed October 8, 2014, www.bu.edu/econ/ files/2012/11/Impacts_paper_v3_Mar202014.pdf.

30. Macrae, "India Faces Huge Job in Giving Bank Accounts to All."

31. Ellen Berry, "After Farmers Commit Suicide, Debts Fall on Families in India," *New York Times*, February 22, 2014, accessed on May 2, 2014, www.nytimes.com/2014/02/23/ world/asia/after-farmers-commit-suicide-debts-fall-on-families-in-india.html; Sonora Jha, "How Suicide and Politics Mix in India," *New York Times*, April 24, 2014, accessed on May 2, 2014, www.nytimes.com/2014/04/25/opinion/how-suicide-and-politics-mix -in-india.html.

32. FAO, "The State of Food Insecurity in the World: How Does International Price Volatility Affect Domestic Economies and Food Security?" (Rome, FAO: 2011), 29, accessed February 24, 2014, www.fao.org/docrep/014/i2330e/i2330e.pdf.

33. Calculated from "Fertilizer Policy: Subsidy outgo on P&K fertilizers and urea during the last 10 years," Ministry of Chemicals and Fertilizers, Department of Fertilizers, Govern- ment of India, accessed February 11, 2015, fert.nic.in/page/fertilizer-policy; "Key Features of Budget 2013–2014," Union Budget of India, accessed February 11, 2014, indiabudget .nic.in/budget2013-2014/ub2013-14/bh/bh1.pdf; Institute for Defense Studies and Analy- sis and Laxman K Behera, "India's Defence Budget 2013–14: A Bumpy Road Ahead," March 4, 2013, accessed February 11, 2015, www.idsa.in/idsacomments/IndiasDefence Budget2013-14_lkbehera_040313.

34. African Centre for Biosafety, "Running to Stand Still: Small-Scale Farmers and the Green Revolution in Malawi," September 2014, vi–xvii, accessed December 1, 2014, www .acbio.org.za/images/stories/dmdocuments/Malawi-running-to-stand-still.pdf.

35. Ibid., vi, xv; African Centre for Biosafety, "Resources transferred from small-scale farmers to multinational agribusinesses in Malawi's Green Revolution," October 6, 2014, accessed December 1, 2014, www.acbio.org.za/index.php/media/64-media-releases/468 -resources-transferred-from-small-scale-farmers-to-multinational-agribusinesses-in -malawis-green-revolution."

36. "Desertification, Land Degradation, and Drought (DLDD)—Some Global Facts and Figures," United Nations Convention to Combat Desertification, accessed March 6, 2014, www.unccd.int/Lists/SiteDocumentLibrary/WDCD/DLDD%20Facts.pdf; David Pimentel and Michael Burgess, "Soil Erosion Threatens Food Production," *Agriculture* 3, no. 3 (2013): 447–448, accessed May 2, 2014, doi: 10.3390/agriculture3030443.

37. P. R. Hepperly, D. Douds, Jr., and R. Seidel, "The Rodale Institute Farming Systems Trial 1981 to 2005: Long-term analysis of organic and conventional maize and soybean cropping systems," in *Long-term Field Experiments in Organic Farming*, ed J. Raupp et al. (2006): 15–31; United Nations Environmental Programme, Reynaldo Victoria, et al., "The Benefits of Soil Organic Carbon: Managing soils for multiple economic, societal and environmental benefits," UNEP Year Book 2012, accessed January 6, 2015, www.unep .org/yearbook/2012/pdfs/UYB_2012_CH_2.pdf; U.S. Department of Agriculture, Natural Resources Conservation Service, "Healthy Soils Are: High in Organic Matter," 1, accessed July 30, 2014, nrcspad.sc.egov.usda.gov/DistributionCenter/pdf.aspx?productID=1024.

38. United Nations Convention to Combat Desertification, "Zero Net Land Degradation: A Sustainable Development Goal for Rio+20," UNCCD Secretariat Policy Brief, May 2012, 11, accessed March 17, 2015, www.unccd.int/Lists/SiteDocumentLibrary/Rio+20/ UNCCD_PolicyBrief_ZeroNetLandDegradation.pdf.

39. This calculation assumes that a full-bed pickup truck can hold 2.5 cubic yards of soil and that one cubic yard of soil weighs approximately 2,200 pounds. The calculation assumes a world population of 7.2 billion people.

40. David Pimentel, "Soil Erosion: A Food and Environmental Threat," *Journal of the Environment, Development and Sustainability* 8 (2006): 119, 123, accessed March 19, 2015, doi: 10.1007/s10668-005-1262-8.

41. David Pimentel et al., "Environmental and Economic Costs of Soil Erosion and Conservation Benefits," *Science* 267, no. 5201 (1995): 1117–1123, accessed January 9, 2015, www.sciencemag.org/content/suppl/2003/11/19/302.5649.1356.DC1/267-5201 -1117.pdf.

42. Calculated from FAOSTAT [Inputs, Fertilizers, Country: United States, Element: Consumption in Nutrients, Items: Nitrogen Fertilizers (N total nutrients), Phosphate fertilizers (P205 total nutrients), and Potash fertilizers (K20 total nutrients), Year: 2012; accessed January 10, 2015], faostat3.fao.org/download/R/RF/E.

43. R. L. Mulvaney, S. A. Khan, and T. R. Ellsworth, "Synthetic nitrogen fertilizers deplete soil nitrogen: A global dilemma for sustainable cereal production," *Journal of Environmental Quality* 38 (2009): 2295–2314, accessed May 7, 2014, doi:10.2134/jeq2008.0527.

44. Saeed Khan et al., "The Myth of Nitrogen Fertilization for Soil Carbon Sequestration," *Journal of Environmental Quality* 36, no. 6 (2007): 1821-1832.

45. James J. Hoorman and Rafiq Islam, "Understanding soil microbes and nutrient recycling," Ohio State University Extension, 2010, 1, accessed May 27, 2015, ohioline .osu.edu/sag-fact/pdf/0016.pdf.

46. Dr. Kristine Nichols, Chief Scientist, Rodale Institute, former Research Soil Microbiologist, U.S. Department of Agriculture, personal communication with the author, December, 2014.

47. Phillip Barak, "Long term effects of nitrogen fertilizers on soil acidity," University of Wisconsin–Madison Department of Soil Science, accessed December 16, 2014, www.soils .wisc.edu/extension/wcmc/proceedings/2A.barak.pdf; J. L. Schroder et al., "Soil acidification from long-term use of nitrogen fertilizers on winter wheat," *Soil Science Society of America Journal* 75, no. 3, (2011): 957–964.

48. "Fact Sheets: Soil Acidity," Soilquality.org, accessed December 16, 2014, soilquality. org.au/factsheets/soil-acidity; "Soil Susceptibility to Compaction," European Commission Joint Research Centre, last updated April 19, 2012, accessed July 31, 2014, eusoils.jrc .ec.europa.eu/library/themes/compaction/susceptibility.html; U.S. Department of Agriculture, Natural Resources Conservation Service, "Soil Compaction: Detection, Prevention, and Alleviation," June 2003, accessed July 31, 2014, www.nrcs.usda.gov/Internet/ FSE_DOCUMENTS/nrcs142p2_053258.pdf.

49. "Threats to Soil Quality," Scottish Environment Protection Agency, accessed July 31, 2014, www.sepa.org.uk/land/soil/threats_to_soil_quality.aspx.

50. West et al., "Leverage points for improving global food security and the environment," 326; Vaclav Smil, "Nitrogen in crop production: An account of global flows," *Global Geochemical Cycles* 13, no. 2 (1999): 647, accessed January 10, 2015, www.vaclavsmil. com/uploads/smil-article-global-biogeochemical-cycles.1999.pdf.

51. L. E. Drinkwater, P. Wagoner, and M. Sarrantonio, "Legume-based cropping systems have reduced carbon and nitrogen losses," *Nature* 396 (1998): 262–264, doi:10.1038/24376.

52. Robert J. Diaz and Rutger Rosenberg, "Spreading Dead Zones and Consequences for Marine Ecosystems," *Science* 321, no. 5891 (2008): 926, doi:10.1126/science.1156401.

53. Zofia E. Taranu et al., "Acceleration of cyanobacterial dominance in north temperate-subarctic lakes during the Anthropocene," *Ecology Letters* (2015): 1, doi: 10.1111/ele.12420.

54. World Health Organization, "Nitrate and nitrite in drinking water: Background document for development of WHO Guidelines for drinking-water quality," 2011, 3, accessed May 12, 2014, www.who.int/water_sanitation_health/dwq/chemicals/ nitratenitrite2ndadd.pdf.

55. Neil M. Dubrovsky, U.S. Geological Survey, personal communication with author, December 02, 2014; Leslie A. DeSimone, Pixie A. Hamilton, and Robert J. Gilliom,"Quality of Water from Domestic Wells in Principal Aquifers of the United States, 1991–2004, Overview of Major Findings," *U.S. Geological Survey Circular* 1332, 2009, 25–26, December 5, 2014, pubs.usgs.gov/circ/circ1332/; U.S. Geological Survey, Matthew C. Larsen, Pixie A. Hamilton, and William H. Workheiser, "Water Quality Status and Trends in the United States," *Monitoring Water Quality* (2013): 24, accessed May 9, 2014, www.usgs .gov/climate_landuse/contacts/presents/Larsen_wq_2013.pdf.

56. "Water-related diseases: Methaemoglobinemia," World Health Organization, 2014, accessed May 9, 2014, www.who.int/water_sanitation_health/diseases/methaemoglob/ en/; Cristina M. Villanueva et al., "Assessing Exposure and Health Consequences of Chemicals in Drinking Water: Current State of Knowledge and Research Needs," *Environmental Health Perspectives* (2014): 12, accessed May 12, 2014, dx.doi.org/10.1289/ehp.1206229; Mary H. Ward et al., "Workgroup Report: Drinking-Water Nitrate and Health—Recent Findings and Research Needs," *Environmental Health Perspectives* 113, no. 11 (2005): 1607, accessed May 12, 2014, doi: 10.1289/ehp.8043.

57. IPCC, WGI, AR5, Myhre et al., "Chapter 8," 714.

58. R. W. Portmann, J. S. Daniel, and A. R. Ravishankara, "Stratospheric ozone depletion due to nitrous oxide: Influences of other gases," *Philosophical Transactions of the Royal Society B* 367 (2012): 1256–1264, accessed May 8, 2014, doi:10.1098/rstb.2011.0377.

59. D. Procházková et al., "Effects of exogenous nitric oxide on photosynthesis," *Photosynthetica* 51, no. 4 (2013): 487, accessed May 8, 2014, doi:10.1007/s11099-013-0053-y; B. Felzer et al., "Future effects of ozone on carbon sequestration and climate change policy using a global biogeochemical model," *Climatic Change* 73 (2005): 345, accessed May 8, 2014, doi:10.1007/s10584-005-6776-4.

60. Calculated from IFADATA (Activity: Consumption, Region: World, Year: 1961–2012, Product: Grand Total Nitrogen; accessed April 2, 2015), ifadata.fertilizer.org/ucResult.aspx?temp=20150402085438.

61. Calculated from United Nations Department of Economic and Social Affairs (UNDESA), "World Population Prospects: The 2012 Revision, Volume I: Comprehensive Tables" (New York: United Nations, 2013), accessed August 6, 2014, esa.un.org/wpp/Documentation/pdf/WPP2012_Volume-I_Comprehensive-Tables.pdf.

62. Peter Mahaffy, "The Human Element: Chemistry Education's Contribution to Our Global Future," in *The Chemical Element: Chemistry's Contribution to Our Global Future*, ed. Javier Garcia-Martinez and Elena Serrano (Weinheim: John Wiley & Sons, 2011).

63. Scott Fields, "Global Nitrogen: Cycling Out of Control," *Environmental Health Perspectives* 112, no. 10 (2004): A557, www.ncbi.nlm.nih.gov/pmc/articles/PMC1247398/pdf/ehp0112-a00556.pdf.

64. James N. Galloway et al., "Nitrogen footprints: Past, present and future," *Environmental Research Letters* 9, no. 11 (2014): 2, accessed December 17, 2014, doi:10.1088/1748-9326/9/11/115003; Fields, "Global Nitrogen," A560–A562.

65. Galloway et al., "Nitrogen Footprints," 2.

66. Calculated from U.S. Environmental Protection Agency, EPA Science Advisory Board, "Reactive Nitrogen in the United States: An Analysis of Inputs, Flows, Consequences, and Management Options," EPA-SAB-11-013, August 2011, ES-5, accessed January 8, 2015, yosemite.epa.gov/sab/sabproduct.nsf/WebBOARD/INCFullReport/$File/Final%20INC%20Report_8_19_11(without%20signatures).pdf.

67. Ibid.

68. Drinkwater, Wagoner, and Sarrantonio, "Legume-based cropping systems," 262–264; Sasha B. Kramer et al., "Reduced nitrate leaching and enhanced denitrifier activity and efficiency in organically fertilized soils," *Proceedings of the National Academy of Sciences* 103, no. 12 (2006): 4522, accessed March 11, 2015, doi: 10.1073/pnas.0600359103; Jennifer B. Gardner and Laurie E. Drinkwater, "The Fate of Nitrogen in Grain Cropping Systems: A Meta-Analysis of [15]N Field Experiments," *Ecological Applications* 19, no. 8 (2009), 2175, 2182, www.jstor.org/stable/40346320.

69. Calculated from FAOSTAT [Production, Crops, Region: World + (Total), Element: Production quantity, Items: Soybeans, Years: 1961–2013; accessed January 7, 2015], faostat3.fao.org/download/Q/QC/E.

70. Calculated from IFADATA (Activity: Consumption, Region: World, Year: 1961–2012, Product: Grand Total P205; accessed April 2, 2015), ifadata.fertilizer.org/ucResult.aspx?temp=20150402085438.

71. Graham K. MacDonald et al., "Agronomic phosphorus imbalances across the world's croplands," *Proceedings of the National Academy of Sciences* 108, no. 7 (2011): 3086, www.pnas.org/cgi/doi/10.1073/pnas.1010808108; Satish Serchan and Daniel Jones, "The

Baltimore Ecosystem Study Virtual Tour Total Phosphorus," Cary Institute of Ecosystem Studies, 2009, accessed January 6, 2015, www.beslter.org/virtual_tour/Total_Phosphorous .html.

72. Seema B. Sharma et al., "Phosphate solubilizing microbes: Sustainable approach for managing phosphorus deficiency in agricultural soils," *Biomedical and Life Sciences* 2 (2013): 587, accessed January 6, 2015, www.springerplus.com/content/2/1/587.

73. Christian J. Peters, Jennifer L. Wilkins, and Gary W. Fick, "Testing a complete-diet model for estimating the land resource requirements of food consumption and agricultural carrying capacity: The New York State example," *Renewable Agriculture and Food Systems* 22, no. 2 (2006): 145, doi:10.1017/S1742170507001767.

74. David A. Vaccari, "Phosphorus: A Looming Crisis," *Scientific American* 300, no. 6 (2009): 54–59.

75. David A. Vaccari, personal communication with the authors, January 11, 2015.

76. Dana Cordell and Stuart White, "Life's Bottleneck: Sustaining the World's Phosphorus for a Food Secure Future," *Annual Review Environment and Resources* 16 (2014): Figure 4, 168, doi: 10.1146/annurev-environ-010213-113300.

77. Ibid., 172.

78. Larisa Epatko, "Regional Instability Threatens Already Tense Western Sahara," PBS, January 18, 2013, accessed January 7, 2015, www.pbs.org/newshour/updates/world-jan -june13-wsahara_01-18/; Kirsten McTighe, "The world's food supply depends on Morocco. Here's why," *Global Post,* November 21, 2013, accessed January 7, 2015, www.globalpost .com/dispatch/news/business/global-economy/131120/phosphate-world-food-supply -morocco-western-sahara.

79. Calculated from U.S. Department of the Interior, U.S. Geological Survey, "Mineral Commodity Summaries 2014," 2014, 119, accessed January 21, 2015, minerals.usgs.gov/ minerals/pubs/mcs/2014/mcs2014.pdf.

80. Dana Cordell and Stuart White, "Peak Phosphorus: Clarifying the Key Issues of a Vigorous Debate About Long-Term Phosphorus Security," *Sustainability* 3 (2011): 2040–2041, accessed January 6, 2015, doi:10.3390/su3102027.

81. "Harmful Algal Blooms," U.S. Environmental Protection Agency, last modified July 21, 2014, accessed August 4, 2014, www2.epa.gov/nutrientpollution/harmful-algal-blooms; "Sources and Solutions," U.S. Environmental Protection Agency, last modified March 16, 2014, accessed August 4, 2014, www2.epa.gov/nutrientpollution/sources-and-solutions.

82. Jane J. Lee, "Driven by Climate Change, Algae Blooms Behind Ohio Water Scare Are New Normal," *National Geographic,* August 6, 2014, accessed May 5, 2015, news .nationalgeographic.com/news/2014/08/140804-harmful-algal-bloom-lake-erie-climate -change-science.

83. Cordell and White, "Life's Bottleneck," 163.

84. Vaclav Smil, "Nitrogen cycle and world food production," *World Agriculture* 2 (2011): 9–11, accessed October 15, 2014, www.vaclavsmil.com/wp-content/uploads/docs/smil -article-worldagriculture.pdf; Jennifer A. Burney, Steven J. Davis, and David B. Lobnell, "Greenhouse Gas Mitigation by Agricultural Intensification," *Proceedings of the National Academy of Sciences* 107, no. 6 (2010): 12052–12057, doi:10.1073/pnas.0914216107.

85. Dale G. Bottrell and Kenneth G. Schoenly, "Resurrecting the ghost of green revolutions past: The brown planthopper as a recurring threat to high-yielding rice production in tropical Asia," *Journal of Asia-Pacific Entomology* 15, no. 1 (2012): 122, doi:1016/j. aspen.2011.09.004.

86. Steve Tally, "Jumping Off the Pesticide Treadmill," *Agricultures Magazine*, Winter 2002, accessed April 24, 2014, www.agriculture.purdue.edu/agricultures/past/winter2002/features/feature_02.html.

87. Union of Concerned Scientists, Doug Gurian-Sherman, and Margaret Mellon, "The Rise of Superweeds and What to Do About It," Policy Brief, 2013, 2, accessed May 2, 2014, www.ucsusa.org/assets/documents/food_and_agriculture/rise-of-superweeds.pdf.

88. Associated Press, "E.P.A. Accepts New Version of Weed Killer for Farming Use," *New York Times*, October 15, 2014, accessed October 16, 2014, www.nytimes.com/2014/10/16/business/energy-environment/epa-accepts-new-version-of-weed-killer-for-farming-use.html?module=Search& mabReward=relbias%3As.

89. U.S. Department of Agriculture, APHIS, and Sid Abel, "Dow AgroSciences Petitions (09-233-01p, 09-349-01p, and 11-234-01p) for Determinations of Nonregulated Status for 2,4-D-Resistant Corn and Soybean Varieties," Draft Environmental Impact Statement, 2013, ix, accessed February 23, 2015, www.aphis.usda.gov/brs/aphisdocs/24d_deis.pdf.

90. David Pimentel et al., "Environmental and economic effects of reducing pesticide use in agriculture," *Agriculture, Ecosystems and Environment* 46, no. 1 (1993): 276, accessed March 19, 2015, doi:10.1016/0167-8809(93)90030-S.

91. "Pesticide Use Peaked in 1981, Then Trended Downward, Driven by Technological Innovations and Other Factors," Economic Research Service, U.S. Department of Agriculture, last modified June 2, 2014, accessed January 9, 2015, www.ers.usda.gov/amber-waves/2014-june/pesticide-use-peaked-in-1981,-then-trended-downward,-driven-by-technological-innovations-and-other-factors.aspx#.VK_uEivF-So.

92. Calculated from CropLife Foundation, "The Role of Seed Treatment in Modern U.S. Crop Production: A Review of Benefits," December 16, 2013, 4, accessed October 16, 2014, www.motherjones.com/files/seedtreatment.pdf; "The Rise of Seed Treatments," SeedWorld, December 2, 2010, accessed January 6, 2015, seedworld.com/the-rise-of-seed-treatments/.

93. "Honey Bees and Colony Collapse Disorder: Why Should the Public Care What Happens to Honey Bees?" United States Department of Agriculture Agricultural Resource Service, accessed August 4, 2014, www.ars.usda.gov/News/docs.htm?docid=15572#public; Chensheng Lu, Kenneth M. Warchol, and Richard A. Callahan, "In situ replication of honey bee colony collapse disorder," *Bulletin of Insectology* 65, no. 1 (2012): 99, accessed July 31, 2014; Chensheng Lu, Kenneth M. Warchol, and Richard A. Callahan, "Sub-lethal exposure to neonicotinoids impaired honey bees winterization before proceeding to colony collapse disorder," *Bulletin of Insectology* 67, no. 1 (2014): 125, accessed July 31, 2014, www.bulletinofinsectology.org/pdfarticles/vol67-2014-125-130lu.pdf; Elizabeth Grossman, "Declining Bee Populations Pose A Threat to Global Agriculture," *Yale Environment 360*, April 30, 2013, accessed March 26, 2015, e360.yale.edu/feature/declining_bee_populations_pose_a_threat_to_global_agriculture/2645/.

94. "Toxic Hazards," World Health Organization, The Health and Environmental Linkages Initiative, 2014, accessed February 20, 2014, www.who.int/heli/risks/toxics/chemicals/en/; "The Impact of Pesticides on Health," World Health Organization, June 2004, accessed February 20, 2014, www.who.int/mental_health/prevention/suicide/en/PesticidesHealth2.pdf.

95. Kristin S. Schafer et al., "Chemical Trespass: Pesticides in Our Bodies and Corporate Accountability," Pesticide Action Network North America, May 2004, 7, accessed March 25, 2014, www.panna.org/sites/default/files/ChemTresMain(screen).pdf.

96. "Findings from Exposure Studies," Center for Environmental Research and Children's Health, accessed January 12, 2015, cerch.org/research-programs/environmental-exposure-studies/exposure-studies-findings/#StudiesinChildren.

97. María Teresa Muñoz-Quezada et al., "Neurodevelopmental effects in children associated with exposure to organophosphate pesticides: A systematic review," *NeuroToxicology* 39 (2013): 158, doi:10.1016/j.neuro.2013.09.003.

98. U.S. President's Cancer Panel and Suzanne H. Reuben, "Reducing Environmental Cancer Risk: What We Can Do Now," 2008–09 Annual Report, 45, accessed May 5, 2014, deainfo.nci.nih.gov/advisory/pcp/annual Reports/pcp08-09rpt/PCP_Report_08-09_508.pdf.

99. K. L. Bassil et al., "Cancer health effects of pesticides: Systematic review," *Canadian Family Physician* 53, no. 10 (2007): 1708, accessed February 20, 2014, www.ncbi.nlm.nih.gov/pmc/articles/PMC2231435/; M.

100. Rachel Carson, *Silent Spring* (Boston: Houghton Mifflin, 1962).

101. Environmental Defense Fund, "25 Years After DDT Ban, Bald Eagles, Osprey Numbers Soar," June 13, 1997, accessed October 27, 2014, www.edf.org/news/25-years-after-ddt-ban-bald-eagles-osprey-numbers-soar.

102. E. G. Valliantos with McKay Jenkins, *Poison Spring: The Secret Hisory of Pollution and the EPA* (New York: Bloomsbury, 2015), 3, 6. See also: Northwest Coalition for Alternatives to Pesticides (NCAP) and Caroline Cox, "Pesticide Registration: No Guarantee of Safety," *Journal of Pesticide Reform* 17 no. 2 (2009): 2-9, www.pesticide.org/get-the-facts/ncap-publications-and-reports/general-reports-and-publications/journal-of-pesticide-reform/journal-of-pesticide-reform-articles/eparegis.pdf.

103. James W. Jaeger, Ian H. Carlson, and Warren P. Porter, "Endocrine, immune, and behavioral effects of aldicarb (carbamate), atrazine (triazine) and nitrate (fertilizer) mixtures at groundwater concentrations," *Toxicology and Industrial Health* 15 (1999): 133–151, accessed May 1, 2014, doi:10.1177/074823379901500111.

104. President's Cancer Panel and Reuben, "Reducing Environmental Cancer Risk," 45.

105. USEPA, Gruber, et al., "Pesticides Industry Sales and Usage," 14.

106. Anthony Samsel and Stephanie Seneff, "Glyphosate's Suppression of Cytochrome P450 Enzymes and Amino Acid Biosynthesis by the Gut Microbiome: Pathways to Modern Diseases," *Entropy* 15 (2013): 1416, 1441, accessed May 2, 2014, doi:10.3390/e15041416; Lea Schinasi and Maria E. Leon, "Non-Hodgkin Lymphoma and Occupational Exposure to Agricultural Pesticide Chemical Groups and Active Ingredients: A Systematic Review and Meta-Analysis," *International Journal of Environmental Research and Public Health* 11, no. 4 (2014): 4449, accessed August 6, 2014, doi:10.3390/ijerph110404449.

107. "§180.364 Glyphosate; tolerances for residues," Electronic Code of Federal Regulations, US GPO, last modified May 4, 2015, accessed May 6, 2015, www.ecfr.gov/cgi-bin/retrieveECFR?gp=1&SID=b5b22f91002a2080f81b0f26387d314e&ty=HTML&h=L&mc=true&r=SECTION&n=se40.24.180_1364; EU Pesticides Database (pesticide: glyphosate, product: all, search current MRL), accessed July 29, 2014, ec.europa.eu/sanco_pesticides/public/?event=substance.selection; "Pesticide Residues in Food and Feed: Glyphosate," FAO/WHO Standards Codex Alimentarious, accessed July 29, 2014, www.codexalimentarius.net/pestres/data/pesticides/details.html?d-16497-o=2&d-16497-s=3&id=158&print=true.

108. International Agency for Research on Cancer, World Health Organization, "IARC Monographs Volume 112: Evaluation of five organophosphate insecticides and herbicides," March 20, 2015, www.iarc.fr/en/media-centre/iarcnews/pdf/MonographVolume112.pdf.

109. J. B. Sass and A. Colangelo, "European Union bans atrazine, while United States negotiates continued use," *International Journal of Occupational and Environmental Health* 12, no. 3 (2006): 260, accessed May 2, 2014, doi:dx.doi.org/10.1179/oeh.2006.12.3.260; "PDP Annual Summary Report: 2012 Summary," Science and Laboratories, Agricultural Marketing Service, U.S. Department of Agriculture, February 2014, Appendices E and F, accessed April 7, 2014, www.ams.usda.gov/AMSv1.0/ams.fetchTemplateData.do?template=Template G&navID=WaterNav2Link3&rightNav1=WaterNav2Link3&topNav=&leftNav=Science andLaboratories&page= PDPDownloadData/Reports&resultType=&acct=pestcddataprg.

110. Tyrone B. Hayes et al., "Hermaphroditic, demasculinized frogs after exposure to the herbicide atrazine at low ecologically relevant doses," *Proceedings of the National Academy of Sciences* 99, no. 8 (2002): 5476–5480, accessed May 8, 2014, doi:10.1073/pnas.082121499; Tyrone Hayes et al., "Atrazine-Induced Hermaphroditism at 0.1 ppb in American Leopard Frogs (*Rana pipiens*): Laboratory and Field Evidence," *Environmental Health Perspectives* 111, no. 4 (2003): 568–575, accessed May 8, 2014, doi:1289/ehp.5932.

111. Lori A. Cragin et al., "Menstrual cycle characteristics and reproductive hormone levels in women exposed to atrazine in drinking water," *Environmental Research* 111, no. 8 (2011): 1293–1301, accessed August 6, 2014, doi: 10.1016/j.envres.2011.09.009; WuQuiang Fan et al., "Atrazine–Induced Aromatase Expression Is SF-1 Dependent: Implications for Endocrine Disruption in Wildlife and Reproductive Cancers in Humans," *Environmental Health Perspectives* 115, no. 5 (2007): 720–727, accessed May 2, 2014, doi:10.1289/ehp.9758.

112. Rachel Aviv, "A Valuable Reputation," *The New Yorker*, February 10, 2014, accessed February 13, 2014, www.newyorker.com/reporting/2014/02/10/140210fa_fact_aviv?currentPage=all.

113. "Meat on Drugs," *Consumer Reports*, June 2012, 2, accessed January 8, 2015, www.consumerreports.org/content/dam/cro/news_articles/health/CR%20Meat%20On%20Drugs%20Report%2006-12.pdf.

114. U.S. Department of Health and Human Services, Centers for Disease Control and Prevention, "Antibiotic Resistance Threats in the United States, 2013," April 2013, 6, 11–12, accessed April 24, 2014, www.cdc.gov/drugresistance/threat-report-2013/pdf/ar-threats-2013-508.pdf#page=6.

115. Food & Water Watch, "Antibiotic Resistance 101: How Antibiotic Misuse on Factory Farms Can Make You Sick," September 2012, 3, accessed August 4, 2014, documents.foodandwaterwatch.org/doc/AntibioticResistance.pdf.

116. "FDA's Strategy on Antimicrobial Resistance—Questions and Answers," U.S. Food and Drug Administration, last modified March 28, 2014, accessed January 30, 2015, www.fda.gov/AnimalVeterinary/GuidanceComplianceEnforcement/GuidanceforIndustry/ucm216939.htm#question8; Rick Young, "Transcript: The Trouble with Antibiotics," Frontline, accessed January 30, 2015, www.pbs.org/wgbh/pages/frontline/health-science-technology/trouble-with-antibiotics/transcript-69/.

117. "Water Trivia Facts," U.S. Environmental Protection Agency, last modified December 16, 2013, accessed January 6, 2015, water.epa.gov/learn/kids/drinkingwater/water_trivia_facts.cfm.

118. United Nations Environment Programme, "Groundwater and Its Susceptibility to Degradation: A global assessment of the problem and options for management" (Nairobi: United Nations, 2003), accessed August 7, 2014, www.unep.org/dewa/water/groundwater/pdfs/Groundwater_INC_cover.pdf.

119. "Groundwater Depletion," U.S. Geological Survey, last modified March 17, 2014, accessed January 7, 2015, water.usgs.gov/edu/wuir.html; "The rate at which groundwater

reserves are being depleted is increasing," *Science Daily*, July 17, 2014, accessed January 7, 2015, www.sciencedaily.com/releases/2014/07/140717094824.htm.

120. Bridget R. Scanlon et al., "Groundwater depletion and sustainability of irrigation in the US High Plains and Central Valley," *Proceedings of the National Academy of Sciences* (2013): 9320, accessed August 7, 2014, doi:10.1073/pnas.1200311109.

121. David R. Steward et al., "Tapping unsustainable groundwater stores for agricultural production in the High Plains Aquifer of Kansas, projections to 2110," *Proceedings of the National Academy of Sciences* (2013): E3477, accessed August 7, 2014, doi:10.1073/pnas.1220351110.

122. Miguel A. Altieri and C. I. Nicholls, "Agroecology Scaling Up for Food Sovereignty and Resiliency," *Sustainable Agriculture Reviews* 11, ed. Eric Lichtfouse (Springer Science+Business Media Dordrecht, 2012), accessed May 2, 2014, usc-canada.org/UserFiles/File/scaling-up-agroecology.pdf; FAO, "Building on Gender, Agrobiodiversity and Local Knowledge: A Training Manual" (Rome: FAO, 2005), 3, accessed December 16, 2014, www.eldis.org/vfile/upload/1/document/0803/ID2678.pdf.

123. Center for Food Safety and Save Our Seeds, and Debbie Barker, "Seed Giants vs. U.S. Farmers: A Report by the Center for Food Safety and Save Our Seeds," 2013, 20, accessed May 13, 2014, www.centerforfoodsafety.org/files/seed-giants_final_04424.pdf.

124. GRAIN, "The end of farm-saved seed? Industry's wish list for the next revision of UPOV," February 16, 2007, accessed October 15, 2014, www.grain.org/article/entries/58-the-end-of-farm-saved-seed-industry-s-wish-list-for-the-next-revision-of-upov.

125. A. J. Ullstrup, "The impacts of the southern corn leaf blight epidemics of 1970–1971," *Annual Review of Phytopathology* 10, no. 1 (1972): 41–44.

126. Colin K. Khoury et al., "Increasing homogeneity in global food supplies and the implications for food security," *Proceedings of the National Academy of Sciences* 111, no. 11 (2014): 4001–4006, accessed December 17, 2014, doi: 10.1073/pnas.1313490111.

127. FAO, "Building on Gender, Agrobiodiversity and Local Knowledge," 3.

128. FAO, "State of Food and Agriculture 2013," ix.

129. "Micronutrient Deficiencies: Iron Deficiency Anemia," World Health Organization, 2014, accessed April 24, 2014, www.who.int/nutrition/topics/ida/en/.

130. Jill Reedy and Susan Krebs-Smith, "Dietary Sources of Energy, Solid Fats, and Added Sugars Among Children and Adolescents in the United States," *Journal of the American Dietetic Association* 110, no. 10 (2010): 1477, accessed August 4, 2014, doi: 10.1016/j.jada.2010.07.010.

131. Doug Gurian-Sherman, "CAFOs Uncovered: The Untold Costs of Confined Animal Feeding Operations," Union of Concerned Scientists, 2008, 2, accessed May 5, 2015, www.ucsusa.org/sites/default/files/legacy/assets/documents/food_and_agriculture/cafos-uncovered.pdf.

132. FAO, Agriculture and Consumer Protection Department, "Livestock impacts on the environment," *Spotlight,* 2006, accessed November 24, 2014, www.fao.org/ag/magazine/0612sp1.htm; "Rising Number of Farm Animals Poses Environmental and Public Health Risks," Worldwatch Institute, accessed November 24, 2014, www.worldwatch.org/rising-number-farm-animals-poses-environmental-and-public-health-risks-0.

133. FAO, "Livestock in the Balance," in *The State of Food and Agriculture* (Rome: FAO, 2009), 27, accessed November 24, 2014, www.fao.org/docrep/012/i0680e/i0680e02.pdf.

134. American Veterinary Medical Association, Animal Welfare Division, Gail C. Golab, "Animal Welfare Assessment and Application to Pregnant Sow Housing," accessed December 2, 2014, www.fmi.org/docs/animal-welfare/gail_golab.pdf?sfvrsn=3; Gurian-Sherman,

"CAFOs Uncovered"; Humane Society of the United States, "Cage-Free vs. Battery-Cage Eggs: Comparison of animal welfare in both methods," September 1, 2009, November 24, 2014, www.humanesociety.org/issues/confinement_farm/facts/cage-free_vs_battery-cage.html.

135. Animal Legal and Historical Center, Elizabeth Overcash, "Detailed Discussion of Concentrated Animal Feeding Operations: Concerns and Current Legislation Affecting Animal Welfare," 2011, accessed November 7, 2014, www.animallaw.info/article/detailed-discussion-concentrated-animal-feeding-operations; Pew Commission on Industrial Farm Animal Production, "Putting Meat on the Table: Industrial Farm Animal Production in America," April 2008, 38, accessed November 24, 2014, www.ncifap.org/_images/PCIFAPFin.pdf.

136. European Commission, "The Welfare of Intensively Kept Pigs," 91; AVMA and Golab, "Animal Welfare Assessment," 2; Mench et al., "The Welfare of Animals in Concentrated Animal Feeding Operations," Report to the Pew Commission on Industrial Farm Animal Production (Washington, DC, 2008), 2, accessed November 24, 2014, www.fao.org/fileadmin/user_upload/animalwelfare/Welfare%20of%20Animals%20in%20Concentrated%20Animal%20Feeding%20Operations1.doc.

137. Daniel Imhoff, "Introduction," in *CAFO: The Tragedy of Industrial Animal Factories*, ed. Daniel Imhoff (San Rafael: Foundation for Deep Ecology & Earth Aware, 2010), xii, accessed November 24, 2014, www.cafothebook.org/thebook_essays_2.htm#up; Pew Commission, "Putting Meat on the Table," 38; Mench et al., "Welfare of Animals," 2.

138. Overcash, "Detailed Discussion of Concentrated Animal Feeding Operations"; Knowles et al., "Leg Disorders in Broiler Chickens: Prevalence, Risk Factors and Prevention," *PLoS ONE* 3, no. 2 (2008): e1545, doi: 10.1371/journal.pone.0001545.

139. Overcash, "Detailed Discussion of Concentrated Animal Feeding Operations," 52.

140. Pew Commission, "Putting Meat on the Table," 13.

141. Wendell Berry, *Bringing It to the Table: On Farming and Food* (Berkeley: Counterpoint Press, 2009), 11.

142. "Animal Factories and Public Health," Center for Food Safety, accessed November 7, 2014, www.centerforfoodsafety.org/issues/307/animal-factories/animal-factories-and-public-health.

143. John A. Painter et al., "Attribution of Foodborne Illnesses, Hospitalizations, and Deaths to Food Commodities by Using Outbreak Data, United States, 1998–2009," *Emerging Infectious Diseases* 9, no. 3 (2013): 409, accessed November 24, 2014, doi: 10.3201/eid1903.111866.

144. Food and Water Watch, "Factory Farm Nation: How America Turned Its Livestock Farms into Factories," November 2010, 2, accessed November 4, 2014, documents.foodandwaterwatch.org/doc/FactoryFarmNation-web.pdf#_ga=1.211590291.723558207.1415132995.

145. Pew Commission, "Putting Meat on the Table," 16–17.

146. National Association of Local Boards of Health and Carrie Hribar, "Understanding Concentrated Animal Feeding Operations and Their Impact on Communities," 2010, 5, accessed on November 24, 2014, www.cdc.gov/nceh/ehs/Docs/Understanding_CAFOs_NALBOH.pdf.

147. Government Accountability Office, "Concentrated Animal Feeding Operations: EPA Needs More Information and a Clearly Defined Strategy to Protect Air and Water Quality from Pollutants of Concern," GAO-08-944, September 4, 2008, Recommendations Page, www.gao.gov/products/GAO-08-944.

148. Hribar, "Understanding Concentrated Animal Feeding Operations," 3.

149. GAO, "Concentrated Animal Feeding Operations," Highlights Page.

150. FAO, "Livestock in the Balance," 4.

151. Cassidy et al., "Redefining Agricultural Yields," 3, Table 1.

152. Punjab State Council for Science and Technology, N. S. Tiwana, et al., "State of Environment Punjab—2007," 2007, v, accessed May 2, 2014, envfor.nic.in/soer/state/SoE%20report%20of%20Punjab.pdf.

153. "Chemical Generation: Punjabis Are Poisoning Themselves," *The Economist*, September 24, 2007, accessed May 2, 2014, 112, www.economist.com/node/9856023?story_id=9856023.

154. Punjab State Council, Tiwana, et al., "State of Environment Punjab," 37.

155. "Punjab's Cancer Cases Exceed National Average," *Times of India*, January 29, 2013, accessed May 2, 2014, articles.timesofindia.indiatimes.com/2013-01-29/chandigarh/36615390_1_cancer-cases-cancer-patients-rich-doaba-region.

156. Punjab State Council, Tiwana, et al., "State of Environment Punjab," 128–129,157–159.

157. David H. Freedman, "Are Engineered Foods Evil?" *Scientific American* 309 (2013): 84, doi:10.1038/scientificamerican0913-80.

158. Henry W. Lane et al., *International Management Behavior: Leading with a Global Mindset,* 6th ed. (United Kingdom: Wiley, 2009), 131.

159. Kristina Hubbard, *Out of Hand: Farmers Face Consequences of a Consolidated Seed Industry*, Farmer to Farmer Campaign on Genetic Engineering, December 2009, 20, accessed February 27, 2015, farmertofarmercampaign.com/Out%20of%20Hand.FullReport.pdf.

160. Emily Marden, "Risk and Regulation: U.S. Regulatory Policy on Genetically Modified Food and Agriculture," *Boston College Law Review* 44, no. 3 (2003): 747, 761, accessed December 5, 2014, lawdigitalcommons.bc.edu/bclr/vol44/iss3/2/; "Key FDA Documents Revealing Hazards of Genetically Engineered Foods and Flaws with How the Agency Made Its Decisions," Alliance for Bio-Integrity, accessed September 26, 2013, www.biointegrity.org/; Editors, "Do Seed Companies Control GM Crop Research?" *Scientific American,* August 2009, accessed March 4, 2015, www.scientificamerican.com/article/do-seed-companies-control-gm-crop-research/.

161. William Freese and David Schubert, "Safety Testing and Regulation of Genetically Engineered Foods," *Biotechnology and Genetic Engineering Review* 21 (2004): 304, accessed September 26, 2013, doi: 10.1080/02648725.2004.10648060.

162. Roni Caryn Rabin, "The Consumer: Information Not on the Label," *New York Times*, May 26, 2014, accessed December 5, 2014, well.blogs.nytimes.com/2014/05/26/information-not-on-the-label/?_r=0.

163. Center for Science in the Public Interest and Doug Gurian-Sherman, "Holes in the Biotech Safety Net: FDA Policy Does Assure the Safety of Genetically Engineered Foods," 2009, 1–2, accessed September 26, 2013, www.cspinet.org/new/pdf/fda_report_final.pdf; Freese and Schubert, "Safety Testing and Regulation," 303–304.

164. National Research Council, *Environmental Effects of Transgenic Plants: Scope and Adequacy of Regulation* (Washington, DC: National Academy Press, 2002), 19; Doug Gurian-Sherman, "A Contrary Perspective on the AAAS Board Statement Against Labeling of Engineering Foods," *The Equation* (blog), Union of Concerned Scientists, November 2, 2012, accessed September 26, 2013, blog.ucsusa.org/a-contrary-perspective-on-the-aaas-board-statement-against-labeling-of-engineered-foods.

165. Elizabeth Grossman, "Banned in Europe, Safe in the U.S.," *Ensia,* June 9, 2014, accessed December 16, 2014, ensia.com/features/banned-in-europe-safe-in-the-u-s/;

Rachael Bale, "5 Pesticides used in the US are banned in other countries," Center for Investigative Reporting, October 23, 2014, accessed December 5, 2014, beta.cironline.org/reports/5-pesticides-used-in-us-are-banned-in-other-countries/print/; "Colony Collapse Disorder: European Bans on Neonicotinoid Pesticides," About Pesticides, U.S. Environmental Protection Agency, last updated August 15, 2013, accessed December 16, 2014, www.epa.gov/pesticides/about/intheworks/ccd-european-ban.html.

166. European Food Safety Authority, "Explanatory statement for the applicability of the Guidance of the EFSA Scientific Committee on conducting repeated-dose 90-day oral toxicity study in rodents on whole food/feed for GMO risk assessment," *EFSA Journal* 12, no. 10 (2014), accessed December 16, 2014, www.efsa.europa.eu/en/efsajournal/doc/3871.pdf; "Commission Implementing Regulation (EU) No 503/2013 of 3 April 2013 on applications for authorisation of genetically modified food and feed in accordance with Regulation (EC) No. 1829/2003 of the European Parliament and of the Council and amending Commission Regulations (EC) No 641/2004 and (EC) No 1981/2006," Official Journal of the European Union, accessed March 4, 2015, eur-lex.europa.eu/LexUriServ/LexUriServ.do?uri=OJ:L:2013:157:0001:0048:EN:PDF.

167. Gurian-Sherman, "Holes in the Biotech Safety Net"; Marden, "Risk"; "Key FDA," Alliance for Bio-integrity.

168. Jack A. Heinemann, "Hope Not Hype: The Future of Agriculture Guided by the International Assessment of Agricultural Knowledge, Science and Technology for Development" (Penang, Malaysia: Third World Network, 2009), 39, accessed May 6, 2014, www.twnside.org.sg/title2/books/Hope.not.Hype.htm; U.S. Department of Agriculture, Office of the Inspector General Southwest Region, "Animal and Plant Health Inspection Service Controls over Issuance of Genetically Engineered Organism Release Permits," Audit 50601-8-Te, December 2005, ii, accessed March 5, 2015, www.usda.gov/oig/webdocs/50601-08-TE.pdf.

169. Freese and Schubert, "Safety Testing and Regulation," 306.

170. Warren Leary, "Genetic Engineering of Crops Can Spread Allergies, Study Shows," *New York Times,* March 14, 1996, accessed December 5, 2014, www.nytimes.com/1996/03/14/us/genetic-engineering-of-crops-can-spread-allergies-study-shows.html.

171. Giles-Eric Séralini et al., "Genetically Modified Crops Safety Assessments: Present Limits and Possible Improvements," *Environmental Sciences Europe* 23:10 (2011), accessed September 26, 2013, doi: 10.1186/2190-4715-23-10; Judy A. Carman et al., "A long-term toxicology study on pigs fed a combined genetically modified (GM) soy and GM maize diet," *Journal of Organic Systems* 8, no. 1 (2013): 46, accessed May 6, 2014, www.organic-systems.org/journal/81/8106.pdf; A. Sagstad et al., "Evaluation of stress- and immune-response biomarkers in Atlantic salmon, *Salmo salar* L., fed different levels of genetically modified maize (Bt maize), compared with its near-isogenic parental line and a commercial suprex maize," *Journal of Fish Diseases* 30, no. 4 (2007): 201–212, accessed March 17, 2015, doi: 10.1111/j.1365-2761.2007.00808.x.

172. Carman et al., "A long-term toxicology study on pigs fed a combined genetically modified (GM) soy and GM maize diet"; "Lynas attack on GMO feed study shredded," GMWatch, 2013 Articles, last modified June 15, 2013, accessed May 6, 2014, www.gmwatch.org/index.php/news/archive/2013/14813-lynas-attack-on-gmo-feed-study-shredded.

173. Independent Science News, "Seralini and Science: An Open Letter," last modified October 2, 2012, www.independentsciencenews.org/health/seralini-and-science-nk603-rat-study-roundup/#more-1087; "Controversial Seralini study linking GM to cancer in rats

is republished," *The Guardian,* June 24, 2014, www.theguardian.com/environment/2014/jun/24/controversial-seralini-study-gm-cancer-rats-republished.

174. Séralini et al., "Republished study: Long-term toxicity of a Roundup herbicide and a Roundup-tolerant genetically modified maize," *Environmental Sciences Europe* 26, no. 14 (2014): 1, accessed December 5, 2014, www.enveurope.com/content/pdf/s12302-014-0014-5.pdf; "Controversial Seralini study," *The Guardian.*

175. "Controversial Seralini study," *The Guardian*; Kate Kelland, "Science journal urged to retract Monsanto GM study," Reuters UK, November 30, 2012, accessed December 2, 2014, uk.reuters.com/article/2012/11/30/us-science-gm-journal-idUKBRE8AT10920121130.

176. "Statement: No Scientific Consensus on GMO Safety," ENSSER, www.ensser.org/increasing-public-information/no-scientific-consensus-on-gmo-safety; "ENSSER Comments on the Retraction of the Séralini et al. 2012 Study," ENSSER, www.ensser.org/democratising-science-decision-making/ensser-comments-on-the-retraction-of-the-seralini-et-al-2012-study.

177. "Elsevier Announces Article Retraction from Journal Food and Chemical Toxicology," Elsevier, November 28, 2013, accessed December 5, 2014, www.elsevier.com/about/press-releases/research-and-journals/elsevier-announces-article-retraction-from-journal-food-and-chemical-toxicology; "Publishing Guidelines: Policies—Article Withdrawal," Elsevier, accessed December 5, 2014, www.elsevier.com/about/publishing-guidelines/policies/article-withdrawal.

178. "Republication of the Séralini study: Science speaks for itself," Sustainable Pulse, June 24, 2014, accessed December 2, 2014, www.gmoseralini.org/republication-seralini-study-science-speaks/.

179. Johan Diels et al., "Association of financial or professional conflict of interest to research outcomes on health risks or nutritional assessment studies of genetically modified products," *Food Policy* 36 (2011): Table 1, 200, doi:10.1016/j.foodpol.2010.11.016.

180. Alden M. Monzon, "UN official says questions remain on GMO health impact, business practices," Business World Online, February 27, 2015, accessed March 2, 2015, www.bworldonline.com/content.php?section=Economy&title=un-official-says-questions-remain-on-gmo-health-impact-business-practices&id=103492.

181. Ian F. Pryme and Rolf Lembcke, "*In vivo* studies on possible health consequences of genetically modified food and feed—with particular regard to ingredients consisting of genetically modified plant materials," *Nutrition and Health* 17 (2003): 7, accessed May 6, 2014, www.combat-monsanto.org/docs/doc%20scan/OGM/Pryme.pdf; Doug Gurian-Sherman, "Are GMOs Worth the Trouble?" MIT Technology Review, March 27, 2014, accessed May 8, 2014, www.technologyreview.com/view/525931/are-gmos-worth-the-trouble/.

182. U.S. Department of Agriculture, Economic Resource Service, Jorge Fernandez-Cornejo, et al., "Genetically Engineered Crops in the United States," ERR-162, February 2014, accessed July 29, 2014, 12, www.ers.usda.gov/publications/err-economic-research-report/err162.aspx#.U9fa9yjyC8A; Doug Gurian-Sherman, "Failure to Yield: Evaluating the Performance of Genetically Engineered Crops," Union of Concerned Scientists, April 2009, 1, accessed January 9, 2014, www.ucsusa.org/sites/default/files/legacy/assets/documents/food_and_agriculture/failure-to-yield.pdf.

183. "Glossary," Monsanto, accessed January 9, 2015, www.monsanto.com/newsviews/pages/glossary.aspx#bt.

184. USEPA, Gruber, et al., "Pesticides Industry Sales and Usage," 26–30; Charles Benbrook, "Impacts of genetically engineered crops on pesticide use in the U.S.—the first

sixteen years," *Environmental Sciences Europe* 24, no. 24 (2012): 8, accessed May 2, 2014, doi:10.1186/2190-4715-24-24.

185. FAO, "Technical Consultation on Low Levels of Genetically Modified (GM) Crops in International Food and Feed Trade," Technical Background Paper No. 2, March 2014, 6, accessed December 4, 2014, www.fao.org/fileadmin/user_upload/agns/topics/LLP/AGD803_3_Final_En.pdf.

186. Food Policy Institute, William K. Hallman, et al., "Public Perceptions of Genetically Modified Foods: A National Study of American Knowledge and Opinion," RR-1003-004 (New Brunswick, 2003), 6, accessed May 2, 2014, core.kmi.open.ac.uk/download/pdf/6435317 .pdf; California Department of Food and Agriculture, "A Food Foresight Analysis of Agricultural Biotechnology: A Report to the Legislature," January 1, 2003, 3, accessed September 26, 2013, www.cdfa.ca.gov/files/pdf/ag_biotech_report_03.pdf.

187. Allison Kopicki, "Strong Support for Labeling Modified Foods," *New York Times*, July 27, 2013, accessed February 13, 2014, www.nytimes.com/2013/07/28/science/strong -support-for-labeling-modified-foods.html?_r=0; "International Labeling Laws," Center for Food Safety, 2014, accessed February 13, 2014, www.centerforfoodsafety.org/issues/976/ ge-food-labeling/international-labeling-laws.

188. Eliza Barclay and Jeremy Bernfeld, "Bracing for a Battle, Vermont Passes GMO Labeling Bill," *The Salt,* National Public Radio, April 24, 2104, accessed September 2, 2014, www.npr.org/blogs/thesalt/2014/04/24/306442972/bracing-for-a-battle-vermont-passes-gmo-labeling-bill; Reid Wilson, "Maine becomes second state to require GMO labels," *Washington Post*, January 10, 2014, accessed September 2, 2104, www .washingtonpost.com/blogs/govbeat/wp/2014/01/10/maine-becomes-second-state -to-require-gmo-labels/.

189. "I-522: GMO Labeling," Voter's Edge, last modified September 30, 2013, accessed September 2, 2014, votersedge.org/washington/ballot-measures/2013/november/i-522/ funding?jurisdictions=28.26.28-upper-wa.26#.VAYJb_ldURo; "Prop. 37: Genetically Engineered Foods," Voter's Edge, last modified November 6, 2012, accessed May 9, 2014, votersedge.org/california/ballot-measures/2012/november/prop-37/funding# .VAYPXPldURo.

190. USDA ERS, Fernandez-Cornejo, et al., "Genetically Engineered Crops," 12.

191. "'Golden Rice' GM Vitamin-A Rice," CBAN Factsheet, January 2014, 1, accessed January 9, 2015, pdf access from www.cban.ca/Resources/Topics/GE-Crops-and-Foods -Not-on-the-Market/Rice/Golden-Rice-GM-Vitamin-A-Rice; Amy Saltzman et al., "Biofortification: Progress toward a more nourishing future," *Global Food Security* 2, no. 1 (2013): 11, accessed March 31, 2014, dx.doi.org/10.1016/j.gfs.2012.12.003.

192. Michael Morris, Greg Edmeades, and Eija Pehu, "The Need for Global Breeding Capacity: What Roles for Public and Private Sectors?" *Horticultural Science* 41, no.1 (2006): 30, accessed December 5, 2014, hortsci.ashspublications.org/content/41/1/30 .full.pdf; Rural Advancement Foundation International, "Proceedings of 2014 Summit on Seeds and Breeds for 21st Century Agriculture," March 2014, accessed December 5, 2014, rafiusa.org/docs/2014SummitProceedings.pdf; Food and Water Watch, "Public Research, Private Gain: Corporate Influence over University Agricultural Research," April 2012, 2-5, accessed December 5, 2014, documents.foodandwaterwatch.org/doc/ PublicResearchPrivateGain.pdf#_ga=1.153402616.1536159999.1417631748.

193. Jack Kloppenburg, personal communication with the authors, February 20, 2015.

194. Marc Lappé and Britt Bailey, *Against the Grain: Biotechnology and the Corporate*

Takeover of Your Food (Monroe: Common Courage Press, 1998). Find the account of events here: "Monsanto and the First Amendment," Center for Ethics and Toxics (now closed): environmentalcommons.org/cetos/articles/monsantoamend.html. CETOS papers are hosted at: environmentalcommons.org/cetos/index.html. It was ultimately published by Common Courage Press, 1998. Monsanto has never denied its threat to the book's original publisher. Dr. Lappé and Ms. Bailey sought dialogue with Monsanto scientists prior to completing their book, but Monsanto did not respond.

195. Food and Water Watch, "Biotech Ambassadors: How the U.S. State Department Promotes the Seed Industry's Global Agenda," May 14, 2013, accessed April 7, 2014, 10–12, www.foodandwaterwatch.org/reports/biotech-ambassadors/; United States Secretary of State, "FY 2008 Biotechnology Outreach Strategy and Department Resources," Cablegate's Cables, Reference ID: 07STATE160639, November 27, 2007, last updated September 1, 2011, accessed January 7, 2015, cablegatesearch.wikileaks.org/search.php?q=07STATE160639& qo=0&qc=0&qto=2010-02-28.

196. Royal Society, "Reaping the benefits: Science and the sustainable intensification of global agriculture," 2009, ix, 7–8, 46, accessed August 4, 2014, royalsociety.org/~/media/ Royal_Society_Content/policy/publications/2009/4294967719.pdf; Thalif Deen, "Climate-Smart Agriculture Is Corporate Green-Washing, Warn NGOs," Inter Press Service News Agency, September 24, 2014, accessed January 12, 2015, www.ipsnews.net/2014/09/ climate-smart-agriculture-is-corporate-green-washing-warn-ngos/.

197. Feed the Future, "Program for Sustainable Intensification," November 2013, 1–2, 4, accessed August 6, 2014, feedthefuture.gov/sites/default/files/resource/files/ftf_ factsheet_fsicsustainableint_nov2013.pdf; Friends of the Earth International, "A Wolf in Sheep's Clothing? An analysis of the sustainable intensification of agriculture, Summary," October 2012, 2-3, accessed August 1, 2014, www.foei.org/wp-content/uploads/2013/12/ Wolf-in-Sheep%E2%80%99s-Clothing-summary.pdf; Emmanuel Tumusiime and Edmund Matotay, "Sustainable and inclusive investments in agriculture: Lessons on the Feed the Future Initiative in Tanzania," Oxfam America, Research Backgrounder, February 2014, March 26, 2015, www.oxfamamerica.org/static/media/files/Tanzania_-_Sustainable_and_ Inclusive_Investments.pdf.

198. Claudia Dreifus, "An Advocate for Science Diplomacy," *New York Times,* August 18, 2008, accessed May 6, 2015, www.nytimes.com/2008/08/19/science/19conv.html.

MYTH 4: Organic & Ecological Farming Can't Feed a Hungry World

1. Lauren C. Ponisio et al., "Diversification practices reduce organic to conventional yield gap," *Proceedings of the Royal Society B* 282, no. 1800 (2014): 1, accessed January 5, 2015, doi: 10.1098/rspb.2014.1396; Verena Seufert, Navin Ramankutty, and Jonathan A. Foley, "Comparing the yields of organic and conventional agriculture," *Nature* 485 (May 2012): 230, accessed May 13, 2014, doi:10.1038/nature11069.

2. Catherine Badgley et al., "Organic agriculture and the global food supply," *Renewable Agriculture and Food Systems* 22, no. 2 (2007): 86–108, accessed January 5, 2015, doi:10.1017/S1742170507001640.

3. Altieri and Nicholls, "Agroecology scaling up for food sovereignty and resiliency," 9–11.

4. "Organic Market Overview," Economic Research Service, U.S. Department of Agriculture, last updated April 7, 2014, accessed August 5, 2014, www.ers.usda.gov/ topics/natural-resources-environment/organic-agriculture/organic-market-overview

.aspx#.U-ENgfldVK0; "American appetite for organic products breaks through $35 billion mark," PRNewswire, May 15, 2014, accessed April 14, 2015, www.prnewswire.com/news-releases/american-appetite-for-organic-products-breaks-through-35-billion-mark-259327061.html; "Table 3. Certified organic and total U.S. acreage, selected crops and livestock, 1995–2011," Economic Research Service, U.S. Department of Agriculture, last updated October 24, 2013, accessed August 5, 2014, www.ers.usda.gov/data-products/organic-production.aspx#.U-DpxPldVK1.

5. John Paull, "The Uptake of Organic Agriculture: A Decade of Worldwide Development," *Journal of Social and Development Sciences*, no. 3 (2011): 111–120, accessed April 1, 2014, orgprints.org/19517/1/Paull2011DecadeJSDS.pdf.

6. FiBL and IFOAM, "The World of Organic Agriculture: Statistics and Emerging Trends 2014," February 2014, 25, accessed October 29, 2014, www.fibl.org/fileadmin/documents/shop/1636-organic-world-2014.pdf.

7. Andre Leu, President, IFOAM, personal communication with author, December 2014.

8. FAO, "AGP—Integrated Production and Pest Management Programme in West Africa," 2011, accessed January 8, 2015, www.fao.org/agriculture/crops/thematic-sitemap/theme/pests/ipm/ipmwestafrica/en/.

9. FAO, Food and Nutrition Division, "Rice and Human Nutrition" (Rome: FAO, 2004), accessed January 29, 2015, www.fao.org/rice2004/en/f-sheet/factsheet3.pdf; "SRI Rice," SRI International Network and Resource Center, Cornell University College of Agriculture and Life Sciences, accessed April 24, 2014, sri.ciifad.cornell.edu/.

10. Union of Concerned Scientists, "The Healthy Farm: A Vision for U.S. Agriculture," Policy Brief, April 2013, 1, accessed May 14, 2014, www.ucsusa.org/assets/documents/food_and_agriculture/The-Healthy-Farm-A-Vision-for-US-Agriculture.pdf.

11. U.S. Department of Agriculture, Natural Resources Conservation Service, "Healthy Soils Are: Full of Life," 1, accessed July 30, 2014, www.nrcs.usda.gov/Internet/FSE_DOCUMENTS/stelprdb1193147.pdf.

12. U.S. Department of Agriculture, Natural Resources Conservation Service California, "Before Spring Planting Expert Says, 'Dig a little. Learn a lot,'" News Release, April 3, 2013, accessed July 30, 2014, www.nrcs.usda.gov/wps/portal/nrcs/detail/ca/people/employees/?cid=nrcs144p2_064312.

13. USDA, NRCS, "Healthy Soils Are: Full of Life," 1.

14. U.S. Department of Agriculture, Natural Resources Conservation Service, "Healthy Soils Are: High in Organic Matter," 1, accessed July 30, 2014, nrcspad.sc.egov.usda.gov/DistributionCenter/pdf.aspx?productID=1024.

15. FAO, Alexandra Bot, and José Benites, "The Importance of Organic Matter: Key to drought-resistant soil and sustained food production," FAO Soils Bulletin 80 (Rome: FAO, 2005), ix, accessed December 2, 2014, www.fao.org/docrep/009/a0100e/a0100e.pdf.

16. USDA, NRCS, "Healthy Soils Are: High in Organic Matter," 1.

17. Abraham et al., "The System of Crop Intensification: Agroecological Innovations," 7–8.

18. Han Olff et al., "Parallel ecological network in ecosystems," *Philosophical Transactions of the Royal Society B: Biological Sciences* 364, no. 1524 (2009): 1755–79, doi:10.1098/rstb.2008.0222.

19. USDA, NRCS, "Healthy Soils Are: High in Organic Matter," 1.

20. U.S. Department of Agriculture, Natural Resources Conservation Service, "Healthy Soils Are: Covered All the Time," 1, accessed July 30, 2014, www.nrcs.usda.gov/wps/portal/nrcs/detail/national/soils/health/?cid=stelprdb1193043.

21. U.S. Department of Agriculture, Natural Resources Conservation Service, "Unlock

Your Farm's Potential: Do Not Disturb," September 2012, 1, accessed July 30, 2014, www .nrcs.usda.gov/Internet/FSE_DOCUMENTS/stelprdb1049424.pdf.

22. U.S. Department of Agriculture, Natural Resources Conservation Service, "Healthy Soils Are: Well Structured," 1, accessed July 30, 2014, www.nrcs.usda.gov/Internet/FSE_DOCUMENTS/stelprdb1193171.pdf.

23. USDA, NRCS, "Healthy Soils Are: High in Organic Matter," 1.

24. Grace Gershuny, personal communication with the author, November 2014.

25. FAO, "Conservation tillage: The end of the plough?" News and Highlights, May 3, 2000, accessed October 29, 2014, www.fao.org/News/2000/000501-e.htm.

26. Rolf Derpsch et al., "Current status of adoption of no-till farming in the world and some of its main benefits," *International Journal of Agricultural & Biological Engineering* 3, no. 1 (2010): 1, accessed October 29, 2014, doi: 10.3965/j.issn.1934-6344.2010.01.001-025.

27. Theodor Friedrich and Amir Kassam, "No-Till Farming and the Environment: Do No-Till Systems Require More Chemicals?" *Outlooks on Pest Management* 23, no. 4 (2012): 156, accessed November 17, 2014, doi: 10.1564/23aug02.

28. Norman Uphoff, *Agroecological Innovations: Increasing Food Production with Participatory Development* (UK, USA: Earthscan Publications Limited, 2002), Chapter 15.

29. Land Institute, "Annual Report Fall 2014," 2014, 1, accessed January 2, 2015, www. landinstitute.org/wp-content/uploads/2014/11/FY2014_Annual_Report.pdf; Frank Lessiter, "29 Reasons Why Many Growers Are Harvesting Higher No-Till Yields in Their Fields Than Some University Scientists Find in Research Plots," No-Till Farmer, January 1, 2015, accessed March 11, 2015, www.no-tillfarmer.com/articles/4038-reasons-why -many-growers-are-harvesting-higher-no-till-yields-in-their-fields-than-some-university -scientists-find-in-research-plots; communication with authors, April 2015.

30. Thomas S. Cox et al., "Prospects for Developing Perennial Grain Crops," *BioScience* 56, no. 8 (2006): 649–659, accessed January 5, 2015, doi: 10.1641/0006-3568(2006)56[649.

31. USDA, NRCS, "Healthy Soils Are: Covered All the Time," 1.

32. Marianne Sarrantonio, "Building Soil Fertility: Building Soil Fertility and Tilth with Cover Crops," in *Managing Cover Crops Profitably*, 3rd ed. (Sustainable Agriculture and Research Education, 2012), 18–19, accessed May 13, 2014, www.sare.org/Learning-Center/Books/Managing-Cover-Crops-Profitably-3rd-Edition/Text-Version/Building-Soil-Fertility.

33. C. Tonitto, M. B. David, and L. E. Drinkwater, "Replacing bare fallows with cover crops in fertilizer-intensive cropping systems: A meta-analysis of crop yield and N dynamics," *Agriculture, Ecosystems and Environment* 112 (2006): 58, 66, doi:10.1016/j.agee.2005.07.003.

34. Stephen R. Gliessman, *Field and Laboratory Investigations in Agroecology* (Boca Raton: CRC Press, 2007), 217.

35. Perfecto, Vandermeer, and Wright, *Nature's Matrix,* 7.

36. Miguel A. Altieri and Clara Ines Nicholls, "Ecologically based pest management: a key pathway to achieving agroecosystem health," in *Managing for Healthy Ecosystems*, ed. William L. Lasley et al. (Boca Raton: CRC Press, 2002), 999–1010.

37. "A Platform Technology for Improving Livelihoods of Resource Poor Farmers," African Insect Science for Food and Health, last modified 2014, accessed July 31, 2014, www.push-pull.net/3.shtml.

38. Miguel A. Altieri and Clara Ines Nicholls, "Agroecology in Action," last updated July 30, 2000, accessed January 2, 2015, nature.berkeley.edu/~miguel-alt/applying_agroecological_concepts.html.

39. Smil, "Nitrogen cycle and world food production," 12.

40. Badgley et al., "Organic agriculture and the global food supply," 92.

41. David A. Vaccari, "Manure as Low-Hanging Fruit," Scope Newsletter No. 6, 28–29, August 2014, www.phosphorusplatform.eu/images/download/SCOPE%20Vision%20texts%20received%205-2014.pdf.

42. Vaccari, "Phosphorus," 55; Vaccari, "Manure as Low-Hanging Fruit"; David Vaccari, personal communication with the authors, January 11, 2015.

43. "Soil Erosion on Cropland 2007," U.S. Department of Agriculture, Natural Resources Conservation Service, 2007, accessed January 7, 2015, www.nrcs.usda.gov/wps/portal/nrcs/detail/national/technical/?cid=stelprdb1041887.

44. Leach et al., "Nitrogen footprint model," 43-56; Galloway et al., "Nitrogen footprints," 115003.

45. Geneviève S. Metson, Elena M. Bennett, and James J. Elser, "The role of diet in phosphorus demand," *Environmental Research Letters* 7, no. 4 (2012): 1–10, doi:10.1088/1748-9326/7/4/044043.

46. Cordell and White, "Life's Bottleneck," 164.

47. Gerald Ondrey, "P-Recovery on the Move," *Chemical Engineering News*, February 2013, 17. www.ostara.com/sites/default/files/pdfs/feb13_Chemical-Engineering-News.pdf.

48. Cordell and White, "Life's Bottleneck," 180.

49. "Land Application and Composting of Biosolids," Water Environment Federation, accessed February 27, 2015, www.wef.org/WEF_LandAppl_Compost_FactSheet_May2010; Zero Waste Committee, "Land Application of Sewage Sludge: Guidance," Sierra Club Conservation Policies, Feburary 18, 2008, accessed February 27, 2015, www.sierraclub.org/sites/www.sierraclub.org/files/uploads-wysiwig/LandApplicationSewageSludge.pdf.

50. Phillip Barak and Alysa Stafford, "Struvite: A Recovered and Recycled Phosphorus Fertilizer," *Proceedings of the 2006 Wisconsin Fertilizer, Aglime & Pest Management Conference*, 45 (2006): 199, www.soils.wisc.edu/extension/wcmc/2006/pap/Barak.pdf.

51. Jessica VanEgeren, "Madison Sewerage District's New Technology to Turn Phosphorus into Fertilizer Pellets," *The Cap Times*, June 2, 2014, accessed January 28, 2015, host.madison.com/news/local/writers/jessica_vanegeren/madison-sewerage-district-s-new-technology-to-turn-phosphorus-into/article_50215ee6-e9ed-11e3-b8d3-0019bb2963f4.html#ixzz3LtsikHb9; Jianbo Shen et al., "Phosphorus Dynamics: From Soil to Plant," *Plant Physiology* 156, no. 3 (2011): 1000–1001, doi: dx.doi.org/10.1104/pp.111.175232; Barak and Stafford, "Struvite," 199; Sharif Ahmed et al., "Imaging the interaction of roots and phosphate fertiliser granules using 4D X-ray tomography," *Plant and Soil* (2015): 6–7, accessed March 27, 2015, doi:10.1007/s11104-015-2425-5.

52. "Scope Newsletter," No. 74, October 2009, 4, accessed March 27, 2015, www.ceep-phosphates.org/Files/Newsletter/Scope74%20Vancouver%20Nutrient%20Recovery%20Conference.pdf.

53. Philip Abrary, President and Chief Executive Officer, Ostara, telephone interview with the author, February 17, 2015.

54. World Health Organization and Caroline Schönning, "Urine diversion—hygienic risks and microbial guidelines for reuse," 2001, 2, 3, 9, accessed January 2, 2015, www.who.int/water_sanitation_health/wastewater/urineguidelines.pdf.

55. Samantha Larson, "Is 'Peecycling' the Next Wave in Sustainable Living?" *National Geographic*, February 2, 2014, accessed January 2, 2015, news.nationalgeographic.com/news/2014/02/140202-peecycling-urine-human-waste-compost-fertilizer/.

56. Dana Cordell et al., "Toward global phosphorus security: A systems framework for

phosphorus recovery and rescue options," *Chemosphere* 84 (2011): 753, accessed January 28, 2015, doi: 10.1016/j.chemosphere.2011.02.032.

57. Paolo D'Odorico et al., "Feeding humanity through global food trade," *Earth's Future* 2, no. 9 (2014): 458–469, accessed October 23, 2014, doi:10.1002/2014EF000250; FAO, "Food, Agriculture and Cities: Challenges of food and nutrition security, agriculture and ecosystem management in an urbanizing world," FAO Food for the Cities multidisciplinary initiative position paper, October 2011, accessed April 30, 2015, www.fao.org/fileadmin/templates/FCIT/PDF/FoodAgriCities_Oct2011.pdf.

58. FAO, "Growing Greener Cities," 2, 4, 16; Christina Ergas, "Cuban Urban Agriculture as a Strategy for Food Sovereignty," *Monthly Review* 64, no. 10 (March 2013), accessed October 27, 2014, monthlyreview.org/2013/03/01/cuban-urban-agriculture-as-a-strategy-for-food-sovereignty/.

59. FAO, "Women, Agriculture, and Food Security," accessed January 8, 2015, www.fao.org/worldfoodsummit/english/fsheets/women.pdf.

60. Courtney Gallaher, Antoinette WinklerPrins, Mary Njena, and Nancy Karanja, "Creating Space: Sack Gardening as a Livelihood Strategy in the Kibera Slums of Nairobi, Kenya," *Journal of Agriculture, Food Systems, and Community Development,* February 16, 2015, 33, accessed April 17, 2015,. www.researchgate.net/publication/272421146_Creating_Space_Sack_gardening_as_a_livelihood_strategy_in_the_Kibera_slums_of_Nairobi_Kenya.

61. Interviews by Courtney Gallaher, Department of Geography, Northern Illinois University, DeKalb.

62. Courtney Gallagher, personal communication with authors, February 11, 2015.

63. Hailu Araya and Sue Edwards, *The Tigray Experience: A Success Story in Sustainable Agriculture* (Penang: Third World Network, 2006).

64. Sue Edwards, e-mail communication with author, August 5, 2014.

65. Araya and Edwards, *Tigray Experience.*

66. Andre Leu, "Achieving Food Security with High Yielding Organic Agriculture in Mountain Systems," Paper presented at IFOAM Conference on Organic and Ecological Agriculture in Mountain Ecosystems, Thimphu, Bhutan, March 5–8, 2014, 4.

67. Sue Edwards, e-mail communication with author, August 5, 2014.

68. World Bank, "Carbon Sequestration in Agricultural Soils," 3, Table 1.1.

69. Araya and Edwards, *Tigray Experience*, 13.

70. Environmental Protection Agency of Ethiopia, trans. Tewolde Berhan Gebre Egziabher, "A Manual for the Preparation of Woreda and Local Community Plans for Environmental Management for Sustainable Development," in *Climate Change and Food Systems Resilience in Sub-Saharan Africa*, ed. Lim Li Ching, Sue Edwards, and Nadia El-Hage Scialabba (Rome: FAO, 2011), 216–224, accessed March 27, 2014, www.fao.org/docrep/014/i2230e/i2230e.pdf.

71. Sue Edwards, e-mail communication with author, August 5, 2014.

72. Sue Edwards, Tewolde Berhan Gebre Egziabher, and Hailu Araya, "Successes and Challenges in Ecological Agriculture: Experiences from Tigray, Ethiopia," in *Climate Change and Food Systems Resilience in Sub-Saharan Africa*, ed. Lim Li Ching, Sue Edwards, and Nadia El-Hage Scialabba (Rome: FAO, 2011), 286, accessed March 27, 2014, www.fao.org/docrep/014/i2230e/i2230e.pdf; Calculated from Sue Edwards, "The Impact of Compost Use on Crop Yields in Tigray, Ethiopia," Presentation given at International Conference on Organic Agriculture and Food Security, FAO, Rome, May 3, 2007, 16, accessed January 28, 2015, harep.org/agriculture/ards.pdf.

73. Araya and Edwards, *Tigray Experience*, 15.

74. Swedish Society for Nature Conservation, Jakob Lundberg and Fredrik Moberg, "Ecological in Ethiopia: Farming with Nature Increases Profitability and Reduces Vulnerability," 2008, 24, accessed January 28, 2015, knowledgebase.terrafrica.org/fileadmin/user_upload/terrafrica/docs/Report_international_Ethiopia.pdf; Araya and Edwards, *Tigray Experience*, 24.

75. United Nations Environmental Program, Reynaldo Victoria, et al., "The Benefits of Soil Carbon," 2012, 19, accessed January 28, 2015, www.unep.org/yearbook/2012/pdfs/UYB_2012_CH_2.pdf.

76. FAO, "Organic Agriculture and Climate Change."

77. S. Siebert et al., "Groundwater use for irrigation—a global inventory," *Hydrology and Earth Systems Sciences* 7 (2010): 3978, accessed April 25, 2014, doi:10.5194/hessd-7-3977-2010.

78. Tim J. LaSalle and Paul Hepperly, "Regenerative Organic Farming: A Solution to Global Warming," Rodale Institute, 2008, 4, accessed May 7, 2014, grist.files.wordpress.com/2009/06/rodale_research_paper-07_30_08.pdf; David Pimentel et al., "Environmental, Energetic, and Economic Comparisons of Organic and Conventional Farming Systems," *Bioscience* 55, no. 7 (2005): 575, accessed May 7, 2014, doi:10.1641/0006-3568(2005)055[0573:EEAECO]2.0.CO.

79. Lotter, Seidel, and Liebhardt, "Performance of organic and conventional cropping systems," 1; Abraham et al., "The System of Crop Intensification: Agroecological innovations," 60–62.

80. "What Is SRI?" SRI International Network and Resources Center, Cornell University, accessed May 7, 2015, sri.ciifad.cornell.edu/; "SRI Methodologies," SRI International Network and Resources Center, Cornell University, accessed May 7, 2015, sri.ciifad.cornell.edu/aboutsri/methods/index.html.

81. Erika Styger et al., "The System of Rice Intensification as a Sustainable Agricultural Innovation: Introducing, Adapting and Scaling Up a System of Rice Intensification Practices in the Timbuktu Region of Mali," *International Journal of Agricultural Sustainability* 9, no. 1 (2011): 71, accessed March 27, 2014, www.tandfonline.com/doi/pdf/10.3763/ijas.2010.0549.

82. Devon Ericksen, "'The Man Who Stopped the Desert': What Yacouba Did Next," Nourishing the Planet, January 28, 2013, blogs.worldwatch.org/nourishingtheplanet/the-man-who-stopped-the-desert-what-yacouba-did-next/#more-18238; Araya and Edwards, Tigray Experience, 13-14, 54.

83. United Nations Environmental Programme and Chris Reij, "Regreening the Sahel," September 2011, 22, accessed November 4, 2014, www.unep.org/pdf/op_sept_2011/EN/OP-2011-09-EN-ARTICLE6.pdf.

84. "About Save Our Seeds," Center for Food Safety, accessed July 30, 2014, www.centerforfoodsafety.org/issues/303/seeds/about-save-our-seeds.

85. ETC Group, "Who Will Control the Green Economy?" November 2011, 22, accessed August 12, 2014, www.etcgroup.org/sites/www.etcgroup.org/files/publication/pdf_file/ETC_wwctge_4web_Dec2011.pdf.

86. Jane Goodall, *Seeds of Hope: Wisdom and Wonder for the World of Plants* (New York: Grand Central Publishing, 2014), 109, 117; "Update on the world's 15 largest seed banks," Ag Professional, last modified August 1, 2013, accessed May 9, 2014, www.agprofessional.com/news/Update-on-the-worlds-15-largest-seed-banks-217990631.html; David J. Merritt

and Kingsley W. Dixon, "Restoration Seed Banks—A Matter of Scale," *Science Magazine* 332, no. 6028 (2011): 424–425, accessed May 9, 2014, doi:10.1126/science.1203083; "About the Millennium Seed Bank Partnership," Kew Royal Botanical Gardens, accessed January 11, 2015, www.kew.org/science-conservation/millennium-seed-bank-partnership/about-millennium-seed-bank-partnership.

87. Andrew Kimbrell, "Seed Banks," Center for Food Safety, accessed October 16, 2014, V=www.centerforfoodsafety.org/issues/303/seeds/seed-banks.

88. Navdanya, accessed May 15, 2014, www.navdanya.org/.

89. Merritt and Dixon, "Restoration Seed Banks," 424.

90. Jack Kloppenburg, University of Wisconsin, personal communication with the authors, January 12, 2015.

91. Ibid.; Open Source Seed Initiative, accessed November 25, 2014, www.opensource seedinitiative.org/about/.

92. Marcin Baranski et al., "Higher antioxidant and lower cadmium concentrations and lower incidence of pesticide residues in organically grown crops: a systematic literature review and meta-analyses," *British Journal of Nutrition* 112 (2014): 794, accessed August 5, 2014, doi:10.1017/S0007114514001366.

93. Ibid.

94. "Hunger: Who Are the Hungry?" World Food Programme, 2014, accessed October 29, 2014, www.wfp.org/hunger/who-are.

95. World Bank, T. Vijay Kumara, et al., "Ecologically Sound, Economically Viable: Community Managed Sustainable Agriculture in Andhra Pradesh, India," 2009, 20, accessed April 3, 2014, www-wds.worldbank.org/external/default/WDSContentServer/WDSP/IB/201 3/03/13/000333037_20130313114055/Rendered/PDF/759610WP0P118800agriculture 0AP02009.pdf.

96. World Bank, "Carbon Sequestration in Agricultural Soils," xxv, Figure E2.

97. World Bank, Kumara et al., "Ecologically Sound, Economically Viable," 20–25.

98. Eric Holt-Giménez, a founder of Campesino a Campesino and executive director of Oakland-based Food First, e-mail communication with the authors, November 12, 2014.

99. FAO, Agricultural Development Economics Division, SOFA Team and Cheryl Doss, "The role of women in agriculture," ESA Working Paper No. 11-02 (Rome: FAO, 2011), 4, accessed January 28, 2015, www.fao.org/docrep/013/am307e/am307e00.pdf.

100. Gerry Marten and Donna Glee Williams, "Getting Clean: Recovering from Pesticide Addiction," *The Ecologist* (December 2006/January2007): 50–53, accessed May 14, 2014, www.ecotippingpoints.org/resources/download-pdf/publication-the-ecologist.pdf; "Buying local worth 400 per cent more," New Economics Foundation, March 7, 2005, accessed May 13, 2014, www.neweconomics.org/press/entry/buying-local-worth-400-per-cent-more.

101. "Biopesticides," Environmental Protection Agency, last updated April 8, 2013, accessed January 5, 2015, www.epa.gov/agriculture/tbio.html; Opender Koul, "Neem: A Global Perspective," in *Neem: Today and in the New Millennium,* ed. Opender Koul and Seema Wahab (New York: Kluwer Academic Publishers, 2004), 13.

102. Marten and Williams, "Getting Clean," 52–53.

103. Altieri and Nicholls, "Agroecology scaling up for food sovereignty and resiliency," 2, 10–11.

104. Jules Pretty et al., "Resource-Conserving Agriculture Increases Yields in Developing Countries," *Environmental Science & Technology* 40, no. 4 (2006): 1115, Supplemental Information Table B, accessed January 12, 2015, doi:10.1021/es051670d.

105. Badgley et al., "Organic agriculture and the global food supply," 86, 88.

106. "EverGreen Agriculture: Re-greening Africa's landscape" (Nairobi: World Agroforestry Center, 2013), 1–6, www.ard-europe.org/fileadmin/SITE_MASTER/content/eiard/Documents/Impact_case_studies_2013/ICRAF_-_EverGreen_agriculture.pdf; FOA, "Climate-Smart Agriculture," last updated June 6, 2011, accessed January 2, 2015, www.fao.org/climatechange/climatesmartpub/66248/en/.

107. "Agriculture at a Crossroads: International Assessment of Agricultural Knowledge, Science and Technology for Development" (Washington, DC: IAASTD, 2009), 538, accessed July 31, 2014, www.unep.org/dewa/agassessment/reports/IAASTD/EN/Agriculture%20at%20a%20Crossroads_Global%20Report%20(English).pdf.

108. Ponisio et al., "Diversification practices reduce organic to conventional yield gap," 1.

109. "Desertification, Land Degradation, and Drought," UNCCD.

110. C. Reij, G. Tappan, and A. Belemvire, "Changing land management practices and vegetation on the Central Plateau of Burkina Faso (1968–2002)," *Journal of Arid Environments* 63, no. 3 (2005): 642–659, accessed July 31, 2014, doi: 10.1016/j.jaridenv.2005.03.010.

111. Perry, "Land of Hope."

112. Inter-American Development Bank, Environmental Safeguards Unit (VPS/ESG), and Susanna B. Hecht, "The Natures of Progress: Land Use Dynamics and Forest Trends in Latin America and the Caribbean," Technical Notes IDB-TN-387, February 2012, 4, accessed January 2, 2015, publications.iadb.org/bitstream/handle/11319/5405/ESG-TN-387%3a%20The%20Natures%20of%20Progress%20Land%20Use%20Dynamics%20and%20Forest%20Trends%20in%20LAC.pdf?sequence=1.

113. P. B. Behere and A. P. Behere, "Farmer suicide in Vidarbha region of Maharashtra state: A myth or reality?" *Indian Journal of Psychiatry* 50, no. 2 (2008): 124–127, accessed January 28, 2015, doi: 10.4103/0019-5545.42401.

114. Pimentel et al., "Environmental, Energetic, and Economic Comparisons," 576.

115. Marcel Mazoyer, "Pour en finir avec la crise alimentaire et agricole mondiale," presented at Conférence mondiale des agronomes, Quebec, September 2012, accessed on April 1, 2014, www.chairedi.fsaa.ulaval.ca/fileadmin/fichiers/fichiersCHAIREDI/7-Evenements/Grandes_conferences/M.Mazoyer_presentation.pdf.

116. Watershed Support and Services Activities Network, "System of Rice Intensification, Weeders: A Reference Compendium," 2006, 12, accessed August 1, 2014, wassan.org/sri/documents/Weeders_Manual_Book.pdf.

117. Calculated from U.S. Department of Agriculture, Center for Nutrition Policy and Promotion, "Eating Healthy on a Budget: The Consumer Economics Perspective," September 2011, accessed May 12, 2014, www.choosemyplate.gov/food-groups/downloads/consumereconomicsperspective.pdf; Cheryl Brown and Mark Sperow, "Examining the cost of an all organic diet," *Journal of Food Distribution Research* 36, no. 1 (2005): 1, accessed January 12, 2015, ageconsearch.umn.edu/bitstream/26759/1/36010020.pdf.

118. Carmen DeNavas-Walt, Bernadette D. Proctor, and Jessica C. Smith, "Income, Poverty, and Health Insurance Coverage in the United States: 2012," U.S Census Bureau, September 2013, 33, Table A-1, accessed October 16, 2014, www.census.gov/prod/2013pubs/p60-245.pdf.

119. Urs Niggli, "Sustainability of organic food production: Challenges and innovations," *Proceedings of the Nutrition Society* (2014): 86, accessed January 2, 2015, doi:10.1017/S0029665114001438.

120. Paolo D'Odorico et al., "Feeding humanity through global food trade," *Earth's Future* 2, no. 9 (2014): 458–469, accessed October 23, 2014, doi: 10.1002/2014EF000250.

121. GRAIN, "Hungry for Land: Small Farmers Feed the World with Less Than a Quarter of all Farmland," Report, May 2014, 3, pdf accessed April 8, 2015, from www.grain .org/article/entries/4952-media-release-hungry-for-land; International Fund for Agricultural Development and the United Nations Environment Programme, Matt Walpole, et al., "Smallholders, Food Security, and the Environment," 2013, 6, 28, accessed April 4, 2014, www.unep.org/pdf/SmallholderReport_WEB.pdf; FAO, "Investing in Smallholder Agriculture for Food Security: A Report by the High Level Panel of Experts on Food Security and Nutrition," HLPE Report 6 (Rome: FAO, June 2013), 26, 28, accessed January 28, 2015, www.fao.org/fileadmin/user_upload/hlpe/hlpe_documents/HLPE_Reports/HLPE-Report -6_Investing_in_smallholder_agriculture.pdf.

MYTH 5: Greater Fairness or More Production? We Have to Choose

1. Calculated from Instituto Brasileiro de Geografia e Estatística,"Censo Agropecuário" (Brazil: IBGE, 2006), Tabela 1.1, www.ibge.gov.br/home/estatistica/economia/ agropecuaria/censoagro/2006/tabela1_1.pdf.

2. Michael Lipton, *Land Reform in Developing Countries: Property Rights and Property Wrongs* (New York: Routledge, 2009), 72; Rehman Sobhan, *Agrarian Reform and Social Transformation* (London: Zed, 1993), 78; William C. Thiesenhusen, *Broken Promises: Agrarian Reform and the Latin American Campesino* (Boulder: Westview Press, 1995), 8, 10, 12, 13, 26, 64, 76, 81, 155.

3. Lipton, *Land Reform in Developing Countries*, 69–70.

4. Amartya Sen, "An Aspect of Indian Agriculture," *The Economic Weekly* 14 (1962): 243–246, accessed August 28, 2014, www.epw.in/system/files/pdf/1962_14/4-5-6/an_ aspect_of_indian_agriculture.pdf; Lipton, *Land Reform in Developing Countries*, 69–70; Peter Rosset, "The Multiple Functions and Benefits of Small Farm Agriculture: In the Context of Global Trade Negotiations," Food First and Transnational Institute, September 1999, accessed March 25, 2015, foodfirst.org/wp-content/uploads/2013/12/PB4-The-Multiple -Functions-and-Benefits-of-Small-Farm-Agriculture_Rosset.pdf.

5. Fatma Gül Ünal, "Small Is Beautiful: Evidence of Inverse Size Yield Relationship in Rural Turkey," October 2006, 20, accessed January 29, 2015, www.policyinnovations.org/ ideas/policy_library/data/01382/_res/id=sa_File1/Unal_paper_updated.pdf.

6. GRAIN, "Hungry for Land," May 2014, 11, accessed October 9, 2014, www.grain.org/ fr/article/entries/4929-hungry-for-land-small-farmers-feed-the-world-with-less-than-a -quarter-of-all-farmland; Mike Wilson, "Brazil's Small Farms Create Pipeline for Profits," *Farm Futures,* May 5, 2011, accessed October 9, 2014, farmfutures.com/blogs-brazils -small-farms-create-pipeline-for-profits-2274.

7. Lipton, *Land Reform in Developing Countries*, 65–69; Ben McKay, "A Socially Inclusive Pathway to Food Security: The Agroecological Alternative," IPC-UNDP Research Brief No. 23, June 2012, 2, accessed January 29, 2015, www.ipc-undp.org/pub/IPCPolicy ResearchBrief23.pdf.

8. Lipton, *Land Reform in Developing Countries*, 70–71.

9. La Vía Campesina, "Peasant and Family Farm-based Sustainable Agriculture Can Feed the World," Via Campesina Views Paper No. 6 (Jakarta: La Vía Campesina, 2010), 7–8, accessed August 28, 2014, viacampesina.org/downloads/pdf/en/paper6-EN.pdf; Robert Netting, *Smallholders, Householders* (Stanford: Stanford University Press, 1993); Gene Wilken, *Good Farmers: Traditional Agricultural Resource Management in Mexico and*

Central America (Berkeley: University of California Press, 1987); Miguel A. Altieri, *Agroecology: The Science of Sustainable Agriculture,* 2nd edition (Boulder: Westview Press, 1995).

10. Ross M. Welch and Robin D. Graham, "Agriculture: the real nexus for enhancing bioavailable micronutrients in food crops," *Journal of Trace Elements in Medicine and Biology* 18, no. 4 (2005): 299–307, accessed October 2, 2014 doi: 10.1016/j.jtemb.2005.03.001.

11. FAO, "Women and Rural Employment: Fighting Poverty by Redefining Gender Roles," Economic and Social Perspectives, Policy Brief No. 5, August 2009, accessed March 12, 2015, www.fao.org/3/a-ak485e.pdf; United Nations Development Programme, "Gender, Climate Change and Food Security," 2012, 2, accessed December 12, 2014, www.undp.org/content/dam/undp/library/gender/Gender%20and%20Environment/PB4_Africa_Gender-ClimateChange-Food-Security.pdf.

12. Rekha Mehra and Mary Hill Rojas, "Women, Food Security and Agriculture in a Global Marketplace," International Center for Research on Women, 2008, 7, accessed October 7, 2014, www.icrw.org/files/publications/A-Significant-Shift-Women-Food%20Security-and-Agriculture-in-a-Global-Marketplace.pdf.

13. FAO, "The State of Food and Agriculture: Closing the Gender Gap for Development" (Rome: FAO, 2011), 5, 34, accessed January 29, 2015, www.fao.org/docrep/013/i2050e/i2050e.pdf.

14. Michael Lipton, "Successes in Anti-Poverty," Issues in Development Discussion Paper No. 8 (Geneva: International Labour Office, 1996), 1996, 62–69, accessed January 29, 2015, www.ilo.int/wcmsp5/groups/public/@ed_emp/documents/publication/wcms_123434.pdf.

15. Martin C. Heller and Gregory A. Keoleian, "Life Cycle–Based Sustainability Indicators for Assessment of the U.S. Food System," Center for Sustainable Systems, Report No. 2000, 42, accessed January 29, 2015, css.snre.umich.edu/css_doc/CSS00-04.pdf; D. Pimentel and M. Pimentel, "Energy Use in Fruit, Vegetable, and Forage Production," in *Food, Energy, and Society*, ed. D. Pimentel and M. Pimentel, revised edition (Niwot, CO: University Press of Colorado, 1996), 131–147.

16. Miguel A. Altieri, Fernando R. Funes-Monzote, and Paulo Petersen, "Agroecologically efficient agricultural systems for smallholder farmers: Contributions to food sovereignty," *Agronomy for Sustainable Development* 32 (2012): 2, accessed January 29, 2015, doi: 10.1007/s13593-011-0065-6.

17. Foley et al., "Solutions for a cultivated planet," 340.

18. "Farmer depend [*sic*] on private moneylenders," *Times of India,* August 2, 2012, accessed October 7, 2014, timesofindia.indiatimes.com/city/nagpur/Farmer-depend-on-private-moneylenders/articleshow/15321269.cms; Pushkar Maitra et al., "Financing Smallholder Agriculture: An Experiment with Agent-Intermediated Microloans in India," March 2014, 9, accessed October 8, 2014, www.bu.edu/econ/files/2012/11/Impacts_paper_v3_Mar202014.pdf.

19. Environmental Working Group, EWG Farm Subsidies Database (Subtotal, Farming Subsidies, Commodity Program Payment, Concentration 2012; accessed January 9, 2015), farm.ewg.org/progdetail.php?fips=&progcode=totalfarm&page=conc&yr=2012&regionname=.

20. FAO, Livelihood Support Programme, Tim Hanstad, Robin Nielsen, and Jennifer Brown, "Land and livelihoods: Making land rights real for India's rural poor," LSP Working Paper 12, May 2004, 9, accessed January 29, 2015, ftp://ftp.fao.org/docrep/fao/007/J2602E/J2602E00.pdf.

21. Lipton, *Land Reform in Developing Countries*, 45; M. Aminul Islam Akanda, Hiroshi Isoda, and Shoichi Ito, "Problem of Sharecrop Tenancy System in Rice Farming in Bangladesh: A case study on Alinapara village in Sherpur district," *Journal of International Farm Management* 4, no. 2 (2008): 1, www.ifmaonline.org/pdf/journals/Vol4_Ed2_Akanda_etal.pdf.

22. Altieri, Funes-Monzote, and Petersen, "Agroecologically efficient agricultural systems for smallholder farmers," 6; Peter M. Rosset, "Alternative Agriculture and Crisis," *Technology and Society Magazine* 16, no. 2 (1997): 19–25, accessed January 29, 2015, doi: 10.1109/44.592253; Oxfam America, "Reforming Cuban Agriculture," *Cuba: Going Against the Grain* 6 (2001), accessed October 14, 2014, www.oxfamamerica.org/static/oa4/OA-CubaGoingAgainstGrain_ReformingGag.pdf.

23. Radha Sinha, *Landlessness: A Growing Problem* (Rome: FAO, 1982), 73.

24. International Fund for Agricultural Development, "Report and Recommendations of the President of the Executive Board on a Proposed Loan to the People's Republic of Bangladesh for the Sunamganj Community-Based Resource Management Project," Executive Board—Seventy-Third Session, Rome, September 12–13, 2001, 2, accessed January 29, 2015, www.ifad.org/gbdocs/eb/73/e/EB-2001-73-R-19.pdf.

25. James K. Boyce, *Agrarian Impasse in Bengal: Institutional Constraints to Technological Change* (Oxford University Press: New York, 1987), 236.

26. Norwood C. Thornton, "Pesticides in Tropical Agriculture," *Advances in Chemistry* 13 (2009): 71–75, accessed January 3, 2015, doi: 10.1021/ba-1955-0013.ch010.

27. Violette Geissen et al., "Soil and Water Pollution in a Banana Production Region in Tropical Mexico," *Bulletin of Environmental Contamination and Toxicology* 85, no. 4 (2010): 407, accessed January 29, 2015, doi: 10.1007/s00128-010-0077-y.

28. Paul B. C. Grant, Million B. Woudneh, and Peter S. Ross, "Pesticides in blood from spectacled caiman (Caiman crocodilus) downstream of banana plantations in Costa Rica," *Environmental Toxicology and Chemistry* 32, no. 11 (2013): 2578, accessed January 29, 2015, doi:10.1002/etc.2358.

29. José Graziano da Silva, "The family farming revolution," FAO, accessed December 21, 2014, www.fao.org/about/who-we-are/director-gen/faodg-opinionarticles/detail/en/c/212364/.

30. "Chapter 6: Transnational Companies in the World Banana Economy," in *The World Banana Economy: 1985–2002* (Rome: FAO, 2003), www.fao.org/docrep/007/y5102e/y5102e09.htm#bm09.2; Sam Cage, "Chiquita and Fyffes Join to Make World's Biggest Banana Firm," Reuters, March 10, 2014, accessed January 29, 2015, www.reuters.com/article/2014/03/10/fyffes-chiquita-brands-idUSL3N0M724Y20140310.

31. James Wiley, *The Banana: Empires, Trade Wars, and Globalization* (Lincoln: University of Nebraska Press, 2008), 19.

32. Human Rights Watch, "Tainted Harvest: Labor and Obstacles to Organizing on Ecuador's Banana Plantations," April 2002, 65, accessed October 9, 2014, www.hrw.org/reports/2002/ecuador/2002ecuador.pdf.

33. Ward Anseeuw et al., "Land Rights and the Rush for Land: Findings of the Global Commercial Pressures on Land Research Project" (Rome: International Land Coalition, 2012), 4, accessed March 17, 2015, www.landcoalition.org/sites/default/files/publication/1205/ILC%20GSR%20report_ENG.pdf.

34. "When others are grabbing their land," *The Economist,* May 5, 2011, accessed January 29, 2015, www.economist.com/node/18648855.

35. Michael Kugelman, "The Global Farmland Rush," *New York Times*, February 5, 2013, accessed January 27, 2014, www.nytimes.com/2013/02/06/opinion/the-global -farmland-rush.html?_r=0.

36. Dietrich Vollrath, "Land Distribution and International Agricultural Productivity," *American Journal of Agricultural Economics* 89, no. 1 (2007): 202–216, accessed January 29, 2015, www.jstor.org/stable/4123573.

37. Frank F. K. Byamugisha, *Securing Africa's Land for Shared Prosperity: A Program to Scale Up Reforms and Investments* (Washington, DC: International Bank for Reconstruction and Development / World Bank, 2013), 4, 78–80, openknowledge.worldbank.org/bitstream/ handle/10986/13837/780850PUB0EPI00LIC00pubdate05024013.pdf?sequence=1.

38. Tania Krutscha, "Brazil's Large Landowners Brace to Resist Reform," *Latin America Press,* September 19, 1985, 1.

39. Sandra Praxedes, "The Landless Workers Movement: Towards Social Transforma- tion," Information Services Latin America, accessed December 21, 2014, isla.igc.org/ Features/Brazil/braz1.html.

40. Joseph Collins and Nick Allen, *Nicaragua: What Difference Could a Revolution Make? Food and Farming in the New Nicaragua* (New York: Grove Atlantic, 1986); William C. Thiesenhusen, "Broken Promises: Agrarian Reform and the Latin American Campesino," *Journal of Latin American Studies* 28, no. 2 (1996): 517–519, www.jstor.org/stable/157639.

41. Lipton, *Land Reform in Developing Countries,* 342; Sobhan, *Agrarian Reform and Social Transformation.*

42. Cho Seok Gon and Park Tae Gyun, "Suggestions for New Perspectives on the Land Reform in South Korea," *Seoul Journal of Korean Studies* 26, no. 1 (2013): 1–21, accessed February 3, 2015, muse.jhu.edu/journals/seo/summary/v026/26.1.cho.html; Yoong-Deok Jeon and Young-Yong Kim, "Land Reform, Income Redistribution, and Agricultural Pro- duction in Korea," *Economic Development and Cultural Change* 48, no. 2 (2000): 255–256, accessed February 3, 2015, www.jstor.org/stable/10.1086/452457; Sobhan, *Agrarian Reform and Social Transformations;* Jeffrey D. Sachs, "Trade and Exchange Rate Policies in Growth-Oriented Adjustment Programs," Working Paper No. 2226, National Bureau of Eco- nomic Research, April 1987, accessed February 3, 2015, www.nber.org/papers/w2226.pdf.

43. Chun-chieh Huang, *Taiwan in Transformation, 1895-2005: The Challenge of a New Democracy to an Old Civilization* (New Brunswick, NJ: Transaction Publishers, 2006), 35.

44. Lipton, *Land Reform in Developing Countries*; Michael Lipton, personal communica- tion with authors, June 2014; Sobhan, *Agrarian Reform and Social Transformations,* 88–89; Sachs, "Trade and Exchange Rate Policies."

45. World Bank, "From Poor Areas to Poor People: China's Evolving Poverty Reduction Agenda: An Assessment of Poverty and Inequality in China," March 2009, 78, accessed Decem- ber 1, 2014, siteresources.worldbank.org/CHINAEXTN/Resources/318949-1239096143906/ China_PA_Report_March_2009_eng.pdf.

46. Calculated from FAOSTAT [Production, Production Indices, Country: China, Years: 1978–84, Item: Agriculture (PIN) + (Total), Element: Net Per Capita Production Index Number (2004–06 = 100); accessed February 3, 2015], faostat3.fao.org/download/Q/QI/E; International Fund for Agricultural Development, Communications and Public Affairs Unit, "Reducing Pov- erty Through Land Reform," accessed January 30, 2014, www.ifad.org/media/pack/land.htm.

47. FAO, "State of Food Insecurity in the World," 2015, 40, 42.

48. Ronald Herring, "Explaining Anomalies in Agrarian Reform: Lessons from South India," in *Agrarian Reform and Grassroots Development: Ten Case Studies*, ed. Roy L.

Prosterman, Mary N. Temple, and Timothey M. Hanstad (Boulder: Lynne Rienner Publishers, 1990), 73.

49. Richard Franke and Barbara Chasin, *Kerala: Radical Reform as Development in an Indian State* (Oakland: Food First Books, 1994), 58.

50. Maitreesh Ghatak and Sanchari Roy, "Land Reform and Agricultural Productivity in India: A review of evidence," *Oxford Review of Economic Policy* 23, no. 2 (2007): 251, accessed February 3, 2015, doi: 10.1093/oxrep/grm017.

51. Purnima Menon, Anil Deolalikar, and Anjor Bhaskar, "India State Hunger Index: Comparisons of Hunger Across States" (Washington, DC, and Bonn: Riverside, 2009), 15, 20, 25.

52. James K. Boyce, Peter Rosset, and Elizabeth A. Stanton, "Land Reform and Sustainable Development," Political Economy Research Institute, University of Massachusetts Amherst, Working Paper No. 98, 2005, accessed January 30, 2014, scholarworks.umass.edu/cgi/viewcontent.cgi?article=1079&context=peri_workingpapers; World Bank and Oxford University Press, *Land Policies for Growth and Poverty Reduction* (Washington, DC: World Bank and Oxford University Press, 2003); IFAD, "Reducing Poverty Through Land Reform."

53. Lipton, "Successes in Anti-poverty," 62 (emphasis in source).

54. Ibid., 62–69; Sobhan, *Agrarian Reform and Social Transformations*, 49.

55. A. Knoll and I. Reyes, "From Fire to Autonomy: Zapatistas, 20 Years of Walking Slowly," *Truth-Out,* January 25, 2014.

56. Richard Stahler-Sholk, "Resisting Neoliberal Homogenization: The Zapatista Autonomy Movement," *Latin American Perspectives* 34, no. 2 (March, 2007): 48-63, doi: 10.1177/0094582X06298747; "EZLN Governs 250,000 Indigenous Mexicans," dorsetchia passolidarity.wordpress.com/2014/02/19/ezln-governs-250000-indigenous-mexicans/; Laura Carlsen, "Zapatista Communities Celebrate 20 Years of Self-Government," *Yes! Magazine,* January 23, 2014; "The EZLN—A Look at Its History: The Guerrilla Nucleus," Parts 1, 2, 3, Upside Down World, accessed May 6, 2015, upsidedownworld.org/main/mexico-archives-79/4611-the-ezln-a-look-at-its-history-part-1-.

57. Friends of the MST, www.mstbrazil.org/content/what-mst; Adam Raney and Chad Heeter, "Brazil: Cutting the Wire: Witnessing a land occupation," PBS Frontline World, December 13, 2005, accessed January 11, 2015, www.pbs.org/frontlineworld/rough/2005/12/brazil_cutting.html.

58. Lipton, "Successes in Anti-Poverty," 62–66.

MYTH 6: The Free Market Can End Hunger

1. Saeed Kamali Dehghan, "Kidneys for Sale: Poor Iranians compete to sell their organs," *The Guardian*, May 27, 2012, accessed August 14, 2014, www.theguardian.com/world/2012/may/27/iran-legal-trade-kidney; Ahad J. Ghods and Shekoufeh Savaj, "Iranian Model of Paid and Regulated Living-Unrelated Kidney Donation," *Clinical Journal of the American Society of Nephrology* 1, no. 6 (2006): 1136, accessed August 14, 2014, doi: 10.2215, cjasn.asnjournals.org/content/1/6/1136.full.pdf.

2. Anti-Slavery International, www.antislavery.org.

3. "Clinton at UN: Food, energy, financial woes linked," FAO Newsroom, October 24, 2008, accessed August 14, 2014, www.fao.org/newsroom/en/news/2008/1000945/.

4. "International Covenant on Economic, Social and Cultural Rights," United Nations, accessed May 8, 2015, treaties.un.org/pages/viewdetails.aspx?chapter=4&lang=en&mtdsg_no=iv-3&src=treaty.

5. Margret Vidar, Yoon Jee Kim, and Luisa Cruz, "Legal Developments in the Progressive Realization of the Right to Food, Right to Food Thematic Study 3," FAO, 2014, 2–3, accessed May 12, 2014, www.fao.org/3/a-i3892e.pdf.

6. Samuel Bowles, Richard Edward, and Frank Roosevelt, *Understanding Capitalism* (New York: Oxford University Press, 1985).

7. World Bank [Data, Indicators, GINI index (World Bank estimate), Countries: Turkey, United States, and India, Years: 2008–22; accessed August 12, 2014], data.worldbank. org/indicator/SI.POV.GINI; Giles Keating et al., "Global Wealth Report 2013" (Zurich: Credit Suisse Ag, 2013), 21, accessed August 12, 2014, publications.credit-suisse.com/ tasks/render/file/?fileID=BCDB1364-A105-0560-1332EC9100FF5C83.

8. Branko Milanovik, "Global Inequality by the Numbers: In History and Now," Policy Research Working Paper No. 6259, World Bank, November 2012, 12, accessed February 3, 2015, elibrary.worldbank.org/doi/pdf/10.1596/1813-9450-6259.

9. U.S. Department of Agriculture, Economic Research Service, Jason P. Brown, and Jeremy G. Weber, "The Off-Farm Occupations of U.S. Farm Operators and Their Spouses," Economic Information Bulletin No. 117, September 2013, ii, accessed August 13, 2014, www.ers.usda.gov/media/1187209/eib-117.pdf.

10. Klaus Schwab, "The Global Competitiveness Report 2013–14," World Economic Forum, Insight Report (Geneva, 2013), 15, accessed August 13, 2014, www3.weforum. org/docs/WEF_GlobalCompetitivenessReport_2013-14.pdf; Heritage Foundation, "2014 Index of Economic Freedom: Finland," accessed August 14, 2014, www.heritage.org/ index/country/finland; Heritage Foundation, "2014 Index of Economic Freedom: Switzerland," accessed August 19, 2014, www.heritage.org/index/country/switzerland#limited -government; Heritage Foundation, "2014 Index of Economic Freedom: Germany," accessed August 19, 2014, www.heritage.org/index/country/germany#limited-government; Heritage Foundation, "2014 Index of Economic Freedom: United States," last modified 2014, accessed August 14, 2014, www.heritage.org/index/country/unitedstates.

11. Lew Daly, "Our Mismeasured Economy," *New York Times*, July 6, 2014, accessed August 13, 2014, www.nytimes.com/2014/07/07/opinion/our-mismeasured-economy. html?_r=1; Alan H. Lockwood, "How the Clean Air Act Has Saved $22 Trillion in Health-Care Costs," *Atlantic*, September 7, 2012, accessed January 30, 2015, www.theatlantic .com/health/archive/2012/09/how-the-clean-air-act-has-saved-22-trillion-in-health -care-costs/262071/.

12. Calculated from Michael Carolan, *Cheaponomics: The High Cost of Low Prices* (New York: Earthscan, 2014), 112.

13. Friedman and Friedman, *Free to Choose*, 65–66.

14. "Soil Erosion on Cropland 2007," Natural Resource Conservation Service, South Carolina, U.S. Department of Agriculture, accessed March 31, 2014, www.nrcs.usda.gov/ wps/portal/nrcs/detail/sc/technical/dma/nri/?cid=stelprdb1041887.

15. Kevin Kruse, "A Christian Nation, Since When?" *New York Times*, Sunday Review, March 15, 2015. www.nytimes.com/2015/03/15/opinion/sunday/a-christian-nation -since-when.html?_r=0.

16. Karl Polanyi, *The Great Transformation: The Political and Economic Origins of Our Time* (Boston: Beacon Press, 1944).

17. Francis X. Sutton et al., *The American Business Creed* (Cambridge: Cambridge University Press, 1956), 64–65.

18. Clyde Prestowitz, *The Betrayal of American Prosperity* (New York: Free Press, 2010), 193.

19. Appalachian Coals, Inc. v. United States, 288 U.S. 344 (1933), accessed August 12, 2014, supreme.justia.com/cases/federal/us/288/344/case.html.

20. United States v. Columbia Steel Co. et al., 334 U.S. 495, 534 (1948).

21. Sophia Murphy, David Burch, and Jennifer Clapp, "Cereal Secrets: The world's largest grain traders and global agriculture," Oxfam Research Reports, August 2010, 3, accessed August 15, 2014, www.oxfam.org/sites/www.oxfam.org/files/rr-cereal-secrets-grain -traders-agriculture-30082012-en.pdf.

22. Thomas A. Lyson and Annalisa Lewis Raymer, "Stalking the wily multinational: Power and control in the US food system," *Agriculture and Human Values* 17, no. 2 (2000): 199, accessed February 3, 2015, doi: 10.1023/A:1007613219447.

23. Joseph Stiglitz, "Inequality Is a Choice," *New York Times,* October 13, 2013, accessed February 3, 2015, opinionator.blogs.nytimes.com/2013/10/13/inequality-is-a-choice/.

24. "Top Ten Most Ridiculous Luxury Items," ThinkAdvisor, last modified December 1, 2013, accessed August 13, 2014, www.thinkadvisor.com/2013/12/01/top-10-most -ridiculous-luxury-items-2013?page_all=1.

25. Hurni et al., "Key Implications of Land Conversions in Agriculture," 221.

26. Calculated from FAOSTAT [Production, Crops, Regions: Africa + (Total), Element: Production Quantity, Items: Cocoa, beans, Year: 1980–2012; accessed August 13, 2014], faostat3.fao.org/faostat-gateway/go/to/browse/Q/QC/E.

27. Elizabeth Kimani, "The Nutrition Paradox in Kenya: What are we doing?" *Global Nutrition Report*, July 18, 2014, accessed January 5, 2015, globalnutritionreport .org/2014/07/18/the-nutrition-paradox-in-kenya/; Food and Water Watch and the Council of Canadians, "Lake Naivasha, Withering Under the Assault of International Flower Vendors," 2008, 1, accessed August 13, 2014, documents.foodandwaterwatch .org/doc/NaivashaReport.pdf#_ga=1.121963787.1468985300.1393100138.

28. U.S. Department of Commerce, Bureau of Economic Analysis, National data, National Income and Product Account Tables, Section 6—Income and Employment by Industry, Table 6.16B and 6.16D Corporate Profits by Industry, Years: 1985 and 2013: Quarter: I, accessed February 3, 2015, www.bea.gov/itable/iTable.cfm?ReqID=9&step=1# reqid=9&step=1&isuri=1.

29. For overview see Charles H. Ferguson, *Predator Nation: Corporate Criminals, Political Corruption, and the Hijacking of America* (New York: Crown, 2012).

30. Ryan Isakson, "Financialization and the Transformation of Agro-food Supply Chains: A Political Economy," Conference Paper No. 9, Food Sovereignty: A Critical Dialogue, International Conference, Yale University, September 14–15, 2013, 7, accessed February 3, 2015, www.yale.edu/agrarianstudies/foodsovereignty/pprs/9_Isakson_2013.pdf.

31. "Lobbying Database," OpenSecrets.org (accessed January 15, 2014,www.open secrets.org/lobby/.

32. Christopher Ketcham, "Monopoly Is Theft," *Harper's Magazine*, October 19, 2012, accessed August 12, 2014, harpers.org/blog/2012/10/monopoly-is-theft/.

33. Calculated from FAOSTAT (Population, Annual population, Country: Malawi, Element: Total population—both sexes, Item: Population—est. & proj., Years: 1995 and 2013; accessed August 19, 2014), faostat3.fao.org/faostat-gateway/go/to/download/O/OA/E; "End of mission statement by the Special Rapporteur on the right to food, Malawi 12 to 22 July 2013," Office of the High Commissioner on Human Rights, United Nations Human Rights, July 22, 2013, accessed February 3, 2015, www.ohchr.org/EN/NewsEvents/Pages/ DisplayNews.aspx?NewsID=13567&.

34. Structural Adjustment Participatory Review International Network, "The Policy Roots of Economic Crisis and Poverty: A Multi-Country Participatory Assessment of Structural Adjustment," 1st edition, April 2002, 173, accessed April 7, 2015, www.saprin.org/SAPRIN_Findings.pdf.

35. Calculated from FAOSTAT [Trade, Crops and Livestock Products, Special Groups: Least Developed Countries + (Total), Elements: Import Value, Items Aggregated: Agricult. Products, Total + (Total), Year: 1985, 2011; accessed February 3, 2015], faostat3.fao.org/download/T/TP/E.

36. USDA, ERS, Coleman-Jensen, Gregory, and Singh, "Household Food Insecurity," 6.

37. FAO, "Smallholders and Family Farmers," Sustainability Pathways Factsheets, 2012, accessed August 13, 2014, www.fao.org/fileadmin/templates/nr/sustainability_pathways/docs/Factsheet_SMALLHOLDERS.pdf; International Fund for Agricultural Development, "Smallholders Can Feed the World," February 2011, 2, accessed March 31, 2015, www.ifad.org/pub/viewpoint/smallholder.pdf.

38. U.S. Department of Agriculture, Economic Research Service, and Jorge Fernandez-Cornejo, "The Seed Industry in U.S. Agriculture: An Exploration of Data and Information on Crop Seed Markets, Regulation, Industry Structure, and Research and Development," Agriculture Information Bulletin No. 786, February 2004, 36–37, accessed December 30, 2014, www.ers.usda.gov/media/260729/aib786_1_.pdf.

39. Philip H. Howard, "Visualizing Consolidation in the Global Seed Industry: 1996–2008,"Sustainability 1 (2009): 1271, doi: 10.3390/su1041266; ETC Group, "Who Will Control the Green Economy," 22.

40. U.S. Department of Agriculture and Jorge Fernandez-Cornejo, "The Seed Industry in U.S. Agriculture," Agriculture Information Bulletin No. 786 (Washington, DC, January 2004), Table 13, 27, accessed August 12, 2014, www.ers.usda.gov/media/260729/aib786_1_.pdf.

41. Ibid.

42. "Why Does Monsanto Sue Farmers Who Save Seeds?" Newsroom, Monsanto, accessed August 19, 2014, www.monsanto.com/newsviews/pages/why-does-monsanto-sue-farmers-who-save-seeds.aspx.

43. Calculated from U.S. Department of Agriculture, Economic Research Service (Data, Commodity Costs and Returns, Recent Costs and Returns: Corn and Soybeans, U.S.: 1996–2000, 2001–04, 2005–09, 2010–13, accessed August 14, 2014), www.ers.usda.gov/data-products/commodity-costs-and-returns.aspx.

44. Timothy Wise, "Agribusiness and the Food Crisis: A new thrust at anti-trust," *GDAE Globalization Commentary,* from *Triple Crisis Blog,* March 22, 2010, accessed August 13, 2014, triplecrisis.com/agribusiness-and-the-food-crisis-a-new-thrust-at-anti-trust/.

45. Chart data from "Farm Household Income (Historical)," Farm Household Well-being, Economic Research Service, U.S. Department of Agriculture, last modified May 22, 2015, accessed June 5, 2015, www.ers.usda.gov/topics/farm-economy/farm-household-well-being/farm-household-income-(historical).aspx.

46. Kumar et al., "Ecologically Sound, Economically Viable," 6–7.

47. Thomas Jefferson, *The Writings of Thomas Jefferson*, collected and edited by Paul Leicester Ford, Volume X, 1816–26 (New York and London: G. P. Putnam's Sons, 1899), 69, accessed August 19, 2014, babel.hathitrust.org/cgi/pt?id=uc2.ark:/13960/t3pv6bn68;view=1up;seq=93.

48. FAO, "Price Volatility from a Global Perspective," Technical background document for the high-level event on: "Food price volatility and the role of speculation," Rome, Italy, July

6, 2012, 5, accessed August 13, 2014, www.fao.org/fileadmin/templates/est/meetings/price_volatility/Price_volatility_TechPaper_V3_clean.pdf.

49. Chris Callieri et al., "Rattling Supply Chains: The Effect of Environmental Trends on Input Costs for the Fast-Moving Consumer Goods Industry," World Resources Institute and A. T. Kearney, 2008, 3, accessed August 12, 2014, pdf.wri.org/rattling_supply_chains.pdf; "Food Staple," National Geographic Encyclopedia, accessed January 22, 2014, education.nationalgeographic.com/education/encyclopedia/food-staple/?ar_a=1; Ankie Scott-Joseph, "The Nature of Rising Food Prices in the Eastern Caribbean: An Analysis of Food Inflation During the Period 2005–08 in a Context of Household Poverty," UNICEF Office for Barbados and the Eastern Caribbean, 2009, accessed October 28, 2014, www.unicef.org/easterncaribbean/The_Nature_of_Rising_Food_Prices_in_the_Eastern_Caribbean.pdf.

50. Phil McMichael, "Historicizing Food Sovereignty: A Food Regime Perspective," Conference Paper No. 13, Food Sovereignty: A Critical Dialogue International Conference, Yale University, September 14–15, 2013, 19, accessed August 13, 2014, www.yale.edu/agrarianstudies/foodsovereignty/pprs/13_McMichael_2013.pdf.

51. FOA, "World Food Situation, FAO Food Price Index," accessed March 31, 2015, www.fao.org/worldfoodsituation/foodpricesindex/en/.

52. Thomas L. Friedman, "The Scary Hidden Stressor," *New York Times*, March 2, 2013, accessed August 12, 2014, www.nytimes.com/2013/03/03/opinion/sunday/friedman-the-scary-hidden-stressor.html.

53. FOA, "World Food Situation, FAO Food Price Index."

54. FAOSTAT [Production, Production indices, Regions: World + (Total), Element: Net per capita Production Index Number (2004–06 = 100), Items Aggregated: Food (PIN) + (Total), Years: 2005–2012; accessed August 19, 2014], faostat3.fao.org/faostat-gateway/go/to/download/Q/QI/E.

55. FAOSTAT [Trade, Crops and Livestock Products, Country: China, Element: Import Quantity, Item Aggregated: Cereals + (Total), Years: 2000–10; accessed August 19, 2014], faostat3.fao.org/faostat-gateway/go/to/download/T/TP/E.

56. Kharunya Paramaguru, "Betting on Hunger: Is Financial Speculation to Blame for High Food Prices?" *Time*, December 17, 2012, accessed August 13, 2014, science.time.com/2012/12/17/betting-on-hunger-is-financial-speculation-to-blame-for-high-food-prices/#ixzz2jQSzLY2j.

57. Frederick Kaufman, *Bet the Farm: How Food Stopped Being Food* (Hoboken: John Wiley & Sons, Inc., 2012), Chapter 12.

58. M. W. Masters, Testimony of Michael W. Masters Before the Committee on Homeland Security and Governmental Affairs, United States Senate, May 20, 2008, accessed August 13, 2014, hsgac.senate.gov/public/_files/052008Masters.pdf.

59. International Food Policy Research Institute, Maximo Torero, "Food Prices: Riding the Rollercoaster," 2011 Global Food Policy Report, 2011, 7, accessed August 13, 2014, www.ifpri.org/node/8436.

60. Eugenio S. A. Bobenrieth and Brian D. Wright, "The Food Price Crisis of 2007/2008: Evidence and Implications," Symposium of Value Chains for Oilseeds, Oils and Fats, Grains and Rice: Status and Outlook, Santiago, Chile, November 4–6, 2009, 10, accessed August 19, 2014, www.fao.org/fileadmin/templates/est/meetings/joint_igg_grains/Panel_Discussion_paper_2_English_only.pdf.

61. FAO, "FAO Statistical Yearbook 2012, Food and Agriculture, Part 4" (Rome: FAO, 2012), 316, accessed August 14, 2014, www.fao.org/docrep/015/i2490e/i2490e04d.pdf; "Land grabbing for biofuels must stop," GRAIN, last updated February 21, 2013, accessed

August 14, 2014, www.grain.org/article/entries/4653-land-grabbing-for-biofuels-must-stop#6; United Nations Environmental Programme, Stefan Bringezu, et al., "Towards sustainable production and use of resources: Assesing biofuels," 2009, 21, accessed March 12, 2015, www.unep.org/PDF/Assessing_Biofuels.pdf.

62. Timothy A. Wise and Emily Cole, "Mandating Food Insecurity: The Global Impacts of Rising Biofuel Mandates and Targets," Global Development and Environment Institute Working Paper No. 15-01, February 2015, accessed February 24, 2015, www.ase.tufts.edu/gdae/Pubs/wp/15-01WiseMandates.pdf.

63. Rebekah Kebede, "Oil hits record above $147," Reuters, July 11, 2008, accessed August 15, 2014, www.reuters.com/article/2008/07/11/us-markets-oil-idUST14048520080711; United Nations, Department of Economic and Social Affairs, *The Global Social Crisis: Report on the World Social Situation 2011* (New York: UN, 2011), 68, accessed August 15, 2014, www.un.org/esa/socdev/rwss/docs/2011/rwss2011.pdf.

64. "Annual Lobbying on Oil and Gas," OpenSecrets.org, accessed August 14, 2014, www.opensecrets.org/lobby/indusclient.php?id=E01&year=2013.

65. "Energy Subsidies," International Energy Agency, accessed April 2, 2015, www.worldenergyoutlook.org/resources/energysubsidies/.

66. FAOSTAT [Production, Production Indices, Regions: World + (Total), Items Aggregated: Cereals, Total + (Total), Element: Net Production Index Number (2004–06 = 100), Years 1961–2011; accessed January 7, 2015], faostat3.fao.org/download/Q/QC/E; Full data set downloaded from "FAO Cereal Supply and Demand Brief," FAO; "World Food Situation, FAO Food Price Index," FAO.

67. Full data set downloaded from "Cereal Supply and Demand Data," "FAO cereal production in 2014," FAO.

68. Agricultural Adjustment Act of 1938, Pub. L. No. 75-430, 52 Stat. 31 (1938), accessed August 18, 2014, nationalaglawcenter.org/wp-content/uploads/assets/farmbills/1938.pdf.

69. Robert G. Chambers and William E. Foster, "Participation in the Farmer-Owned Reserve Program: A discrete choice model," *American Journal of Agricultural Economics* 65, no. 1 (1983): 120, accessed February 3, 2015, doi: 10.2307/1240346.

70. Peter Rosset, "Agrofuels, Food Sovereignty, and the Contemporary Food Crisis," *Bulletin of Science, Technology & Society* 29, no. 3 (2009): 190, accessed August 15, 2014, globalalternatives.org/files/RossetAgrofuels.pdf; Frederick Kaufman "How to fight a food crisis: To blunt the ravages of drought and market greed, we need a national grain reserve," *Los Angeles Times*, September 21, 2012, articles.latimes.com/print/2012/sep/21/opinion/la-oe-kaufman-food-hunger-drought-20120921.

71. "Cargill reports fourth-quarter and fiscal 2008 earnings," Cargill, August 2008, accessed October 28, 2014, www.cargill.com/news/releases/2008/NA3007599.jsp.

72. Oxfam, "Our Land, Our Lives: Time out on the global land rush," Oxfam Briefing Note, October 2012, 2, accessed April 16, 2015, www.oxfam.org/sites/www.oxfam.org/files/bn-land-lives-freeze-041012-en_1.pdf; The Land Matrix Partnership, W. Anseeuw, et al., "Transnational Land Deals for Agriculture in the Global South," Analytical Report based on Land Matrix Database No. 1, April 2012, vii, accessed February 3, 2015, www.landcoalition.org/sites/default/files/publication/1254/Analytical%20Report%20Web.pdf.

73. Oxfam, "Our Land, Our Lives," 2.

74. Timothy A. Wise, "Picking up the pieces from a failed land grab project in Tanzania," *Global Post,* June 27, 2013, accessed August 14, 2014, www.globalpost.com/dispatches/globalpost-blogs/rights/picking-the-pieces-failed-land-grab-project-tanzania.

75. Oakland Institute, Shephard Daniel, and Anuradha Mittal, "The Great Land Grab: Rush for World's Farmland Threatens Food Security for the Poor," 2009, 13, accessed August 15, 2014, www.oaklandinstitute.org/sites/oaklandinstitute.org/files/LandGrab_final_web.pdf.

76. "Franklin D. Roosevelt: 59—Message to Congress on Curbing Monopolies," American Presidency Project, accessed February 3, 2015, www.presidency.ucsb.edu/ws/index.php?pid=15637.

77. Adam Liptak, "Justices, 5–4, Reject Corporate Spending Limit," *New York Times,* January 21, 2010, accessed October 29, 2014, www.nytimes.com/2010/01/22/us/politics/22scotus.html?pagewanted=all&_r=0.

78. Roger Blobaum, "Chapter 7: The Worldwide Expansion of Organic Farming: Its Potential Contribution to a Global Transition to Sustainable Agriculture," in *For All Generations: Making World Agriculture More Sustainable*, ed. Patrick Madden and Scott Chaplowe (Madison, WI: Om Pub Consultants, 1997, rogerblobaum.com/737/.

79. FAO, Regional Office for Asia and the Pacific, ed. John Pontius, Russell Dilts, and Andrew Bartlett, "From Farmer Field School to Community IPM: Ten Years of IPM Training in Asia," 2002, 1, accessed February 3, 2015, ftp://ftp.fao.org/docrep/fao/005/ac834e/ac834e00.pdf; International Food Policy Research Institute, Kristin Davis, et al., "Impact of Farmer Field Schools on Agricultural Productivity and Poverty in East Africa," Discussion Paper 00992, June 2010, 13, accessed February 3, 2015, www.ifpri.org/sites/default/files/publications/ifpridp00992.pdf.

80. PBL Netherlands Environmental Assessment Agency, "Evaluation of the Policy Document on Sustainable Crop Protection," February 15, 2012, 2, accessed August 12, 2014, www.pbl.nl/sites/default/files/cms/publicaties/PBL-2012-Duurzame-Gewasbescherming-Engelse-samenvatting.pdf; Pesticides Action Network Europe, Catherine Wattiez, and Stephanie Williamson, "Pesticide Use Reduction Is Working: An assessment of national reduction strategies in Denmark, Sweden, the Netherlands and Norway," December 2003, 6–9, accessed February 3, 2015, www.epha.org/IMG/pdf/Pure_is_Working.pdf.

81. Soil Association, "The Lazy Man of Europe: Wake Up to What Europe Can Teach the UK About Backing Organic Food and Farming," 2011, 7–15, accessed August 12, 2014, www.soilassociation.org/LinkClick.aspx?fileticket=0DLNEXzSlJk%3D&tabid=387.

82. Institute for Agriculture and Trade Policy, and Sophia Murphy, "Strategic Grain Reserves in an Era of Volatility," October 2009, 7, accessed October 29, 2014, www.iatp.org/documents/strategic-grain-reserves-in-an-era-of-volatility.

83. U.S. Department of Agriculture, Economic Resource Service, and Jerry A. Sharples, "An Evaluation of U.S. Grain Reserve Policy, 1977–80," no. 481 (March 1982): 28, accessed January 6, 2015, naldc.nal.usda.gov/naldc/download.xhtml?id=CAT85840194&content=PDF.

84. Mulat Demeke, Guendalina Pangrazio, and Materne Maetz, "Country responses to the food security crisis: Nature and preliminary implications of the policies pursued," FAO, 2009, 17, accessed January 6, 2015, www.fao.org/fileadmin/user_upload/ISFP/pdf_for_site_Country_Response_to_the_Food_Security.pdf.

85. Asean, "Asean Plus Three Emergency Rice Reserve Agreement" 5, October 7, 2011, www.asean.org/images/2012/Economic/AMAF/Agreements/ASEAN%20Plus%20Three%20Emergency%20Rice%20Reserve%20Agreement%2022.pdf; "ASEAN Plus Three Emergency Rice Reserve: Overview," APTERR, accessed October 28, 2014, www.apterr.org/about-us.

86. IATP and Murphy, "Strategic Grain Reserves," 7.

87. Edward Ojulu, "Africa: Economic Projections in the Eyes of the World Bank," All Africa, November 3, 2013, accessed February 3, 2015, m.allafrica.com/stories/201311030020 .html/; "Equity Case Study: Rwanda—One Cow per Poor Family," UNICEF, July 10, 2012, accessed December 31, 2014, www.unicef.org/equity/index_65274.html.

88. FAO, "The Right to Food in Practice, Implementation at the National Level" (Rome: FAO, 2006), 7, accessed August 13, 2014, www.fao.org/docrep/016/ah189e/ah189e.pdf.

89. Deborah Wetzel, "Bolsa Familia: Brazil's Quiet Revolution," World Bank, November 4, 2013, accessed February 3, 2015, www.worldbank.org/en/news/opinion/2013/11/04/ bolsa-familia-Brazil-quiet-revolution.

90. Cecilia Rocha, "Urban Food Security Policy: The Case of Belo Horizonte, Brazil," *Journal for the Study of Food and Society* 5, no. 1 (2001): 37, www.ryerson.ca/content/ dam/foodsecurity/publications/articles/BeloHorizonte.pdf.

91. Adriana Aranha, interview with coauthor Frances Moore Lappé in Belo Horizonte, July 2000; Frances Moore Lappé and Anna Lappe, *Hope's Edge: The Next Diet for a Small Planet* (New York: Tarcher/Penguin, 2002), 96.

92. FAO, "Right to Food: Lessons Learned in Brazil" (Rome: FAO, 2007), 12, accessed May 14, 2014, ftp://ftp.fao.org/docrep/fao/010/a1331e/a1331e.pdf.

93. FAO, "Growing Greener Cities," 75.

94. Cecilia Rocha and Lara Lessa, "Urban Governances for Food Security: The Alternative Food System in Belo Horizonte, Brazil," *International Planning Studies* 14, no. 4 (2009): 389–400, 2014, www.area-net.org/fileadmin/user_upload/Food_Security/Urban_ governance_for_food_security__Belo_Horizonte.pdf; nickel calculation by authors.

95. "The Irish famine: Opening old wounds," *The Economist,* December 12, 2012, accessed April 14, 2015, www.economist.com/blogs/prospero/2012/12/irish-famine.

Myth 7: Free Trade Is the Answer

1. Jeff Madrick, "Our Misplaced Faith in Free Trade," *New York Times* Sunday Review, October 3, 2014, accessed October 23, 2014, www.nytimes.com/2014/10/04/opinion/ sunday/our-misplaced-faith-in-free-trade.html?_r=0; Ha-Joon Chang, "Kicking Away the Ladder: The 'Real' History of Free Trade," December 2003, 2, accessed January 2, 2015, fpif .org/kicking_away_the_ladder_the_real_history_of_free_trade.

2. Arantxa Guerena, "The Soy Mirage," Oxfam Research Reports, August 2013, 3, accessed August 18, 2014, www.oxfam.org/sites/www.oxfam.org/files/rr-soy-mirage-corporate- social-responsibility-paraguay-290813-en.pdf.

3. Ibid.

4. Oxfam, "Smallholders at Risk: Monoculture expansion, land, food and livelihoods in Latin America," Oxfam Briefing Paper No. 180, April 23, 2013, 8, accessed August 18, 2014, www.oxfam.org/sites/www.oxfam.org/files/bp180-smallholders-at-risk-land-food -latin-america-230414-en.pdf.

5. Jeremy Hobbs, "Paraguay's Destructive Soy Boom," *New York Times,* July 2, 2012, accessed August 18, 2014, www.nytimes.com/2012/07/03/opinion/paraguays-destructive -soy-boom.html.

6. Simon Romero, "Boom Times in Paraguay Leave Many Behind," *New York Times*, April 24, 2013, accessed February 11, 2015, www.nytimes.com/2013/04/25/world/americas/ boom-times-in-paraguay-leave-many-behind.html?pagewanted=all&_r=0; Guerena, "Soy Mirage," 38; Hobbs, "Paraguay's Destructive Soy Boom."

7. Guerena, "Soy Mirage," 16.

8. Hobbs, "Paraguay's Destructive Soy Boom."

9. Ibid.

10. "Côte d'Ivoire: Cocoa farmers welcome state-imposed prices," *IRIN*, UN Office for the Coordination of Humanitarian Affairs, November 7, 2012, accessed December 8, 2014, www.irinnews.org/report/96731/cote-d-ivoire-cocoa-farmers-welcome-state -imposed-prices.

11. Calculated from FAOSTAT (Browse Data, Trade, Crop and Livestock Products, Item: Cocoa Bean, Area: Côte d'Ivoire, from Year: 1990, to Year: 2011, Aggregation: Average; accessed October 31, 2014), faostat3.fao.org/browse/T/TP/E.

12. Ibid.

13. FAO, Trade and Markets Division, Manitra A. Rakotoarisoa, Massimo Iafrate, and Marianna Paschali, "Why Has Africa Become a Net Food Importer? Explaining Africa agricultural and food trade deficits" (Rome, 2011), Table 7, 16, accessed August 18, 2014, www.fao.org/docrep/015/i2497e/i2497e00.pdf.

14. U.S. Department of Labor, "2012 Findings on the Worst Forms of Child Labor: Côte d'Ivoire," 2012, 1–2, accessed August 8, 2014, www.dol.gov/ilab/reports/child-labor/ findings/2012TDA/cotedivoire.pdf.

15. Tom Philpott, "Bloody Valentine: Child Slavery in Ivory Coast's Cocoa Fields," *Mother Jones*, February 14, 2012, accessed August 6, 2014, www.motherjones.com/tom -philpott/2012/02/ivory-coast-cocoa-chocolate-child-slavery.

16. International Monetary Fund, "Côte d'Ivoire: Poverty Reduction Strategy Paper," IMF Country Report No. 13/172 (Washington, DC, June 2013), 3, accessed September 29, 2014, www.imf.org/external/pubs/ft/scr/2013/cr13172.pdf; FAO, "State of Food Inse- curity in the World 2014: Strengthening the Enabling Environment" (Rome, 2014), Annex 1, Table A1, 40, accessed October 31, 2014, www.fao.org/3/a-i4030e.pdf; FAOSTAT (Population, Annual Population, Country: Côte d'Ivoire, Elements: Population, both sexes, Item: Population, est. and proj., 1990–2014, accessed January 8, 2015), faostat3 .fao.org/download/O/OA/E.

17. Oxfam, "Smallholders at Risk," 10.

18. Ibid., 15.

19. Ibid., 13.

20. Ibid.

21. Ibid., 10.

22. Ibid.

23. WHO, Global Health Observatory Data Repository, Child malnutrition country esti- mates (WHO global database), "Children aged <5 years stunted data by country," accessed April 16, 2015, apps.who.int/gho/data/node.main.1097?lang=en.

24. Tracy Wilkinson, "Mexico's tomato-farm workers toil in "circle of poverty," *Los Angeles Times*, November 11, 2013, accessed August 6, 2014, www.latimes.com/world/ la-fg-mexico-sinaloa-workers-20131111-story.html#page=1; Barry Estabrook, *Tomatoland* (Kansas City: Andrews McMeel Publishing, 2011), xv.

25. The advertisement was reprinted in "Comparative (Dis) Advantage," *Dollars and Sense* 114 (March 1986): 15.

26. "CBI Product Factsheet: Fresh Pineapples in the European Market," CBI Ministry of Foreign Affairs, 2013, accessed September 22, 2014, www.cbi.eu/sites/default/files/ study/product-factsheet-pineapple-europe-fresh-fruit-vegetables-2014.pdf.

27. Bernard Wideman, "Dominating the Pineapple Trade," *Far Eastern Economic Review*,

July 8, 1974; Duane P. Bartholomew, Richard A. Hawkins, and Johnny A. Lopez, "Hawaii Pineapple: The Rise and Fall of an Industry," *HortScience* 47, no. 10 (2012): 1390–1398, accessed January 9, 2015, hortsci.ashspublications.org/content/47/10/1390.full.

28. *The Case Against the Global Economy: And for a Turn Toward the Local*, ed. Edward Goldsmith and Jerry Mander (San Francisco: Sierra Club Books, 1996), 194, 287.

29. Paolo D'Odorico et al., "Feeding humanity through global food trade," Department of Environmental Sciences, University of Virginia 2, no. 9 (2014): 458, accessed September 22, 2014, doi: 10.1002/2014EF000250 onlinelibrary.wiley.com/doi/10.1002/2014EF000250/abstract.

30. See Walden Bello, with Shea Cunningham and Bill Rau, *Dark Victory: The United States, Structural Adjustment and Global Poverty* (London: Pluto Press/Food First/Transnational Institute, 1994), Chapters 4 and 8.

31. FOA, "Multilateral Trade Negotiations on Agriculture: A Resource Manual," 2000. online www.fao.org/docrep/003/x7352e/X7352E00.htm#TopOfPage.

32. Larry D. Sanders et al., "The GATT Uruguay Round and the World Trade Organization: Opportunities and Impacts for U.S. Agriculture," Southern Agriculture in World Economy, accessed October 23, 2014, www.ces.ncsu.edu/depts/agecon/trade/seven.html.

33. Sophia Murphy, Institute for Agriculture and Trade Policy, personal communication with author, August 24, 2014.

34. "The WTO Agreement on Agriculture," ActionAid, London, accessed October 27, 2014, www.actionaid.org.uk/sites/default/files/doc_lib/51_1_agreement_agriculture.pdf.

35. FAO et al., "Why Has Africa Become a Net Food Importer?"

36. Sophia Murphy, "Land Grabs and Fragile Food Systems: The Role of Globalization," Institute for Agriculture and Trade Policy, February 2013, 5-6, accessed October 27, 2014, www.iatp.org/files/2013_02_14_LandGrabsFoodSystem_SM_0.pdf.

37. World Bank Group, "Voting Powers" (Washington, DC, 2013), web.worldbank.org/WBSITE/EXTERNAL/EXTABOUTUS/ORGANIZATION/BODEXT/0, contentMDK:21429866~menuPK:64020035~pagePK:64020054~piPK:64020408~theSitePK:278036,00.html; "Factsheet: How the IMP Makes Decisions," International Monetary Fund, last modified April 2, 2014, accessed September 19, 2014, www.imf.org/external/np/exr/facts/govern.htm.

38. Aileen Kwa, "Power Politics in the WTO," 2nd edition, January 2003, 5, accessed September 18, 2014, www.citizen.org/documents/powerpoliticsKWA.pdf.

39. Julian M. Alston, Daniel A. Sumner, and Henrich Brunke, "Impacts of Reductions in US Cotton Subsidies on West African Cotton Producers," Oxfam America, 2007, 10, accessed September 18, 2014, www.oxfamamerica.org/static/oa3/files/paying-the-price.pdf.

40. Ideas Centre Geneva, "Cotton Update: US-Brazil Cotton Case Deal of the 5th of April: What does it mean for Brazil, the US, the multilateral trading system and the African countries," Newsletter No. 85 (Geneva, April 9, 2010), 2–5, accessed September 22, 2014, www.ideascentre.ch/wp-content/uploads/2013/09/Newsletter-85-Cotton-Brazil-vs-USA.pdf; Congressional Research Service and Randy Schnepf, "Status of the WTO Brazil-U.S. Cotton Case," R43336, February 21, 2014, accessed September 22, 2014, nationalaglawcenter.org/wp-content/uploads/assets/crs/R43336.pdf.

41. Bill Clinton, "Remarks on the Signing of NAFTA," University of Virginia Miller Center of Public Affairs, December 8, 1993, accessed August 19, 2014, millercenter.org/president/clinton/speeches/speech-3927.

42. Eric Holt-Giménez and Raj Patel, *Food Rebellions: Crisis and the Hunger for Justice* (Oakland, CA: Food First Books, 2009), 56–57.

43. Alicia Puyana, "Mexican Agriculture and NAFTA: A 20-Year Balance Sheet," *Review of Agrarian Studies* 2, no. 1 (2012), 2, accessed August 19, 2014, www.ras.org.in/mexican_agriculture_and_nafta.

44. Ibid.

45. Timothy A. Wise and Betsy Rakocy, "Hogging the Gains from Trade: The Real Winners of U.S. Trade and Agricultural Policies," Global Development and Environment Institute, Tufts University, Policy Brief No. 10-01, January 2010, 3, accessed August 20, 2014, www.ase.tufts.edu/gdae/pubs/rp/pb10-01hogginggainsjan10.pdf.

46. Raj Patel, *Stuffed and Starved: The Hidden Battle for the World Food System* (Brooklyn, NY: Melville House Publishing, 2012), 68.

47. Puyana, "Mexican Agriculture and NAFTA: A 20-Year Balance Sheet," 7; Organization for Economic Co-Operation and Development, OECD StatExtracts, accessed October 2, 2014, stats.oecd.org/index.aspx?querytype=view&queryname=221#.

48. ActionAid International USA, "Biofueling Hunger: How US Corn Ethanol Policy Drives Up Food Prices in Mexico," May 2012, accessed January 9, 2015, www.actionaidusa.org/sites/files/actionaid/biofueling_hunger_aausa.pdf.

49. Patel, *Stuffed and Starved*, 63.

50. Wise and Rackocy, "Hogging the Gains from Trade," 3.

51. Ibid., 1.

52. Ibid., 3; Peter Goldsmith and Philip L. Martin, "Community and Labor Issues in Animal Agriculture," *Choices*, 2006, accessed August 20, 2014, www.choicesmagazine.org/2006-3/animal/2006-3-12.htm.

53. Wise and Rakocy, "Hogging the Gains from Trade," 3.

54. Saeed Azhar and Steven Aldred, "Exclusive: Smithfield's China bidders plan Hong Kong IPO after deal—sources," Reuters, 16 July 2013, accessed August 20, 2014, www.reuters.com/article/2013/07/16/smithfield-shuanghui-idUSL4N0FM20B20130716.

55. Doug Palmer, "U.S. approves Chinese company's purchase of Smithfield," *Politico*, September 6, 2013, accessed August 20, 2014, www.politico.com/story/2013/09/us-china-smithfield-96399.html.

56. Patel, *Stuffed and Starved*.

57. Lori Wallach and Ben Beachy, "Obama's Covert Trade Deal," *New York Times*, June 2, 2013, accessed August 13, 2014, www.nytimes.com/2013/06/03/opinion/obamas-covert-trade-deal.html.

58. Jo Comerford, "TPP=NAFTA on Steroids," MoveOn.Org, January 28, 2015, accessed February 18, 2015, front.moveon.org/tpp-nafta-on-steroids/#.VOSyhlPF_uU/.

59. Public Citizen's Global Trade Watch and Lori Wallach, "TPP Presentation: Washington Joint Legislative Oversight Committee on Trade Policy," November 2012, 11, accessed January 9, 2015, leg.wa.gov/JointCommittees/LOCTP/Documents/2012Nov14/TPP%20 Presentation.pdf.

60. Wallach and Beachy, "Obama's Covert Trade Deal."

61. Madrick, "Our Misplaced Faith in Free Trade."

62. Karen Hansen-Kuhn, "Who's at the Table? Demanding Answers on Agriculture in the Trans-Pacific Partnership," Institute for Agriculture and Trade Policy, February 2013, 2, accessed August 13, 2014, www.iatp.org/files/2013_02_28_TPP_KHK_0.pdf.

63. Joseph Stiglitz, "On the Wrong Side of Globalization," *New York Times*, March 15, 2014, accessed August 13, 2014, opinionator.blogs.nytimes.com/2014/03/15/on-the-wrong-side-of-globalization/.

64. Roy Carrol, "Uruguay Bows to Pressure over Anti-Smoking Amendments," *The Guardian*, July 27, 2010, accessed August 13, 2014, www.theguardian.com/world/2010/jul/27/uruguay-tobacco-smoking-philip-morris.

Myth 8: U.S. Foreign Aid Is the Best Way to Help the Hungry

1. "2013 Survey of Americans on the U.S. Role in Global Health," Henry J. Kaiser Family Foundation, 2013, accessed November 6, 2014, kff.org/global-health-policy/poll-finding/2013-survey-of-americans-on-the-u-s-role-in-global-health.

2. Calculated from U.S. Office of Management and Budget, "Fiscal Year 2014 Budget of the U.S. Government" (Washington, DC: US OMB, 2014), 69, 183, accessed May 7, 2015, www.whitehouse.gov/sites/default/files/omb/budget/fy2014/assets/budget.pdf.

3. "Dollars to Results: Foreign Aid Spending," USAID, accessed November 6, 2014, results.usaid.gov/less-than-one-percent.

4. John Kerry, "Executive Budget Summary Function 150 and Other International Programs: Fiscal Year 2014," U.S. State Department, April 10, 2013, 2, accessed July 10, 2014, www.usaid.gov/sites/default/files/documents/1868/207305.pdf.

5 Pamela Constable, "Robert E. White, who criticized policy on El Salvador as U.S. ambassador, dies at 88," *Washington Post,* January 15, 2014, accessed February 11, 2015, www.washingtonpost.com/world/robert-e-white-who-criticized-policy-on-el-salvador-as-us-ambassador-dies-at-88/2015/01/15/0c504738-9c29-11e4-96cc-e858eba91ced_story.html.

6. Joseph Collins, *Nicaragua: What Difference Could a Revolution Make?* 3rd edition (New York: Grove Press, 1986).

7. Peter Smith, *Talons of the Eagle: Dynamics of US-Latin American Relations* (New York: Oxford, 1996), 153.

8. USAID (EADS, U.S. Overseas Loans and Grants, Foreign Assistance Data, Fast Facts, Fiscal Year: 2012; accessed November 4, 2014), eads.usaid.gov/gbk/data/fast_facts.cfm.

9. "United States Military Aid," World Heritage Encyclopedia, accessed January 5, 2014, www.worldheritage.org/articles/United_States_ military_aid.

10. USAID (EADS, U.S. Overseas Loans and Grants, Foreign Assistance Data, Fast Facts, Fiscal Year: 2012, Top 10 Recipients; accessed November 4, 2014), eads.usaid.gov/gbk/data/fast_facts.cfm; Human Rights Watch, "World Report 2014: Events of 2013," accessed November 4, 2014, www.hrw.org/sites/default/files/wr2014_web_0.pdf.

11. USAID (EADS, U.S. Overseas Loans and Grants, Foreign Assistance Data, Fast Facts, Fiscal Year: 2012, Overview; accessed January 2, 2015), eads.usaid.gov/gbk/data/fast_facts.cfm.

12. FAO, "State of Food Insecurity in the World," 2014, 40–43.

13. Patrick Radden Keefe, "Corruption and Revolt: Does tolerating graft undermine national security?" *The New Yorker,* January 19, 2015, www.newyorker.com/magazine/2015/01/19/corruption-revolt.

14. Rod Nordland, "Afghanistan's Worsening, and Baffling, Hunger Crisis," *New York Times*, January 4, 2014, accessed November 5, 2015, www.nytimes.com/2014/01/05/world/asia/afghanistans-worsening-and-baffling-hunger-crisis.html?_r=0.

15. FAO, "State of Food Insecurity in the World," 2014, 42.

16. World Health Organization, Global Health Observatory Data Repository, Child malnutrition country estimates (WHO global database), "Children aged <5 years stunted data by country," accessed April 16, 2015, apps.who.int/gho/data/node.main.1097?lang=en.

17. Government Accountability Office, "Afghanistan Development: USAID Continues to Face Challenges in Managing and Overseeing U.S. Development Assistance Programs," GAO-10-932T, July 15, 2010, 5, accessed July 10, 2014, www.gao.gov/new.items/d10932t.pdf.

18. "Corruption Perception Index 2014," Transparency International, accessed November 4, 2014, www.transparency.org/cpi2014.

19. World Bank, "Afghanistan in Transition: Looking Beyond 2014," May 2012, Volume 1, 1, accessed July 10, 2014, siteresources.worldbank.org/INTAFGHANISTAN/Resources/Vol1Overview8Maypm.pdf.

20. Anthony H. Cordesman, "Volume II: Afghan Economic and Outside Aid," in *The Afghan War in 2013: Meeting the Challenges of Transition*, Center for Strategic and International Studies (Lanham, MD: Rowman and Littlefield, 2013), 28, accessed July 10, 2014, csis.org/files/publication/130327_afghan_war_in_2013_vol_II.pdf.

21. Radden Keefe, "Corruption and Revolt."

22. "Corruption Perception Index 2014," Transparency International; "Poverty and Development," Transparency International, accessed January 2, 2015, archive.transparency.org/global_priorities/poverty/corruption_aid.

23. USAID (EADS, U.S. Overseas Loans and Grants, Data, Country Profile, Fiscal Year: 2012, Economic Foreign Assistance Profile: Ethiopia; accessed November 4, 2014), eads.usaid.gov/gbk/data/profile.cfm.

24. International Food Policy Research Institute, "Global Hunger Index 2013: The Challenge of Hunger: Building Resilience to Achieve Food and Nutrition Security," 2013, 15, Table 2.1, accessed November 4, 2014, www.ifpri.org/sites/default/files/publications/ghi13.pdf.

25. James Quirin, "Is the Successful Military Resistance to European Colonialism in Late Nineteenth-Century Ethiopia Still Significant Today?" *Journal of African History* 48, no. 2 (2007): 344–345, doi:10.1017/S0021853707003015.

26. John Norris and Connie Veillette, "Country Assistance Profiles," in *Engagement amid Austerity: A Bipartisan Approach to Reorienting the International Affairs Budget*, Center for Global Development and Center for American Progress, 2013, 86, accessed July 10, 2014, www.cgdev.org/doc/Rethinking%20Aid/EAA-Country_Profiles.pdf.

27. World Bank [Data, by Country, Kenya, World Development Indicators, Poverty head count ratio at national poverty line (% of population); accessed May 7, 2015], data.worldbank.org/country/kenya.

28. Joseph Collins and Bill Rau, "AIDS in the Context of Development," Paper No. 4, United Nations Research Institute for Social Development, Programme on Social Policy and Development, December 2000, accessed February 4, 2015, www.unrisd.org/80256B3 C005BCCF9/%28httpAuxPages%29/329E8ACB59F4060580256B61004363FE/$file/collins .pdf (especially see "The Socioeconomic Context Driving the Pandemic"); Paul Farmer, *Infections and Inequalities* (Berkeley, University of California Press, 1999).

29. "Corruption Perception Index 2014," Transparency International.

30. Jake Johnston and Alexander Main, "Breaking Open the Black Box: Increasing Transparency and Accountability in Haiti," Center for Economic and Policy Research, April 2013, 3, accessed January 2, 2015, www.cepr.net/documents/publications/haiti -aid-accountability-2013-04.pdf.

31. "Corruption Perception Index 2014," Transparency International.

32. Alexis Sowa, "Aid to Egypt by the numbers," Center for Global Development (blog), July 19, 2013, accessed January 2, 2015, www.cgdev.org/blog/aid-egypt-numbers.

33. USAID (EADS, U.S. Overseas Loans and Grants, Data, Country Profile, Fiscal Year: 2012, Military Foreign Assistance Profile: Israel; accessed January 5, 2015), eads.usaid.gov/gbk/data/profile.cfm.

34. World Food Programme, "The Status of Poverty and Food Security in Egypt: Analysis and Policy Recommendations," May 2013, 2, 28, accessed July 10, 2014, documents.wfp.org/stellent/groups/public/documents/ena/wfp257467.pdf; World Food Programme Egypt, "Vulnerability and Analysis and Review of Food Subsidy in Egypt," October 2005, ix, accessed January 5, 2014, home.wfp.org/stellent/groups/public/documents/ena/wfp198346.pdf.

35. WFP, "Status of Poverty and Food Security in Egypt," 6, 10.

36. Foreign Assistance Data (Foreign Assistance Summarized by Year, Organizational Unit, and Category, Obligation Data, Fiscal Year: 2012, Category: Health, Organizational Unit: Afghanistan, Pakistan, Ethiopia, Jordan, Iraq, Kenya, Columbia, Haiti, West Bank/Gaza, Tanzania; accessed January 12, 2015), www.foreignassistance.gov/web/ObjectiveView.aspx.

37. Collins and Rau, "AIDS in the Context of Development"; Farmer, *Infections and Inequalities*.

38. "Humanitarian Assistance," Transparency International, accessed January 10, 2015, www.transparency.org/topic/detail/humanitarian_assistance.

39. USAID (EADS, U.S. Overseas Loans and Grants, Greenbook, Standard Country Report, Summary of All Countries; accessed January 10, 2015), eads.usaid.gov/gbk/data/greenbook.cfm.; USAID, "U.S. Overseas Loans and Grants: Obligations and Loan Authorizations July 1, 1945–September 30, 2013," 2013, 1, accessed August 28, 2014, pdf.usaid.gov/pdf_docs/pnaec300.pdf; Congressional Research Service, Charles E. Hanrahan, and Carol Canada, "International Food Aid: U.S. and Other Donor Contributions," November 12, 2013, 2, accessed August 26, 2014, www.fas.org/sgp/crs/misc/RS21279.pdf.

40. Jennifer Clapp, *Hunger in the Balance: The New Politics of International Food Aid* (Ithaca, NY: Cornell University Press, 2012), 69–77.

41. Ibid., 74.

42. U.S. Department of Agriculture and USAID, "U.S. International Food Assistance Report, FY2012," Appendices E, F, G, H, accessed May 12, 2015, www.fas.usda.gov/sites/default/files/2014-07/usda-usaid_fy2012_food_assistance_report.pdf.

43. Forrest Laws, "Rice industry in crisis," *Delta Farm Press*, July 20, 2001, accessed August 28, 2014, deltafarmpress.com/rice-industry-crisis-president-told.

44. Oxfam International, "Food Aid or Hidden Dumping," Oxfam Briefing Paper No. 71, March 2005, 19, accessed August 28, 2014, www.oxfam.org/sites/www.oxfam.org/files/bp71_food_aid.pdf; Ron Nixon, "Obama Administration Seeks to Overhaul International Food Aid," *New York Times,* April 4, 2013, accessed November 5, 2014, www.nytimes.com/2013/04/05/us/politics/white-house-seeks-to-change-international-food-aid.html?_r=0.

45. Edward Clay, Barry Riley, and Ian Urey, "The Development Effectiveness of Food Aid: Does Tying Matter?" OECD, 2006, 9, accessed September 10, 2014, www.odi.org/sites/odi.org.uk/files/odi-assets/publications-opinion-files/3043.pdf.

46. Gawain Kripke, personal communication with the authors, November 5, 2014.

47. Oxfam, "Food Aid or Hidden Dumping," 11–12.

48. Clapp, *Hunger in the Balance*, 73.

49. Ibid., 74.

50. "Food Aid: A Critical Program, Ripe for Reform," Oxfam America, accessed February 18, 2015, www.oxfamamerica.org/take-action/campaign/food-farming-and-hunger/food-aid/.

51. Congressional Research Service, Randy Schnepf, "International Food Aid Programs: Background and Issues," May 2014, 20–21, accessed September 10, 2014, www.fas.org/sgp/crs/misc/R41072.pdf.

52. Clapp, *Hunger in the Balance*, 40.

53. Dan Charles, "A Political War Brews over 'Food for Peace' Aid Program," NPR, April 4, 2013, accessed September 10, 2014, www.npr.org/blogs/thesalt/2013/04/04/176154775/a-political-war-brews-over-food-for-peace-aid-program.

54. Doug Sanders, "Food aid exposes the West's uncharitable charity," *The Globe and Mail*, January 15, 2005, accessed September 10, 2014, www.theglobeandmail.com/news/national/food-aid-exposes-the-wests-uncharitable-charity/article1113134/.

55. Ibid.

56. Clapp, *Hunger in the Balance*, 37.

57. Johnston and Maine, "Breaking Open the Black Box," 9; Congressional Research Service, Rhoda Margesson, and Maureen Taft-Morales, "Haiti Earthquake: Crisis and Response," February 2, 2010, 11, accessed September 10, 2014, fas.org/sgp/crs/row/R41023.pdf.

58. Oxfam, "Planting Now 2nd Edition: Revitalizing agriculture for reconstruction and development in Haiti," Oxfam Briefing Paper No. 162, October 15 2012, 10, accessed September 12, 2014, www.oxfam.org/sites/www.oxfam.org/files/bp162-planting-now-second-edition-haiti-reconstruction-151012-en.pdf.

59. Center for Economic and Policy Research, Mark Weisbrot, Jake Johnston, and Rebecca Ray, "Using Food Aid to Support, Not Harm Haitian Agriculture," CEPR Issue Brief, April 2010, 1, accessed January 9, 2015, www.cepr.net/documents/publications/haiti-2010-04.pdf.

60. Mukhtar Diriye, Abdirizak Nur, and Abdullahi Khalif, "Food Aid and the Challenge of Food Security in Africa," *Development* 56, no. 3 (2014): 396–403, accessed January 11, 2015, doi: 10.1057/dev.2014.15.

61. Carlo del Ninno, Paul A. Dorosh, and Kalanidhi Subbarao, "Food Aid and Food Security in the Short and Long Term: Country Experience from Asia and Sub-Saharan Africa," SP Discussion Paper No. 0538, World Bank Social Safety Nets Primer Series, November 2005, i, accessed September 10, 2014, www.fao.org/fsnforum/sites/default/files/files/Have%20your%20say/Food_aid_in_Africa_and_Asia.pdf.

62. U.S. Government Accountability Office, "International Food Assistance: Funding Development Projects Through the Purchase, Shipment, and Sale of U.S. Commodities Is Inefficient and Can Cause Adverse Market Impacts," GAO-11-636, June 23, 2011, accessed September 10, 2014, www.gao.gov/assets/330/320013.pdf.

63. "Andrew Natsios Extended Interview," PBS Religion and Ethics Newsweekly, February 2010, accessed November 6, 2014, www.pbs.org/wnet/religionandethics/2010/02/19/february-19-2010-andrew-natsios-extended-interview/5720/.

64. Andrew Natsios, "Obama's Promising Food Aid Reforms," *U.S. News and World Report*, May 6, 2013, accessed September 10, 2014, www.usnews.com/opinion/blogs/world-report/2013/05/06/usaid-food-aid-reforms-must-protect-food-for-the-poor.

65. Getaw Tadesse and Gerald Shively, "Food Aid, Food Prices, and Producer Disincentives in Ethiopia," *American Journal of Agricultural Economics* 91, no. 4 (2009): 942–955, accessed February 4, 2015, doi: 10.1111/j.1467-8276.2009.01324.x.

66. Roger Thurow and Scott Kilman, "As U.S. Food Aid Enriches Farmers, Poor Nations Cry Foul," *Wall Street Journal*, September 11, 2003, accessed September 10, 2014, online .wsj.com/news/articles/SB106323428552114700.

67. USAID, "U.S. International Food Assistance Report for FY2012," accessed January 11, 2015, www.fas.usda.gov/sites/default/files/2014-07/usda-usaid_fy2012_food_assistance_report.pdf; USAID [EADS, U.S. Overseas Loans and Grants, Standard Country Report (current dollars), Ethiopia, obligations in millions, current $US; accessed January 11, 2015], eads.usaid.gov/gbk/data/country_report.cfm.

68. Future Agricultures Consortium and Samuel Gebreselassie, "Food Aid and Smallholder Agriculture in Ethiopia," Future Agricultures Policy Brief 003, January 2006, 4, accessed September 10, 2014, www.future-agricultures.org/policy-engagement/policy-briefs/131 -food-aid-and-smallholder-agriculture-in-ethiopia/file.

69. Clapp, *Hunger in the Balance*, 47–48, 61.

70. World Food Programme, "Food Procurement Annual Report, 2013," 2013, 2, accessed January 8, 2015, documents.wfp.org/stellent/groups/public/documents/communications/wfp264134.pdf.

71. Damien Fontaine, World Food Programme, e-mail communication with author, January 8, 2015; World Food Programme, "Purchase for Progress (P4P): Final Consolidated Procurement Report (September 2008-December 2013)," accessed January 11, 2015, 2, documents.wfp.org/stellent/groups/public/documents/reports/wfp270609.pdf.

72. Oxfam, "Food Aid or Hidden Dumping," 6, 12.

73. Gawain Kripke, personal communication with authors, November 3, 2014; Delilah Griswold, "U.S. Food Aid Reform Through Alternative Dispute Resolution," *Sustainable Development Law and Policy* 14, no. 1 (2014): 47–58, digitalcommons.wcl.american.edu/cgi/viewcontent.cgi?article=1547&context=sdlp; Clapp, *Hunger in the Balance*, 84.

74. Feed the Future, "Growing Innovation, Harvesting Results: Progress Report June 2013," June 2013, 12, accessed July 10, 2014, feedthefuture.gov/sites/default/files/resource/files/feed_the_future_progress_report_2013.pdf.

75. Tjada McKenna and Jonathan Shrier, "Five Questions on the New Alliance for Food Security and Nutrition," Feed the Future, May 23, 2012, accessed September 9, 2014, feedthefuture.gov/article/five-questions-about-new-alliance-food-security-and-nutrition; Feed the Future, "Growing Innovation, Harvesting Results," 12.

76. World Bank, "Growing Africa: Unlocking the Potential of Agribusiness," March 2013, 4, accessed July 10, 2014, www-wds.worldbank.org/external/default/WDSContentServer/WDSP/IB/2013/03/13/000350881_20130313100019/Rendered/PDF/759720REPLACEM 0mmary0pub03011013web.pdf.

77. Rebecca Robbins, "With Africa's Private Sector on the Rise, U.S. Businesses Seek to Expand Their Presence," *Washington Post,* August 5, 2014, accessed September 8, 2014, www.washingtonpost.com/business/economy/with-africas-private-sector-on-the-rise -us-businesses-seek-to-expand-their-presence/2014/08/04/c7b17aae-1c1a-11e4-82f9 -2cd6fa8da5c4_story.html.

78. "Countries," Feed the Future, accessed September 8, 2014, feedthefuture.gov/countries.

79. "African Civil Society Organizations to Counter Corporatisation of African Agriculture," Food First, August 8, 2013, accessed July 10, 2014, foodfirst.org/press-releases/african-civil-society-organisations-to-counter-corporatisation-of-african-agriculture/.

80. Feed the Future, "Growing Innovation, Harvesting Results," 12; Claire Provost, Liz Ford, and Mark Tran, "G8 New Alliance condemned as new wave of colonialism

in Africa," *The Guardian*, February 18, 2014, accessed September 8, 2014, www.theguardian.com/global-development/2014/feb/18/g8-new-alliance-condemned-new-colonialism.

81. "The G8 and land grabs in Africa," GRAIN, March 11, 2013, accessed January 9, 2015, 2, www.grain.org/article/entries/4663-the-g8-and-land-grabs-in-africa.

82. Provost, Ford, and Tran, "G8 New Alliance condemned."

83. USAID Press Office, "USAID, Dupont Work with Government of Ethiopia to Improve Food Security," January 24, 2013, accessed July 10, 2014, www.usaid.gov/news-information/press-releases/usaid-dupont-work-government-ethiopia-improve-food-security.

84. "Cooperation Framework to Support the New Alliance for Food Security and Nutrition in Ethiopia," Feed the Future, 2012, Annex 1, 4, accessed July 10, 2014, feedthefuture.gov/sites/default/files/resource/files/Ethiopia_web.pdf.

85. "Cooperation Framework to Support the New Alliance for Food Security and Nutrition in Burkina Faso," Feed the Future, 2012, Annex 1, 17, accessed July 10, 2014, feedthefuture.gov/sites/default/files/resource/files/Burkina%20Faso%20Coop%20Framework%20ENG%20Final%20w.cover_.pdf.

86. "Cooperation Framework to Support the New Alliance for Food Security and Nutrition in Mozambique," Feed the Future, 2012, Annex 1, 4, accessed July 10 2014, feedthefuture.gov/sites/default/files/resource/files/Mozambique%20Coop%20Framework%20ENG%20FINAL%20w.cover%20REVISED.pdf.

87. "New Alliance for Food Security and Nutrition: Part 1," One, December 10, 2012, accessed July 10, 2014, www.one.org/us/policy/policy-brief-on-the-new-alliance/.

88. Provost, Ford, and Tran, "G8 New Alliance Condemned."

89. Glen Ashton, "The Fifth Horseman of the Apocalypse—G8 Corporate Power," GRAIN, June 27, 2012, accessed July 10, 2014, www.grain.org/bulletin_board/entries/4535-the-fifth-horseman-of-the-apocalypse-g8-corporate-power; Joshua Humphreys, Ann Solomon, and Emmanuel Tumusiime, "US Investment in Large-Scale Land Acquisition in Low- and Middle-Income Countries," Oxfam America Research Backgrounder Series, 2013, 28, accessed September 8, 2014, www.oxfamamerica.org/static/media/files/us-land-investment.pdf.

90. World Bank [Data, by Country, Côte d'Ivoire, World Development Indicators, Poverty head count ratio at national poverty line (% of population); accessed September 22, 2014], data.worldbank.org/country/cote-divoire#cp_wdi; "The G8 and land grabs in Africa," GRAIN.

91. "The G8 and land grabs in Africa," GRAIN; U.S. Department of Agriculture, Foreign Agriculture Service (PSD, Custom Query, Commodity: Rice Milled, Data Type: Exports, Country: Côte d'Ivoire, Year: 2010 and 2014; accessed February 4, 2015), apps.fas.usda.gov/psdonline/psdquery.aspx.

92. "Tanzania: Swedes in US$ 550 Million Tanzania Ethanol Venture," All Africa, May 11, 2014, accessed September 8, 2014, allafrica.com/stories/201405121542.html.

93. Gerald Kitabu, "Dubious Land Sale in Bagamoyo Creates Dispute Between Villagers and Investor," *IPPMedia,* March 8, 2013, accessed September 8, 2014, www.ippmedia.com/frontend/?l=52068.

94. "Home," Agro Eco Energy, accessed January 10, 2015, www.ecoenergy.co.tz/home/; "Resettlement Action Plan Programme," Agro EcoEnergy, accessed January 10, 2015, www.ecoenergy.co.tz/sustainability/social-development/rap-programme/; African

Development Bank, "Executive Summary of the Resettlement Action Plan," 3, accessed January 11, 2015, www.afdb.org/fileadmin/uploads/afdb/Documents/Environmental -and-Social-Assessments/Tanzania%20-%20Bagamoyo%20Sugar%20Project% 20-%20 RAP%20Summary.pdf.

95. "The G8 and Land Grabs in Africa," GRAIN.

96. Humphreys, Solomon, and Tumusiime, "U.S. Investment in Large-Scale Land Acquisition," 28.

97. Oxfam, "For Whose Benefit: The G8 New Alliance for Food Security and Nutrition in Burkina Faso," Oxfam Briefing Note, May 22, 2014, 7, accessed September 8, 2014, www.oxfam.org/sites/www.oxfam.org/files/bn-whose-benefit-burkina-faso-g8-new -alliance-220514-en.pdf.

98. Tim Gore, "On Leaving the Leadership Council of the New Alliance for Food Security and Nutrition," Oxfam International, accessed January 10, 2015, blogs.oxfam.org/en/ blogs/14-10-06-leaving-leadership-council-new-alliance-food-security-and-nutrition.

99. Feed the Future, "Growing Innovation, Harvesting Results," 2, 4; "Approach," accessed January 9, 2015, www.feedthefuture.gov/.

100. Scott Higham and Steven Rich, "Whistleblowers say USAID's IG removed critical details from public reports," *Washington Post*, October 22, 2014, accessed January 9, 2015, www.washingtonpost.com/investigations/whistleblowers-say-usaids -ig-removed-critical-details-from-public-reports/ 2014/10/22/68fbc1a0-4031-11e4 -b03f-de718edeb92f_story.html.

101. Emmanuel Tumusiime and Edmunt Matotay, "Sustainable and Inclusive Investments in Agriculture: Lessons on the Feed the Future Initiative in Tanzania," Oxfam America Research Backgrounder Series, 2013, 18, 27, 31, February 4, 2015, www .oxfamamerica.org/static/media/files/Tanzania_-_Sustainable_and_Inclusive_Investments .pdf.

102. Danielle Fuller-Wimbush and Cardyn Fils-Aimé, "Feed the Future Investment in Haiti: Implications for sustainable food security and poverty reduction," Oxfam America Research Backgrounder Series, 2014, 19, 25, 31, 37, February 4, 2015, www.oxfamamerica .org/static/media/files/Haiti_Feed_the_Future_RB.pdf.

103. Gregg Rapaport, "Saving Lives Across Nepal: Female Community Health Volunteers," USAID (blog), July 29, 2011, accessed November 6, 2014, blog.usaid.gov/2011/07/ saving-lives-across-nepal-female-community-health-volunteers/.

104. USAID, "Empowering women, saving lives," *Health Bulletin,* USAID Special Publication, February 26, 2014, 1, accessed November 10, 2014, www.usaid.gov/sites/default/ files/documents/1861/health_bulletin.pdf.

105. USAID Nepal, "Fact Sheet: National Female Community Health Volunteer Program," accessed January 10, 2015, www.healthynewbornnetwork.org/sites/default/files/ resources/USAID%20Nepal%20Factsheet%20-%20Nepal%20Female%20Community%20 Health%20Volunteers.pdf; New ERA, "An Analytical Report on Female Community Health Volunteers of Selected Districts of Nepal," October 2008, 1, accessed November 6, 2014, pdf.usaid.gov/pdf_docs/PNADN017.pdf; "Nepal: Background," UNICEF, February 26, 2003, accessed November 6, 2014, www.unicef.org/infobycountry/nepal_nepal_background .html; World Health Organization, "WHO vaccine-preventable diseases: monitoring system 2014 global summary," July 15, 2014, accessed November 6, 2011, apps.who .int/immunization_monitoring/globalsummary/countries?countrycriteria%5Bcountry% 5D%5B%5D=NPL&commit=OK.

106. Jeffrey Ashe, personal communication with authors, October 31, 2014. Note: Jeffrey Ashe, with Kyla Neilan, wrote *In Their Own Hands: How Savings Groups Are Revolutionizing Development*, and was an earlier pioneer in microfinance.

107. Executive Office of the President's Office of Management and Budget, "Annual Report on United States Contribution to the United Nations," Washington, DC, June 7, 2010, 4, accessed November 6, 2014, www.whitehouse.gov/sites/default/files/omb/assets/ legislative_reports/us_contributions_to_the_un_06112010.pdf.

108. Khalil Sesmou, "The UN Food and Agriculture Organization: An Insider's View," *The Ecologist* 21, no. 2 (1991): 46–56, exacteditions.theecologist.org/browse/307/308/5643/1/1.

109. FAO and K. D. Gallagher, "Global Integrated Production and Pest Management (IPPM) Development," accessed January 7, 2015, www.fao.org/docrep/006/y4751e/y4751e0m.htm.

110. "EverGreen Agriculture: Re-Greening Africa's Landscape," World Agroforestry Centre, last modified February 2013, 2, accessed January 7, 2015, www.ard-europe.org/ fileadmin/SITE_MASTER/content/eiard/Documents/Impact_case_studies_2013/ICRAF_-_ EverGreen_agriculture.pdf.

111. Yacouba Ouedraogo, Project Manager/Agent of Change, Regional Programme in the Sahel and Horn of Africa, World Agroforestry Centre, personal communication with author, January 9, 2015.

112. "Our Investors," World Agroforestry Centre, accessed January 2, 2015, www .worldagroforestrycentre.org/about_us/our_investors.

113. Holly Creighton-Hird, "The future of family faming is in our hands," *The Ecologist,* October 19, 2014, accessed January 11, 2015, www.theecologist.org/campaigning/2599256/ the_future_of_family_farming_is_in_our_hands.html.

114. Eric Holt-Giménez and Raj Patel with Annie Shattuck, *Food Rebellions! Crisis and the Hunger for Justice* (Oakland, CA: Food First Books, with copublishers, 2009), 56.

Myth 9: It's Not Our Problem

1. USDA, ERS, Coleman-Jenson, Gregory, and Singh, "Household Food Insecurity," 6.

2. Calculated from Central Intelligence Agency, The World Factbook (Country Comparison: Infant Mortality Rate, Year: 2014; accessed September 8, 2014), www.cia.gov/ library/publications/the-world-factbook/rankorder/2091rank.html.

3. Hilary Waldron, "Trends in Mortality Differentials and Life Expectancy for Male Social Security Covered Workers, by Average Relative Earnings," U.S. Social Security Administration Office of Policy, 2007. (Based on males covered by Social Security.)

4. Fumiaki Imamura et al., "Dietary quality among men and women in 187 countries in 1990 and 2010: A systemic assessment," *The Lancet* 3 (March 2015): 141, www.thelancet .com/pdfs/journals/langlo/PIIS2214-109X%2814%2970381-X.pdf

5. Mark R. Rank and Thomas A. Hirschl, "The likelihood of poverty across the American adult life span," *Social Work* 44, no. 3 (1999): 206, accessed December 29, 2014, www -personal.umich.edu/~mdover/website/Social%20Welfare%20Policy%20Main%20Folder/ poverty.pdf.

6. UNICEF and Peter Adamson, "Measuring Child Poverty," Report Card No. 10, May 2012, Figure 1b, 3, September 9, 2014, www.unicef-irc.org/publications/pdf/rc10_eng .pdf.

7. U.S. Department of Health and Human Services, "2014 Poverty Guidelines," accessed September 22, 2014, aspe.hhs.gov/poverty/14poverty.cfm.

8. Calculated from "Top Ten Cheapest US Cities to Rent an Apartment," *CBS Money Watch,* July 20, 2013, accessed September 10, 2014, www.cbsnews.com/media/top-10-cheapest-us-cities-to-rent-an-apartment/.

9. World Bank [Data, Indicators, GINI Index (World Bank estimate), India, Liberia, Yemen, United States; accessed January 12, 2015], data.worldbank.org/indicator/SI.POV.GINI.

10. Aviva Aron-Dine and Arloc Sherman, "New CBO Data Show Income Inequality Continues to Widen: After-Tax Income for Top 1 Percent Rose by $146,000 in 2004," Center on Budget and Policy Priorities, January 23, 2007, accessed September 18, 2014, www.cbpp.org/cms/?fa=view&id=957.

11. Brenda Cronin, "Some 95% of 2009–2012 Income Gains Went to Wealthiest 1%," *Wall Street Journal,* 2013, accessed September 9, 2014, blogs.wsj.com/economics/2013/09/10/some-95-of-2009-2012-income-gains-went-to-wealthiest-1/.

12. Calculated from Kerry A. Dolan and Luisa Kroll, "Inside the 2014 Forbes 400: Facts and Figures About America's Wealthiest," *Forbes,* September 29, 2014, accessed March 24, 2015, www.forbes.com/sites/kerryadolan/2014/09/29/inside-the-2014-forbes-400-facts-and-figures-about-americas-wealthiest/; Federal Reserve Statistical Release, Board of Governors of the Federal Reserve System, "Financial Accounts of the United States: Flow of Funds, Balance Sheets, and Integrated Macroeconomic Accounts, Fourth Quarter 2014," March 12, 2015, i, accessed March 24, 2015, www.federalreserve.gov/releases/z1/current/z1.pdf; Congressional Research Service and Linda Levine, "An Analysis of the Distribution of Wealth Across Households, 1989-2010," July 17, 2012, 4, accessed March 24, 2015, www.fas.org/sgp/crs/misc/RL33433.pdf.

13. Richard Fry and Rakesh Kochhar, "America's wealth gap between middle-income and upper-income families is widest on record," Pew Research Center, December 17, 2014, accessed December 29, 2014, www.pewresearch.org/fact-tank/2014/12/17/wealth-gap-upper-middle-income/.

14. Andrew Dugan and Frank Newport, "In U.S., Fewer Believe 'Plenty of Opportunity' to Get Ahead," Gallup Economy, October 25, 2013, accessed August 26, 2014, www.gallup.com/poll/165584/fewer-believe-plenty-opportunity-ahead.aspx.

15. Pew Charitable Trust, "Pursuing the American Dream: Economic Mobility Across Generations," July 2012, 2, accessed September 9, 2014, www.pewtrusts.org/~/media/legacy/uploadedfiles/pcs_assets/2012/Pursuing AmericanDreampdf.pdf.

16. Richard Fry and Paul Taylor, "The Rise of Residential Segregation by Income," Pew Research Center, last modified August 1, 2012, accessed September 9, 2014, www.pewsocialtrends.org/2012/08/01/the-rise-of-residential-segregation-by-income/#fn-14312-1.

17. Institute for the Study of Labor, Markus Jäntti, et al., "American Exceptionalism in a New Light: A Comparison of Intergenerational Earnings Mobility in the Nordic Countries, the United Kingdom and the United States," Discussion Paper No. 1938, 2006, September 9, 2014, ftp.iza.org/dp1938.pdf.

18. International Labour Office, International Programme on the Elimination of Child Labour, Yacouba Diallo, Alex Etienne, and Farhad Mehran, "Global child labour trends 2008 to 2012" (Geneva: ILO, 2013), Table 2, viii, September 9, 2014, www.ilo.org/ipec/Informationresources/WCMS_IPEC_PUB_23015/lang--en/index.htm; Reuters, "Supplier for Samsung and Lenovo Accused of Using Child Labor," *New York Times,* August 28, 2014, accessed September 9, 2014, www.nytimes.com/2014/08/29/technology/supplier-for-samsung-and-lenovo-accused-of-using-child-labor.html.

19. U.S. Department of Labor, Employment Standards Administration Wage and Hour Division, "Child Labor Requirements in Agricultural Occupations Under the Fair Labor Standards Act (Child Labor Bulletin 102)," WH-1295, June 2007, 3, accessed September 9, 2014, www.dol.gov/whd/regs/compliance/childlabor102.pdf.

20. Human Rights Watch, "Fields of Peril: Child Labor in US Agriculture," May 2010, 6, accessed August 27, 2014, www.hrw.org/sites/default/files/reports/crd0510webwcover_1.pdf.

21. Lawrence Lessig, *Republic, Lost: How Money Corrupts Congress—and a Plan to Stop It* (New York: Twelve, 2012).

22. Joseph Stiglitz, "Inequality Is Not Inevitable," *New York Times*, June 27, 2014, accessed August 26, 2014, opinionator.blogs.nytimes.com/2014/06/27/inequality-is -not-inevitable/.

23. Oxfam, "Working for the Few: Political Capture and Economic Inequality," Briefing Paper No. 178, January 2014, 3, accessed September 22, 2014, www.oxfam.org/sites/www .oxfam.org/files/bp-working-for-few-political-capture-economic-inequality-200114-en .pdf.

24. Mamta Bhaurya, "Tax Evasion in India: Causes and Remedies," 2012, www .academia.edu/6810585/TAX_EVASION_IN_INDIA_CAUSES_ AND_REMEDIES.

25. Jacques Leslie, "The True Cost of Hidden Money: A Piketty Protégé's Theory on Tax Havens," *New York Times*, June 15, 2014, accessed July 8, 2014, www.nytimes. com/2014/06/16/opinion/a-piketty-proteges-theory-on-tax-havens.html; Jonathan Weisman, "Senate Passes $3.7 Trillion Budget, Setting Up Contentious Negotiations," *New York Times,* March 23, 2013, accessed September 25, 2014, www.nytimes.com/2013/03/24/ us/politics/senate-passes-3-7-trillion-budget-its-first-in-4-years.html.

26. Leslie, "True Cost of Hidden Money."

27. "Taxing for Some," *The Economist,* May 22, 2013, accessed October 1, 2014, www.economist.com/blogs/graphicdetail/2013/05/daily-chart-14?fsrc=scn%2Ftw%2Fdc %2F&%3Ffsrc%3Dscn%2F=tw%2Fdc.

28. Estimated from Glenn Ruffenach, "Navigating the Dividend Storm," *Wall Street Journal,* January 10, 2013, accessed October 1, 2014, online.wsj.com/news/articles/SB1 0001424127887323689604578219952168695148; Elizabeth Rosen, "Marginal income tax brackets," U.S. Tax Center, September 10, 2013, accessed October 1, 2014, www.irs.com/ articles/marginal-income-tax-brackets.

29. Scott Klinger, Sarah Anderson, and Javier Rojas, "Corporate Tax Dodgers: 10 Companies and Their Tax Loopholes: 2013 Report," Institute for Policy Studies, 2013, accessed September 25, 2014, www.americansfortaxfairness.org/files/Corporate-Tax-Dodgers -Report-Final.pdf.

30. "Statement of Senator Carl Levin (D-Mich) before U.S. Senate Permanent Subcommittee on Investigations on Offshore Profit Shifting and the U.S. Tax Code—Part 2 (Apple Inc.)," Permanent Subcommittee on Investigations, May 21, 2013, 1, accessed September 22, 2014, www.hsgac.senate.gov/subcommittees/investigations/hearings/offshore-profit -shifting-and-the-us-tax-code_-part-2.

31. U.S. Department of Agriculture, Food and Nutrition Service, Program Data Home, Overview: "Summary of Annual Data, FY 2009–2013," data as of September 5, 2014, accessed September 22, 2014, www.fns.usda.gov/pd/overview; "Farm Bill 2014: Latest News," Food Research and Action Center, accessed September 22, 2014, frac.org/leg -act-center/farm-bill-2012/.

32. Calculated from U.S. Census Bureau Retail Sales 2013-2014 and Wal-Mart, "Walmart 2014 Annual Report," accessed February 17, 2015, cdn.corporate.walmart.com/66/e5/9ff9a87445949173fde56316ac5f/2014-annual-report.pdf.

33. Democratic Staff of U.S. House Committee on Education and the Workforce, "The Low-Wage Drag on Our Economy: Wal-Mart's low wages and their effects on taxpayers and economic growth," May 2013, 5, accessed August 28, 2014, democrats.edworkforce.house.gov/sites/democrats.edworkforce.house.gov/files/documents/WalMartReport-May2013.pdf.

34. Robert E. Scott, "Wal-Mart's Reliance on Chinese Imports Costs U.S. Jobs," Economic Policy Institute, June 26, 2007, accessed August 28, 2014, www.epi.org/publication/webfeatures_snapshots_20070627/.

35. "The Economic Case for Raising the Minimum Wage," White House Council of Economic Advisers, February 12, 2014, accessed September 10, 2014, www.whitehouse.gov/blog/2014/02/12/economic-case-raising-minimum-wage; Democratic Staff, "Low-Wage Drag on Our Economy," 5.

36. Robert B. Reich, *Beyond Outrage: What Has Gone Wrong with Our Economy and Our Democracy and How to Fix It* (New York: Vintage Books, 2012), 13, 43, 48–51; Joseph E. Stiglitz, *The Price of Inequality* (New York: W. W. Norton, 2012), 60–64; Robert B. Reich, *Aftershock: The Next Economy and America's Future* (New York: Alfred A. Knopf, 2010), 52–56; Alan Tonelson, *The Race to the Bottom* (Cambridge, MA: Westview Press, 2002), xxii–xxiii, 2–18, 35–51.

37. Carolan, *Cheaponomics,* 200.

38. Clare O'Connor, "Report: Walmart Workers Cost Taxpayers $6.2 Billion In Public Assistance," *Forbes*, April 15, 2014, accessed September 10, 2014, www.forbes.com/sites/clareoconnor/2014/04/15/report-walmart-workers-cost-taxpayers-6-2-billion-in-public-assistance/.

39. UC Berkeley Labor Center, Sylvia A. Allegretto, et al., "Fast Food, Poverty Wages: The Public Cost of Low-Wage Jobs in the Fast-Food Industry," October 2013, 3, September 10, 2014, laborcenter.berkeley.edu/pdf/2013/fast_food_poverty_wages.pdf; United States Census Bureau (People and Households, Families and Living Arrangements, America's Families and Living Arrangements: 2013: Households, Table H1; accessed September 16, 2014), www.census.gov/hhes/families/data/cps2013H.html.

40. Jayne O'Donnell and Alicia McElhaney, "Costco pays more . . . because it can," *USA Today,* January 30, 2014, accessed November 4, 2014, www.usatoday.com/story/money/business/2014/01/29/costco-wages-walmart-federal-minimum-wage-obama/5029211/.

41. Rick Ungar, "Walmart Pays Workers Poorly and Sinks While Costco Pays Workers Well and Sails—Proof That You Get What You Pay For," *Forbes,* April 17, 2013, accessed November 4, 2014, www.forbes.com/sites/rickungar/2013/04/17/walmart-pays-workers-poorly-and-sinks-while-costco-pays-workers-well-and-sails-proof-that-you-get-what-you-pay-for/.

42. International Labour Organization, "Declaration concerning the aims and purposes of the International Labour Organisation (Declaration of Philadelphia)," accessed September 10, 2014, www.ilo.org/wcmsp5/groups/public/---asia/---ro-bangkok/---ilo-islamabad/documents/policy/wcms_142941.pdf.

43. "ILO Constitution," International Labour Organization, accessed April 15, 2015, www.ilo.org/dyn/normlex/en/f?p=NORMLEXPUB: 62:0::NO::P62_LIST_ENTRIE_ID:2453907.

44. "NORMLEX," International Labour Organization, accessed September 10, 2014, www.ilo.org/dyn/normlex/en/f?p=NORMLEXPUB:1:0.

45. "Ratifications of fundamental Conventions and Protocols by country," NORMLEX, International Labour Organization, accessed February 5, 2015, www.ilo.org/dyn/normlex/en/f?p=1000:10011:0::NO:10011:P10011_DISPLAY_BY,P10011_CONVENTION_TYPE_CODE:1,F; "ILO Constitution," International Labour Organization.

46. George I. Long, "Differences between union and nonunion compensation," *Monthly Labor Review*, Bureau of Labor Statistics, April 2013, 16–18. www.bls.gov/opub/mlr/2013/04/art2full.pdf.

47. Gardiner Harris, "'Superbugs' Kill India's Babies and Pose an Overseas Threat," *New York Times*, December 3, 2014, accessed December 17, 2014, www.nytimes.com/2014/12/04/world/asia/superbugs-kill-indias-babies-and-pose-an-overseas-threat.html?_r=0.

48. U.S. Government Accountability Office, "FDA can better oversee food imports by assessing and leveraging other countries' oversight resources," GAO-12-933, September 2012, 1, accessed December 29, 2014, www.gao.gov/assets/650/649010.pdf.

49. David M. Konisky, "Regulatory Competition and Environmental Enforcement: Is There a Race to the Bottom?," *American Journal of Political Science* 51, no. 4 (2007), 853–872, doi: 10.1111/j.1540-5907.2007.00285.x.

50. Suzanne Goldenberg, "CO2 emissions are being 'outsourced' by rich countries to rising economies," *The Guardian,* January 19, 2014; this article is based on Jintai Lin et al., "China's international trade and air pollution in the United States," *Proceedings of the National Academy of Sciences* 111, no. 5 (2014): 1736–1741; Elisabeth Rosenthal, "Lead from Old Batteries Sent to Mexico Raises Risks," *New York Times,* December 8, 2011; Naomi Klein, *This Changes Everything: Capitalism vs. the Climate* (New York: Simon & Schuster, 2014), 19–21, 80–83.

51. "CDC research shows outbreaks linked to imported foods increasing," U.S. Centers for Disease Control and Prevention, March 14, 2012, accessed August 28, 2014, www.cdc.gov/media/releases/2012/p0314_foodborne.html.

52. James Andrews, "Formaldehyde Detected in Supermarket Fish Imported from Asia," Food Safety News, September 11, 2013, accessed August 28, 2014, www.foodsafetynews.com/2013/09/formaldehyde-detected-in-supermarket-fish-imported-from-asia/#.U_88wPldUTR.

53. Nicole Gilbert, "Tainted Seafood Reaching U.S., Food Safety Experts Say," News 21, accessed August 28, 2014, foodsafety.news21.com/2011/imports/seafood/index.html.

54. "CDC research shows outbreaks linked," U.S. CDC.

55. Andrew C. von Eschenbach, "Enhanced Aquaculture and Seafood Inspection—Report to Congress," U.S. Food and Drug Administration, November 20, 2008, accessed March 30, 2015, www.fda.gov/Food/GuidanceRegulation/GuidanceDocumentsRegulatoryInformation/Seafood/ucm150954.htm; U.S. Government Accountability Office, "Food Safety: FDA and USDA Should Strengthen Pesticide Residue Monitoring Programs and Further Disclose Monitoring Limitations," GAO-15-38, October 7, 2014, accessed March 31, 2015, www.gao.gov/products/GAO-15-38.

56. Kaye, *The Fight for the Four Freedoms*. Image of the medal is on the back cover.

57. "1944 State of the Union Address: FDR's Second Bill of Rights or Economic Bill of Rights Speech," Franklin D. Roosevelt Presidential Library and Museum, accessed September 10, 2014, www.fdrlibrary.marist.edu/archives/stateoftheunion.html.

58. "Poverty Overview," World Bank, last modified October 7, 2014, accessed January 22, 2015, www.worldbank.org/en/topic/poverty/overview.

59. Eric Schmitt, Michael R. Gordon, and Helene Cooper, "Destroying ISIS May Take 3 Years, White House Says," *New York Times* (print), September 8, 2014.

60. Tori DeAngelis, "Understanding terrorism," *American Psychological Association* 40, no. 10 (2009), accessed January 22, 2015, www.apa.org/monitor/2009/11/terrorism.aspx.

61. Riaz Hassan, "What Motivates the Suicide Bombers?" Yale Center for the Study of Globalization, September 2009, accessed January 21, 2015, yaleglobal.yale.edu/content/ what-motivates-suicide-bombers-0; Riaz Hassan, *Life as a Weapon: The Global Rise of Suicide Bombing* (London: Routledge, 2010).

62. DeAngelis, "Understanding terrorism."

63. Dr. Thomas F. Lynch III, "Sources of Terrorism and Rational Counters," TRENDS Research and Analyses CVE Paper (2), Future Security of the GCC: Fighting Extremism forum, December 17, 2014, 10, accessed January 21, 2015, trendsinstitution.org/wp -content/uploads/2015/01/Terrorism-Fuel-Counters-By-Dr.-Thomas-F.-Lynch-III.pdf.

64. "G.I. Bill of Rights," State University of Education, accessed September 9, 2014, education.stateuniversity.com/pages/2008/G-I-Bill-Rights.html.

65. Paul N. Van de Water, Arloc Sherman, and Kathy Ruffing, "Social Security Keeps 22 Million Americans Out of Poverty: A State-by-State Analysis," Center on Budget and Policy Priorities, October 2013, accessed October 7, 2014, www.cbpp.org/cms/?fa=view&id=4037.

66. Calculated from DeNavas-Walt, Proctor, and Smith, "Income, Poverty, and Health Insurance Coverage," 13.

67. Robert B. Reich, *Supercapitalism: The Transformation of Business, Democracy, and Everyday Life* (New York: Alfred A. Knopf, 2007), 106.

68. Robert Pollin, *Contours of Descent: U.S. Economic Fractures* (London: Verso, 2003).

69. DeNavas-Walt, Proctor, and Smith, "Income, Poverty, and Health Insurance Coverage," 6.

Myth 10: Power Is Too Concentrated for Real Change—It's Too Late

1. Keating et al., "Global Wealth Report 2013," 4; Zephyr Teachout, *Corruption in America: From Benjamin Franklin's Snuff Box to Citizens United* (Cambridge, MA: Harvard University Press, 2014), 53–54.

2. Capra and Luisi, *Systems View of Life*.

3. Robert A. Dahl, *Democracy in the United States* (Boston: Houghton Mifflin, 1981), 32.

4. Dr. G. V. Ramanjaneyulu, Executive Director, Center for Sustainable Agriculture, Taranka, Secunderabad, Andhra Pradesh, e-mail communication with author, November 13, 2014; "Non Pesticide Management in Andhra Pradesh, India," accessed April 14, 2015, ftp://ftp.fao.org/sd/sda/sdar/sard/GP%20updates/pest_management_India.pdf.

5. "Sustainable Agriculture: a pathway out of poverty for India's poor," Sustainet, 40, accessed July 1, 2012, www.mamud.com/Docs/sustainet_india08_lowres.pdf.

6. Ramanjaneyulu, e-mail communication with author, November 13, 2014; Gerry Marten, "Escaping the Pesticide Trap: Non-Pesticide Management for Agricultural Pests (Andhra Pradesh, India)," EcoTipping Points Project, June 2005, www.ecotippingpoints .org/our-stories/indepth/india-pest-management-nonpesticide-neem.html.

7. Regina Gregory, "Replications of Punukula Example to Other Villages in Andhra Pradesh," EcoTipping Points Project, www.ecotippingpoints.org/our-stories/indepth/india -pest-management-nonpesticide-neem.html.

8. Ted Swagerty, "Why Has the Adoption of Non-Pesticide Management Been More Successful in Some Villages Than Others? An Update on the Dissemination of Non-Pesticide

Management Through Andhra Pradesh, India," EcoTipping Points Project, February 2014, www.ecotippingpoints.org/our-stories/indepth/india-pest-management-nonpesticide -neem.html.

9. "Sustainable Agriculture," Sustainet, 46.

10. Swagerty, "Why Has the Adoption of Non-Pesticide Management."

11. Ibid.

12. Sylvia Rowley, "In India, Profitable Farming with Fewer Chemicals," *New York Times,* Opinionator, April 24, 2015, opinionator.blogs.nytimes.com/2015/04/24/in-india -profitable-farming-with-fewer-chemicals.

13. Swagerty, "Why Has the Adoption of Non-Pesticide Management."

14. "Sustainable Agriculture," Sustainet, 45.

15. Swagerty, "Why Has the Adoption of Non-Pesticide Management"; Government of Andhra Pradesh Planning Department, "Socio- Economic Survey 2013–2014," 4, accessed September 29, 2014, www.aponline.gov.in/apportal/Downloads/Socio_Economic_ Survey_Book_let.pdf.

16. Swagerty, "Why Has the Adoption of Non-Pesticide Management."

17. "Torch bearers for millet seed security," *Deccan Herald,* January 19, 2014, accessed September 29, 2014, www.deccanherald.com/content/381375/torch-bearers-millet-seed -security.html.

18. Vandana Shiva, "Everything I Need to Know I Learned in the Forest," *Yes! Magazine*, December 2012, accessed October 7, 2014, www.yesmagazine .org/issues/what-would -nature-do/vandana-shiva-everything-i-need-to-know-i-learned-in-the-forest.

19. Ibid.

20. See Navdanya.org; "Navdanya: Two Decades of Service to the Earth and Small Farmers," Navdanya, accessed July 1, 2014, www.navdanya.org/attachments/Navdanya.pdf.

21. A. M. Nicolaysen, C. Francis, and G. Lieblein, "Farmer Supported Biodiversity Conservation in Uttarakhand, India," Proceeding of the International Farming Systems Association, 2014, 885, ifsa.boku.ac.at/cms/fileadmin/Proceeding2014/WS_1_9_Nicolaysen.pdf.

22. Binju Abraham et al., "SCI: The System of Crop Intensification: Agroecological Innovations for Improving Agricultural Production, Food Security, and Resilience to Climate Change," SRI International Network and Resources Center and Technical Centre for Agricultural and Rural Cooperation, 2014, 4, accessed October 6, 2014, sri.ciifad.cornell .edu/aboutsri/othercrops/SCImonograph_SRIRice2014.pdf.

23. John Vidal, "India's rice revolution," *The Observer,* February 16, 2013, accessed September 30, 2014, www.theguardian.com/global-development/2013/feb/16/india -rice-farmers-revolution.

24. Ibid.

25. Government of Kerala, Agriculture (P.B.) Department, "Kerala State Organic Farming Policy, Strategy, and Action Plan," G.O.(P) No. 39/2010/Agri., February 10, 2010, 7, accessed September 30, 2014, foodprocessingindia.co.in/state_pdf/Kerala/Kerala -Organic-Farming-Policy_2008.PDF.

26. Sapna E. Thottathil, *India's Organic Farming Revolution: What It Means for the Global Food System*, (Iowa City: University of Iowa Press, 2014), 3.

27. Eric Holt-Giménez, "Scaling Up Sustainable Agriculture—Lessons from the Campesino a Campesino Movement," Agricultures Network, www.agriculturesnetwork .org/magazines/global/lessons-in-scaling-up/scaling-up-sustainable-agriculture, originally published in Spanish in *LEISA Magazine* 17, no. 3 (October 2001), www.leisa-al.org/

web/revista-leisa/30-vol17n3.html#Ampliando_el_impacto; Holt-Giménez and Patel, *Food Rebellions!*; Eric Holt-Giménez, *Campesino a Campesino: Voices from Latin America's Farmer to Farmer Movement for Sustainable Agriculture* (Oakland, CA: Food First Books, 2006); Eric Holt-Giménez, "The Campesino a Campesino Movement: Farmer-Led Sustainable Agriculture," Food First: Institute for Food and Development Policy, Development Report 10, June 1996, accessed February 5, 2015, foodfirst.org/publication/the-campesino-a-campesino-movement/.

28. Holt-Giménez, "Scaling Up Sustainable Agriculture."

29. Daniel Buckles, "Hearing the mucuna story," *ILEIA Newsletter* 8, no. 3 (1992), accessed January 12, 2015, www.agriculturesnetwork.org/magazines/global/livestock-sustaining-livelihoods/hearing-the-mucuna-story.

30. Holt-Giménez, "Scaling Up Sustainable Agriculture"; Eric Holt-Giménez, "La Canasta Metodologica: Metodologías campesinas para la enseñanza agroecologica y el desarrollo de la agricultura sostenible," Rep. No. 28. SIMAS, Managua, 1997.

31. Holt-Giménez, "Scaling Up Sustainable Agriculture."

32. Chris Reij, Gray Tappan, and Melinda Smale, "Re-Greening the Sahel: Farmer-Led innovation in Burkina Faso and Niger," International Food Policy Research Institute, accessed September 29, 2014, www.ifpri.org/sites/default/files/publications/oc64ch07.pdf.

33. Rob Finlayson, "Land tenure and agroforestry at the heart of a sustainable future Earth," World Agroforestry Centre, October 4, 2013, blog.worldagroforestry.org/index.php/2013/10/04/land-tenure-and-agroforestry-at-the-heart-of-a-sustainable-future-earth/.

34. See viacampesina.org/en/ and 2013 Annual Report viacampesina.org/downloads/pdf/en/EN-annual-report-2013.pdf.

35. "Mayan People's Movement Defeats Monsanto Law in Guatemala," La Via Campesina, 2014, accessed October 21, 2014, viacampesina.org/en/index.php/main-issues-mainmenu-27/biodiversity-and-genetic-resources-mainmenu-37/1668-mayan-people-s-movement-defeats-monsanto-law-in-guatemala.

36. "2013 Annual Report," La Via Campesina, 3; David Alire Garcia, "Past and future collide as Mexico fights over GMO corn," Reuters, November 12, 2013, accessed December 3, 2014, www.reuters.com/article/2013/11/12/us-mexico-corn-idUSBRE9AB11Q20131112; Timothy A. Wise, "Monsanto Meets Its Match in the Birthplace of Maize," *Triple Crisis*, May 12, 2014, triplecrisis.com/monsanto-meets-its-match-in-the-birthplace-of-maize/

37. Richard L. Harris and Jorge Nef, eds., *Capital, Power, and Inequality in Latin America and the Caribbean* (Lanham, MD: Rowman and Littlefield, 2008), 265; estimate of prior landless from personal communication with Hannah Wittman; Hannah Wittman, "Agrarian reform and the production of locality: Resettlement and community building in Mato Grosso, Brazil," *Revista Nera* 8, no. 7 (July/December 2005), accessed November 5, 2014, revista.fct.unesp.br/index.php/nera/article/viewFile/1457/1433; Jonathan Watts, "Brazil's Landless Workers Movement renews protest on 30th anniversary," *The Guardian,* February 13, 2014, www.theguardian.com/global-development/2014/feb/13/brazil-landless-workers-movement-mst-protest-30th-anniversary.

38. For this and other quotations from MST members, see Frances Moore Lappé and Anna Lappé, *Hope's Edge: The Next Diet for a Small Planet* (New York: Jeremy P. Tarcher/Putnam, 2002), Chapter 3.

39. "What Is the MST?" Friends of the MST, accessed July 1, 2014, www.mstbrazil
.org/whatismst.

40. "We are millions," *New Internationalist*, December 12, 2009, accessed May 22,
2012, www.newint.org/features/special/2009/12/01/we-are-millions/. (Estimate from
Pastoral Land Commission.)

41. See www.mstbrazil.org; Cassia Bechara, Coletivo de Relacoes Internacionais Secretaria Nacional—MST, personal communication with the author, June 30, 2014; Instituto Nacional de Colonização e Reforma Agrária, "Assentamentos de Trabalhadores(as)
Rurais- Numeros Oficiais," December 31, 2013, accessed November 10, 2014, www.incra
.gov.br/sites/default/files/uploads/reforma-agraria/questao-agraria/reforma
-agraria/02-assentamentos.pdf.

42. "The MST at 30: Far beyond the distribution of land," Friends of the MST, January
28, 2014, accessed June 6, 2015, www.mstbrazil.org/news/mst-30-far-beyond-distribution
-land; Cassia Bechara, National Secretariat, MST, personal communication with authors,
2014–2015.

43. Cassia Bechara, personal communication with the author, September 2014.

44. Community-Wealth.org, a project of the Democracy Collaborative, www.community
-wealth-org.

45. Hannah Wittman, Associate Professor of Food, Nutrition and Health, University
of British Columbia, and specialist on Brazil, personal communication with the author,
November 4, 2014; Sérgio Sauer and Sergio Pereira Leite, "Agrarian structure, foreign
land ownership, and land value in Brazil," paper presented at International Conference
on Global Land Grabbing, Institute of Development Studies (IDS), University of Sussex,
Brighton, UK, April 6–8, 2011, 4, accessed February 5, 2015, www.future-agricultures.org/
publications/search-publications/global-land-grab/conference-papers-2/1281-english
-agrarian-structure-foreign-land-ownership-and-land-value-in-brazil/file.

46. Cassia Bechara, Coletivo de Relacoes Internacionais Secretaria Nacional—MST,
personal communication with the author, June 30, 2014; Jason Mark, "Brazil's MST: Taking
Back the Land," *Multinational Monitor* 22, nos. 1 and 2 (January–February 2001), accessed
October 13, 2014, multinationalmonitor.org/mm2001/01jan-feb/corp2.html; Hannah
Wittman, "Reworking the metabolic rift: La Vía Campesina, agrarian citizenship, and food
sovereignty," *Journal of Peasant Studies* 36, no. 4 (2009): 810, accessed November 4, 2014,
doi: 10.1080/03066150903353991.

47. FAO, "Seed multiplication by resource-limited farmers," FAO Plant Production and
Protection Paper No. 180 (Rome: FAO, 2004), 27, accessed October 8, 2014, ftp://ftp.fao
.org/docrep/fao/009/y5706e/y5706e.pdf; "MST and Agroecology: BioNatur (Organic)
Seeds," Friends of the MST, accessed April 18, 2012, www.mstbrazil.org/?q=seeds.

48. "Brasil: Centro Chico Mendes de Agroecologia: Terra livre de transgênicos e sem
agrotócivos," Biodiversidad en América Latina y El Caribe, June 2, 2005, accessed March
1, 2012, www.biodiversidadla.org/content/view/full/16526.

49. Avery Cohn et al., "Agroecology and the Struggle for Food Sovereignty in the Ameri-
cas," International Institute for Environment and Development, Yale School of Forestry and
Environmental Studies, and IUCN Commission on Environmental, Economic and Social
Policy, 2006, 92, accessed November 10, 2014, foodsecurecanada.org/sites/default/files/
Agroecology_and_the_Struggle_for_FS_in_the_Americas.pdf.

50. Calculated from "Landless in Brazil," WaronWant.org.

51. Clif Welch, "Movement Histories: A Preliminary Historiography of the Brazil's Landless Laborers' Movement (MST)," *Latin American Research Review* 41, no. 1 (2006): 199,

accessed February 5, 2015, doi: 10.1353/lar.2006.0015; "Land Reform Creates Food Security in Brazil," 2009, accessed May 8, 2015, hopebuilding.pbworks.com/w/page/19222594/Land%20reform%20creates%20food%20security%20in%20Brazil.

52. Lappé and Lappé, *Hope's Edge*, 89.

53. "Impact of the Sixth Congress, MST, Interview with Kelli Mafort," Friends of the MST, December 27, 2014, www.mstbrazil.org/news/impact-6th-congress-interview-kelli-mafort.

54. Lappé and Lappé, *Hope's Edge*, 80.

55. E. Hansen and M. Donohoe, "Health Issues of Migrant and Seasonal Farmworkers," *Journal of Health Care for the Poor and Underserved* 14, no. 2 (2003): 153, doi: 10.1353/hpu.2010.0790.

56. Calculated from "Workplace Safety and Health Topics: Agricultural Safety," Centers for Disease Control and Prevention, June 25, 2014, accessed October 7, 2014, www.cdc.gov/niosh/topics/aginjury/.

57. Shane Stephens, Bureau of Labor Statistics, e-mail communication with author, December 29, 2012.

58. Paul K Mills, Jennifer Dodge, and Richard Yang, "Cancer in Migrant and Seasonal Hired Farm Workers," *Journal of Agromedicine* 14 (2009): 185–191, accessed February 5, 2015, doi: 10.1080/10599240902824034.

59. Irma Morales Waugh, "Examining the Sexual Harassment Experiences of Mexican Immigrant Farmworking Women," *Violence Against Women* 16, no. 3 (2010): 237–261, accessed December 29, 2014, doi: 10.1177/1077801209360857; Maria M. Dominguez, "Sex Discrimination and Sexual Harassment in Agricultural Labor," *The American University Journal of Gender & the Law* 6 (1997): 232–260, accessed February 6, 2015, digitalcommons.wcl.american.edu/cgi/viewcontent.cgi?article=1132&context=jgspl.

60. U.S. Department of State, "Trafficking in Persons Report," 10th edition, June 2010, 44, accessed October 6, 2014, www.state.gov/documents/organization/142979.pdf; "Anti-Slavery Campaign," Coalition of Immokalee Workers, 2012, accessed October 6, 2014, ciw-online.org/slavery/.

61. Estabrook, *Tomatoland*, x.

62. "President Bill Clinton, Secretary of State Hillary Clinton Honor CIW with Global Citizen Award!" Coalition of Immokalee Workers, September 22, 2014, accessed October 6, 2014, ciw-online.org/blog/2014/09/bill-clinton/.

63. See ciw-online.org/; Jake Ratner, Just Harvest USA, e-mail communication with authors, December 29, 2014.

64. "NLRB Office of the General Counsel Issues Complaint Against Wal-Mart," Office of Public Affairs, National Labor Relations Board, January 15, 2014, accessed October 24, 2014, www.nlrb.gov/news-outreach/news-story/nlrb-office-general-counsel-issues-complaint-against-walmart.

65. "Press Release: Coalition of Immokalee Workers Announces Wal-Mart to Join Groundbreaking Fair Food Program," Coalition of Immokalee Workers, January 16, 2014, accessed October 24, 2014, ciw-online.org/blog/2014/01/Wal-Mart-press-release/; "Wal-Mart Stores, Inc.," Case Number: 16-CA-096240, National Labor Relations Board, last updated April 11, 2014, accessed October 24, 2014, www.nlrb.gov/case/16-CA-096240?page=1.

66. U.S. Department of Agriculture, Agricultural Marketing Service, Farmers Market and Direct Marketing Research Branch, and Debra Tropp, "Why Local Food Matters: The rising importance of locally grown food in the U.S. food system—a national perspective," March 2, 2014, accessed December 2, 2014, www.ams.usda.gov/AMSv1.0/getfile?dDocName=STELPRDC5105706.

67. "Farmers Markets and Local Food Marketing," Agricultural Marketing Service, U.S. Department of Agriculture, last modified August 14, 2014, accessed December 2, 2014, www.ams.usda.gov/AMSv1.0/ams.fetchTemplateData.do?template=TemplateS& leftNav=WholesaleandFarmersMarkets&page=WFMFarmersMarketGrowth&description =Farmers%20Market%20Growth&acct=frmrdirmkt.

68. John Fisher, Director of Programs and Partnerships at Life Lab, personal communication with author, April 12, 2015.

69. "FAQ," American Community Gardening Association, accessed December 2, 2014, communitygarden.org/resources/faq/; Anne Todd, "Green Thumbs Up: Interest in community gardens sprouting all across America," *Rural Cooperatives* 76, no. 3 (May/June 2009), accessed December 2, 2014, www.rurdev.usda.gov/rbs/pub/may09/green.htm; U.S. Department of Agriculture, Economic Research Service, "Chapter 7: Community Food Projects," in *Access to Affordable and Nutritious Food: Measuring and Understanding Food Deserts and Their Consequences*, ed. USDA, 2009, 99, accessed December 2, 2014, www .ers.usda.gov/media/242618/ap036g_1_.pdf.

70. See www.growingpower.org/; Roger Bybee, "Growing Power in an Urban Food Desert," *Yes! Magazine,* February 13, 2009, accessed December 2, 2014, www.yesmagazine .org/issues/food-for-everyone/growing-power-in-an-urban-food-desert.

71. "International Covenant on Economic, Social and Cultural Rights," United Nations, accessed May 8, 2015, treaties.un.org/pages/viewdetails.aspx?chapter=4&lang=en&mtdsg_ no=iv-3&src=treaty.

72. Margret Vidar, Yoon Jee Kim, and Luisa Cruz, "Legal Developments in the Progressive Realization of the Right to Food, Right to Food Thematic Study 3," FAO, 2014, 2–3, accessed May 12, 2014, www.fao.org/3/a-i3892e.pdf.

73. Standing Committee on Food, Consumer Affairs and Public Distribution (2012–13), Fifteenth Lok Sabha Ministry of Consumer Affairs, Food and Public Distribution Department of Food and Public Distribution, National Food Security Bill, 2013, Twenty-seventh Report, 9, law passed in 2013, 164.100.47.134/lsscommittee/Food,%20Consumer%20Affairs%20 &%20Public%20Distribution/Final%20Report%20on%20NFSB.pdf.

74. "UN Special Rapporteur on the Right to Food," Right to Food, accessed October 7, 2014, www.righttofood.org/work-of-jean-ziegler-at-the-un/un-soecial-rapporteu-on -the-right-to-food/.

75. Kaye, *Fight for the Four Freedoms*.

76. Economic Policy Institute, Lawrence Mishel, et al., *The State of Working America,* 12th edition (Ithaca, NY: Cornell University Press, 2012), 27, Fig. 1J, accessed February 6, 2015, www.stateofworkingamerica.org/files/book/Chapter1-Overview.pdf.

77. Kaye, *Fight for the Four Freedoms*.

78. Larry Rohter, "Picking Butter over Guns, Brazil Puts Off Buying Jets," *New York Times,* January 4, 2003, accessed October 7, 2014, www.nytimes.com/2003/01/04/world/ picking-butter-over-guns-brazil-puts-off-buying-jets.html.

79. Deborah Wetzel, "Bolsa Família: Brazil's Quiet Revolution," *The World Bank News,* November 4, 2013, accessed October 30, 2014, www.worldbank.org/en/news/ opinion/2013/11/04/bolsa-familia-Brazil-quiet-revolution; Kathy Lindert et al., "The Nuts and Bolts of Brazil's Bolsa Família Program: Implementing Conditional Cash Transfers in a Decentralized Context," SP Discussion Paper No. 0709, The World Bank Social Protection, May 2007, 142, accessed October 13, 2014, siteresources.worldbank.org/ SOCIALPROTECTION/Resources/SP-Discussion-papers/Safety-Nets-DP/0709.pdf; Oxfam, "Halving Hunger: Still Possible? Building a rescue package to set the MDGs back on track,"

Oxfam Briefing Paper No. 139, September 2010, 27, accessed October 13, 2014, www.oxfam .org/sites/www.oxfam.org/files/file_attachments/oxfam-halving-hunger-sept-2010_5.pdf.

80. Sarojini Ganju Thakur and Catherine Arnold, "Gender and Social Protection," in *Promoting Pro-Poor Growth: Social Protection* (Organisation for Economic Cooperation and Development, 2009), 170, 173–175, accessed June 30, 2014, www.oecd.org/development/povertyreduction/43280899.pdf; FAO, "The State of Food Insecurity in the World: Economic growth is necessary but not sufficient to accelerate reduction of hunger and malnutrition," 2012, 37, accessed February 6, 2015, www.fao.org/docrep/016/i3027e/i3027e.pdf.

81. Brazil's Conditional Cash Transfer Programme, Bolsa Família, IBSA International Conference on South-South Cooperation "Innovations in Public Employment Programmes and Sustainable Inclusive Growth, New Delhi, March 2012, www.ilo.org/wcmsp5/groups/public/---asia/---ro-bangkok/---sro-new_delhi/documents/presentation/wcms_175274 .pdf.4, doi:10.1038/24376;-and-stats.

82. Jonathan Watts, "Brazil's *bolsa famili*a scheme marks a decade of pioneering poverty relief," *The Guardian*, December 17, 2013, www.theguardian.com/global -development/2013/dec/17/brazil-bolsa-familia-decade-anniversary-poverty-relief.

83. Tina Rosenberg, "To Beat Back Poverty, Pay the Poor," *New York Times,* January 3, 2011, accessed October 30, 2014, opinionator.blogs.nytimes.com/2011/01/03/to-beat -back-poverty-pay-the-poor/?_php=true&_type=blogs&_r=0.

84. Janine Berg, "Laws or Luck? Understanding Rising Formality in Brazil in the 2000s," International Labor Office—Brasilia, 2010, 2, 4, 8, accessed February 6, 2015, www.ase .tufts.edu/gdae/Pubs/rp/BergLaborFormalityBrazil.pdf.

85. Luis F. Lopez-Calva, "Declining Income Inequality in Brazil: The Proud Outlier," in *Inequality in Focus*, World Bank, 2012, 5. siteresources.worldbank.org/EXTPOVERTY/ Resources/Inequality_in_Focus_April2012.pdf.

86. World Bank [Data, Indicators, Mortality rate, under-5 (per 1,000 live births), Brazil, Years: 2000–12; accessed February 6, 2015], data.worldbank.org/indicator/SH.DYN.MORT.

87. International Policy Centre for Inclusive Growth, Ryan Nehring, and Ben McKay, "Sustainable Agriculture: An Assessment of Brazil's Family Farm Programmes in Scaling Up Agroecological Food Production," One Pager No. 246, March 2014, accessed February 6, 2015, www.ipc-undp.org/pub/IPCOnePager246.pdf.

88. FAO, José Graziano da Silva, Mauro Eduardo del Grossi, and Caio Galvão de França, eds., *The Fome Zero (Zero Hunger) Program: The Brazilian Experience* (Brazil: Ministry of Agrarian Development, 2011), 249–252, accessed July 8, 2014, www.fao.org/docrep/016/ i3023e/i3023e.pdf.

89. Bolivia, Final Report, Constitutional Referendum, January 25, 2009, European Union Election Observation Mission, 31; "The Human Right to Food in Bolivia," Report of an International Fact-finding Mission, Rights and Democracy, Montreal, 2010, 26, cesr.org/ downloads/Bolivia_Right_To_Food_eng.pdf.

90. FAO, "State of Food Insecurity in the World," 2014, 20–22.

91. Peeyush Bajpai, Laveesh Bhandari, and Aali Sinha, *Social and Economic Profile of India* (New Delhi: Esha Béteille, 2005), 40.

92. M. Karuna, "Adding Millets to the Basket," India Together, December 2011, www .indiatogether.org/2011/dec/pov-millets.htm; "Honouring Women Farmers as Environmental Saviors—These millet farmers fulfill India's global commitment in CBD," Millet Network of India, June 2013, www.milletindia.org/recentevents.asp; Gardiner Harris, "Study Says Pregnant Women in India Are Gravely Underweight," *New York*

Times, March 2, 2015, accessed March 11, 2015, www.nytimes.com/2015/03/03/world/asia/-pregnant-women-india-dangerously-underweight-study.html?_r=1; Government of India, Ministry of Health and Family Welfare, Adolescent Division, "Guidelines for Control of Iron Deficiency Anemia," 2013, accessed June 2, 2015, www.pbnrhm.org/docs/iron_plus_guidelines.pdf.

93. See www.milletindia.org/.

94. David J. Thompson, "Italy's Emilia Romagna: Clustering Co-op Development," *Cooperative Grocer Magazine* 109 (November–December 2003), accessed February 6, 2015, www.cooperativegrocer.coop/articles/2003-12-02/italys-emilia-romagna.

95. "It's time to make the global food system work for smallholders," Fairtrade Labelling Organizations International, May 10, 2013, accessed December 3, 2014, www.fairtrade.net/single-view+M549f154b502.html.

96. See Fairtrade Labelling Organizations International, www.fairtrade.net/; Fairtrade International, "Monitoring the Scope and Benefits of Fairtrade," 5th edition, 2013, accessed November 17, 2014, www.fairtrade.net/fileadmin/user_upload/content/2009/resources/2013-Fairtrade-Monitoring-Scope-Benefits_web.pdf; "Fairtrade Impact," 2011, info.fairtrade.net/info-impact.0.html.

97. John Paull, "The Fairtrade movement: Six lessons for the organics sector," Proceedings of the Third Scientific Conference of ISOFAR (International Society of Organic Agriculture Research), September 28–October 1, 2011, Namyangju, Korea, accessed February 6, 2015, orgprints.org/19527/1/Paull2011FairtradeISOFAR.pdf.

98. "It's time to make the global food system work for smallholders," Fairtrade Labelling Organizations International.

99. "Cooperatives: Resilient to crises, key to sustainable growth," International Labour Organization, last modified July 6, 2012, accessed July 2, 2014, www.ilo.org/global/about-the-ilo/newsroom/news/WCMS_184777/lang--en/index.htm. The logic of this estimate: The International Labour Organization reports one billion cooperative members worldwide. Considering the combined population of the EU and U.S. (about 830 million in 2013), and even assuming that half of the people in these two regions own corporate shares, and that there are as many as several hundred additional million shareholders in the rest of the world, we can still arrive at less than one billion.

100. "The Divine Story," www.divinechocolate.com/us/about-us/divine-story.

101. Calculated from "Kuapa Kokoo," Divine Chocolate, accessed October 6, 2014, www.divinechocolate.com/uk/about-us/research-resources/divine-story/kuapa-kokoo; Institute of Development Studies and University of Ghana, "Mapping sustainable production in Ghanaian cocoa: Report to Cadbury," 2009, 22, accessed October 15, 2014, www.academia.edu/6190281/Mapping_sustainable_production_in_ghanaian_cocoa.

102. Shashi Kolavalli and Marcella Vigneri, "Cocoa in Ghana: Shaping the Success of an Economy," in Punam Chuhan-Pole and Manka Angwafo, *Yes, Africa Can: Success Stories from a Dynamic Continent* (Washington, DC: World Bank, 2011), 208–209, accessed July 2, 2014, siteresources.worldbank.org/AFRICAEXT/Resources/258643-1271798012256/Ghana-cocoa.pdf.

103. International Food Policy Research Institute, Adam Salifu, Gian Nicola Francesconi, and Shashidhara Kolavalli, "A Review of Collective Action in Rural Ghana," IFPRI Discussion Paper 00998 June 2010, 10, accessed February 6, 2015, www.ifpri.org/sites/default/files/publications/ifpridp00998.pdf.

104. See www.amuldairy.com/.

105. "Cooperatives: Resilient to crises, key to sustainable growth," International Labour Organization, last modified July 6, 2012, accessed July 2, 2014, www.ilo.org/global/about -the-ilo/newsroom/news/WCMS_184777/lang--en/index.htm; "Co-ops Facts and Stats," International Co-operative Alliance, last modified 2011, accessed July 2, 2014, ica.coop/ en/co-op-facts-and-stats.

106. Michel Pimbert et al., "Democratising Agricultural Research for Food Sovereignty in West Africa" (London: International Institute for Environment and Development, 2010), 4–5, pubs.iied.org/pdfs/14603IIED.pdf; Frances Moore Lappé, *EcoMind: Changing the Way We Think to Create the World We Want* (New York: Nation Books, 2011), 163.

107. Michel Pimbert, personal communication with author, May 2014; Diverse Food System, July 2002, accessed October 6, 2014, www.diversefoodsystems.org/prajateerpu/ download/reply%20to%20dfid.pdf.

108. Pimbert et al., "Democratising Agricultural Research,"6–7.

109. Archon Fung and Erik Olin Wright, *Deepening Democracy: Institutional Innovations in Empowered Participatory Governance* (New York: Verso, 2003), 13–14.

110. Ibid., 78–79.

111. World Bank, "World Development Report: Equity and Development," 32204, 2006, 70, accessed November 10, 2014, www-wds.worldbank.org/servlet/WDSContentServer/ WDSP/IB/2005/09/20/000112742_20050920110826/Rendered/PDF/322040World0 Development0Report02006.pdf.

112. Prefeitura BH, "Participatory Budgeting in Belo Horizonte: Fifteen years," December 23, 2008, 29, accessed September 30, 2014, www.pbh.gov.br/comunicacao/pdfs/ publicacoesop/revista_op15anos_ingles.pdf.

113. Ibid., 15.

114. Fung and Wright, *Deepening Democracy*, 50.

115. Clare Fox, "Food Policy Councils: Innovations in Democratic Governance for a Sustainable and Equitable Food System," prepared for Los Angeles Food Policy Task Force, June 2010, 11, accessed December 3, 2014; Mark Bittman et al., "How a national food policy could save millions of American lives," *Washington Post,* November 7, 2014, accessed December 3, 2014, www.washingtonpost.com/opinions/how-a-national -food-policy-could-save-millions-of-american-lives/2014/11/07/89c55e16-637f-11e4 -836c-83bc4f26eb67_story.html.

116. Harvard Law School Food Law and Policy Clinic and Community Food Security Coalition, "Good Laws, Good Food: Putting Local Food Policy to Work for Our Communities," July 2012, accessed December 2, 2014, www.law.harvard.edu/academics/clinical/ lsc/documents/FINAL_LOCAL_TOOLKIT2.pdf.

117. Mark Winne, "Food Policy Councils: A Look Back at 2012," January 8, 2013, accessed December 3, 2014, www.markwinne.com/food-policy-councils-a-look-back -at-2012/.

118. Michael Burgan and Mark Winne, "Doing Food Policy Councils Right: A Guide to Development and Action," September 2012, 8, accessed December 2, 2014, www .markwinne.com/wp-content/uploads/2012/09/FPC-manual.pdf.

119. Minnesota Food Charter, "Minnesota Food Charter Fact Sheet," accessed December 3, 2014, mnfoodcharter.com/wp-content/uploads/2014/10/MNFC-Fact-Sheet.pdf; "What Is the Minnesota Food Charter," Minnesota Food Charter, accessed December 3, 2014, mnfoodcharter.com/the-charter/what-is-the-minnesota-food-charter/.

120. Letter from José Graziano da Silva to the Members of the Facilitating Committee

of IPC, FAO, Rome, May 23, 2014, accessed October 6, 2014, www.foodsovereignty.org/wp-content/uploads/2014/05/director-general.pdf.

121. Nora McKeon, *Food Security Governance: Empowering Communities, Regulating Corporations* (New York: Routledge Critical Security Studies, 2015), Chapter 4.

122. Letter from José Graziano da Silva, FAO.

123. McKeon, *Food Security Governance*, Chapter 4.

124. Ibid., Chapter 6.

125. "World Food Security Committee Puts Corporate Investors Before Human Rights," International Union of Food, Agricultural, Hotel, Restaurant, Catering, Tobacco and Allied Workers' Associations, October 17, 2014, accessed February 6, 2015, www.iuf.org/w/?q=node/3714.

126. McKeon, *Food Security Governance*, Chapter 7.

127. "Lawrence Lessig on 'Tweedism,'" YouTube, 2014. Lessig is a professor of law, Harvard University; "Donor Demographics," opensecrets.org, last modified November 16, 2014, accessed December 17, 2014, www.opensecrets.org/overview/donordemographics.php.

128. Nicholas Confessore, "Koch Brothers' Budget of $889 Million for 2016 Is on Par with Both Parties' Spending," *New York Times,* January 26, 2015, accessed February 11, 2015, www.nytimes.com/2015/01/27/us/politics/kochs-plan-to-spend-900-million-on-2016-campaign.html?_r=0; Walter Hickey, "The Total Cost of the 2012 Presidential Race Was Astounding," *Business Insider*, www.businessinsider.com/election-cost-2-billion-2012-12.

129. "Lawrence Lessig on 'Tweedism'"; film, "Donor Demographics," opensecrets.org, last modified November 16, 2014, accessed December 17, 2014, www.opensecrets.org/overview/donordemographics.php.

130. "PACs, Big Companies, Lobbyists, and Banks and Financial Institutions Seen by Strong Majorities as Having Too Much Power and Influence in DC," Harris Poll, May 29, 2012, accessed September 30, 2014, www.harrisinteractive.com/NewsRoom/HarrisPolls/tabid/447/mid/1508/articleId/1069/ctl/ReadCustom%20Default/Default.aspx.

131. William H. Wiist, "Citizens United, Public Health, and Democracy: The Supreme Court Ruling, Its Implications, and Proposed Action," National Center for Biotechnology Information 2011, accessed October 7, 2014, www.ncbi.nlm.nih.gov/pmc/articles/PMC3110222/#bib43.

132. "Local and State Resolutions," United for the People, 2014, accessed October 6, 2014, www.united4thepeople.org/local.html.

133. Referendum Amending City of Tallahassee Charter, 2014. citizensforethicsreform.org/wp-content/uploads/2014/06/CfER_Petition.pdf; Josh Silver, "One Community Beat Big Money on Election Day: Here's How They Did It," Huffington Post, updated, January 6, 2015. www.huffingtonpost.com/josh-silver/one-community-beat-big-mo_b_6115044.html.

134. Elin Falgura, Samuel Jones, and Magnus Ohman, eds., "Funding of Political Parties and Election Campaigns: A Handbook on Political Finance," International Institute of Democracy and Electoral Assistance, 2014, 215–221, accessed October 7, 2014, www.idea.int/publications/funding-of-political-parties-and-election-campaigns/loader.cfm?csModule=security/getfile&pageID=64347.

135. Pippa Norris, Ferran Martinez, i Coma, and Max Grömping, "The Year in Elections, 2014: Why Elections Fail and What We Can Do About It," Electoral Integrity Project (2014), University of Sydney, Australia, Fig. 4, 8, 9, 19, sites.google.com/site/electoralintegrityproject4/projects/expert-survey-2/the-year-in-elections-2014.

136. U.S. Department of Health and Human Services, Marian F. MacDorman, et al., "National Vital Statistics Reports: "International Comparisons of Infant Mortality and Related Factors: United States and Europe, 2010," *National Vital Statistics Reports* 63, no. 5 (2014): 1, accessed October 21, 2014, www.cdc.gov/nchs/data/nvsr/nvsr63/nvsr63_05 .pdf.

137. United States Department of Agriculture, "WIC at a Glance," www.fns.usda.gov/ wic/about-wic-wic-glance.

Beyond the Myths of Hunger

1. Harry C. Boyte, *Everyday Politics: Reconnecting Citizens and Public Life* (Philadelphia: University of Pennsylvania Press, 2004), 190.

2. Charles Lindblom, *Politics and Markets* (New York: Basic Books, 1977), 49–50.

3. Emmanuel Saez and Gabriel Zucman, "Wealth Inequality in the United States Since 1913: Evidence from Capitalized Income Tax Data," Working Paper No. 20625, National Bureau of Economic Research, October 2014, 3, accessed November 20, 2014, gabriel -zucman.eu/files/SaezZucman2014.pdf.

4. Thomas Jefferson, *Thomas Jefferson on Democracy* (New York: D. Appleton-Century, 1939), 215.

5. Teachout, *Corruption in America*, 53–54.

6. Sorapop Kiatpongsan and Michael I. Norton, "How Much (More) Should CEOs Make? A Universal Desire for More Equal Pay," *Perspectives on Psychological Science* 9, no. 6 (2014): 587–593, accessed December 18, 2014, doi: 10.1177/1745691614549773, pps.sagepub .com/content/9/6/587.abstract.

7. "Franklin D. Roosevelt, XXXII President of the United States: 1933–45: State of the Union Message to Congress," January 11, 1944, American Presidency Project, accessed November 5, 2014, www.presidency.ucsb.edu/ws/index.php?pid=16518.

8. Henry Shue, *Basic Rights: Subsistence, Affluence, and U.S. Foreign Policy* (Princeton, NJ: Princeton University Press, 1980), 24–25.

9. Gregory Claeys, *Thomas Paine: Social and Political Thought* (Boston: Unwin Hyman, 1989), 201. Italics in original.

10. William Cronon, *Changes in the Land: Indians, Colonists, and the Ecology of New England* (New York: Hill and Wang, 1983).

11. Teachout, *Corruption in America,* 38.

12. "Political advertising: Case studies and monitoring," Background paper—Plenary, 23rd EPRA Meeting, May 17–19, 2006, accessed December 30, 2014, www.rtdh.eu/ pdf/20060517_epra_meeting.pdf; Edith Palmer, "Campaign Finance: Germany," Library of Congress, last updated May 2009, accessed December 18, 2014, www.loc.gov/law/ help/campaign-finance/germany.php.

13. Richard Laynard, *Happiness: Lessons from a New Science* (New York: Penguin, 2005).

Index

Index

Index

Index

Index

Index

Index

Index

Index

Index

Index

women (*continued*)
    people-to-people effect and, 28–29
    risk from pressure to lower birthrates,
        29–30
    role in food production, 5–6
    worsening family situation for, 6–7
Women's Empowerment Program in
    Nepal, 260
World Bank, 154–155, 175–176. *See also*
    financial industry
World Food Program (WFP) forecast, 36,
    196, 201, 204

World Trade Organization (WTO),
    177–179

yield
    gains from agroecology, 124–126
    gaps, 45–46
    "high yield" seeds, 72
    not increased by industrial agriculture
        and GMOs, 94–95, 101–102

Zambia, agroforestry in, 61
Zero Hunger campaign, 167, 253, 254